INTRODUCTION TO PETROLEUM ENGINEERING

INTRODUCTION TO PETROLEUM ENGINEERING

JOHN R. FANCHI
and
RICHARD L. CHRISTIANSEN

Copyright © 2017 by John Wiley & Sons, Inc. All rights reserved

Published by John Wiley & Sons, Inc., Hoboken, New Jersey
Published simultaneously in Canada

No part of this publication may be reproduced, stored in a retrieval system, or transmitted in any form or by any means, electronic, mechanical, photocopying, recording, scanning, or otherwise, except as permitted under Section 107 or 108 of the 1976 United States Copyright Act, without either the prior written permission of the Publisher, or authorization through payment of the appropriate per-copy fee to the Copyright Clearance Center, Inc., 222 Rosewood Drive, Danvers, MA 01923, (978) 750-8400, fax (978) 750-4470, or on the web at www.copyright.com. Requests to the Publisher for permission should be addressed to the Permissions Department, John Wiley & Sons, Inc., 111 River Street, Hoboken, NJ 07030, (201) 748-6011, fax (201) 748-6008, or online at http://www.wiley.com/go/permissions.

Limit of Liability/Disclaimer of Warranty: While the publisher and author have used their best efforts in preparing this book, they make no representations or warranties with respect to the accuracy or completeness of the contents of this book and specifically disclaim any implied warranties of merchantability or fitness for a particular purpose. No warranty may be created or extended by sales representatives or written sales materials. The advice and strategies contained herein may not be suitable for your situation. You should consult with a professional where appropriate. Neither the publisher nor author shall be liable for any loss of profit or any other commercial damages, including but not limited to special, incidental, consequential, or other damages.

For general information on our other products and services or for technical support, please contact our Customer Care Department within the United States at (800) 762-2974, outside the United States at (317) 572-3993 or fax (317) 572-4002.

Wiley also publishes its books in a variety of electronic formats. Some content that appears in print may not be available in electronic formats. For more information about Wiley products, visit our web site at www.wiley.com.

Library of Congress Cataloging-in-Publication Data:

Names: Fanchi, John R., author. | Christiansen, Richard L. (Richard Lee), author.
Title: Introduction to petroleum engineering / by John R. Fanchi and Richard L. Christiansen.
Description: Hoboken, New Jersey : John Wiley & Sons, Inc., [2017] | Includes bibliographical references and index.
Identifiers: LCCN 2016019048| ISBN 9781119193449 (cloth) | ISBN 9781119193647 (epdf) | ISBN 9781119193616 (epub)
Subjects: LCSH: Petroleum engineering.
Classification: LCC TN870 .F327 2017 | DDC 622/.3382–dc23
LC record available at https://lccn.loc.gov/2016019048

Printed in the United States of America

V10008956_031919

CONTENTS

ABOUT THE AUTHORS

John R. Fanchi

John R. Fanchi is a professor in the Department of Engineering and Energy Institute at Texas Christian University in Fort Worth, Texas. He holds the Ross B. Matthews Professorship in Petroleum Engineering and teaches courses in energy and engineering. Before this appointment, he taught petroleum and energy engineering courses at the Colorado School of Mines and worked in the technology centers of four energy companies (Chevron, Marathon, Cities Service, and Getty). He is a Distinguished Member of the Society of Petroleum Engineers and coedited the General Engineering volume of the *Petroleum Engineering Handbook* published by the Society of Petroleum Engineers. He is the author of numerous books, including *Energy in the 21st Century*, 3rd Edition (World Scientific, 2013); *Integrated Reservoir Asset Management* (Elsevier, 2010); *Principles of Applied Reservoir Simulation*, 3rd Edition (Elsevier, 2006); *Math Refresher for Scientists and Engineers*, 3rd Edition (Wiley, 2006); *Energy: Technology and Directions for the Future* (Elsevier-Academic Press, 2004); *Shared Earth Modeling* (Elsevier, 2002); *Integrated Flow Modeling* (Elsevier, 2000); and *Parametrized Relativistic Quantum Theory* (Kluwer, 1993).

Richard L. Christiansen

Richard L. Christiansen is an adjunct professor of chemical engineering at the University of Utah in Salt Lake City. There, he teaches a reservoir engineering course as well as an introductory course for petroleum engineering. Previously, he engaged in all aspects of petroleum engineering as the engineer for a small oil and gas exploration company in Utah. As a member of the Petroleum Engineering faculty at the Colorado School of Mines from 1990 until 2006, he taught a variety of courses, including multiphase flow in wells, flow through porous media, enhanced oil recovery, and phase behavior. His research experiences include multiphase flow in rock, fractures, and wells; natural gas hydrates; and high-pressure gas flooding. He is the author of *Two-Phase Flow in Porous Media* (2008) that demonstrates fundamentals of relative permeability and capillary pressure. From 1980 to 1990, he worked on high-pressure gas flooding at the technology center for Marathon Oil Company in Colorado. He earned his Ph.D. in chemical engineering at the University of Wisconsin in 1980.

PREFACE

Introduction to Petroleum Engineering introduces people with technical backgrounds to petroleum engineering. The book presents fundamental terminology and concepts from geology, geophysics, petrophysics, drilling, production, and reservoir engineering. It covers upstream, midstream, and downstream operations. Exercises at the end of each chapter are designed to highlight and reinforce material in the chapter and encourage the reader to develop a deeper understanding of the material.

Introduction to Petroleum Engineering is suitable for science and engineering students, practicing scientists and engineers, continuing education classes, industry short courses, or self-study. The material in *Introduction to Petroleum Engineering* has been used in upper-level undergraduate and introductory graduate-level courses for engineering and earth science majors. It is especially useful for geoscientists and mechanical, electrical, environmental, and chemical engineers who would like to learn more about the engineering technology needed to produce oil and gas.

Our colleagues in industry and academia and students in multidisciplinary classes helped us identify material that should be understood by people with a range of technical backgrounds. We thank Helge Alsleben, Bill Eustes, Jim Gilman, Pradeep Kaul, Don Mims, Wayne Pennington, and Rob Sutton for comments on specific chapters and Kathy Fanchi for helping prepare this manuscript.

John R. Fanchi, Ph.D.
Richard L. Christiansen, Ph.D.
June 2016

ABOUT THE COMPANION WEBSITE

This book is accompanied by a companion website:

www.wiley.com/go/Fanchi/IntroPetroleumEngineering

The website includes:

- Solution manual for instructors only

1

INTRODUCTION

The global economy is based on an infrastructure that depends on the consumption of petroleum (Fanchi and Fanchi, 2016). Petroleum is a mixture of hydrocarbon molecules and inorganic impurities that can exist in the solid, liquid (oil), or gas phase. Our purpose here is to introduce you to the terminology and techniques used in petroleum engineering. Petroleum engineering is concerned with the production of petroleum from subsurface reservoirs. This chapter describes the role of petroleum engineering in the production of oil and gas and provides a view of oil and gas production from the perspective of a decision maker.

1.1 WHAT IS PETROLEUM ENGINEERING?

A typical workflow for designing, implementing, and executing a project to produce hydrocarbons must fulfill several functions. The workflow must make it possible to identify project opportunities; generate and evaluate alternatives; select and design the desired alternative; implement the alternative; operate the alternative over the life of the project, including abandonment; and then evaluate the success of the project so lessons can be learned and applied to future projects. People with skills from many disciplines are involved in the workflow. For example, petroleum geologists and geophysicists use technology to provide a description of hydrocarbon-bearing reservoir rock (Raymond and Leffler, 2006; Hyne, 2012). Petroleum engineers acquire and apply knowledge of the behavior of oil, water, and gas in porous rock to extract hydrocarbons.

Introduction to Petroleum Engineering, First Edition. John R. Fanchi and Richard L. Christiansen.
© 2017 John Wiley & Sons, Inc. Published 2017 by John Wiley & Sons, Inc.
Companion website: www.wiley.com/go/Fanchi/IntroPetroleumEngineering

Some companies form asset management teams composed of people with different backgrounds. The asset management team is assigned primary responsibility for developing and implementing a particular project.

Figure 1.1 illustrates a hydrocarbon production system as a collection of subsystems. Oil, gas, and water are contained in the pore space of reservoir rock. The accumulation of hydrocarbons in rock is a reservoir. Reservoir fluids include the fluids originally contained in the reservoir as well as fluids that may be introduced as part of the reservoir management program. Wells are needed to extract fluids from the reservoir. Each well must be drilled and completed so that fluids can flow from the reservoir to the surface. Well performance in the reservoir depends on the properties of the reservoir rock, the interaction between the rock and fluids, and fluid properties. Well performance also depends on several other properties such as the properties of the fluid flowing through the well; the well length, cross section, and trajectory; and type of completion. The connection between the well and the reservoir is achieved by completing the well so fluid can flow from reservoir rock into the well.

Surface equipment is used to drill, complete, and operate wells. Drilling rigs may be permanently installed or portable. Portable drilling rigs can be moved by vehicles that include trucks, barges, ships, or mobile platforms. Separators are used to separate produced fluids into different phases for transport to storage and processing facilities. Transportation of produced fluids occurs by such means as pipelines, tanker trucks, double-hulled tankers, and liquefied natural gas transport ships. Produced hydrocarbons must be processed into marketable products. Processing typically begins near the well site and continues at refineries. Refined hydrocarbons are used for a variety of purposes, such as natural gas for utilities, gasoline and diesel fuel for transportation, and asphalt for paving.

Petroleum engineers are expected to work in environments ranging from desert climates in the Middle East, stormy offshore environments in the North Sea, and

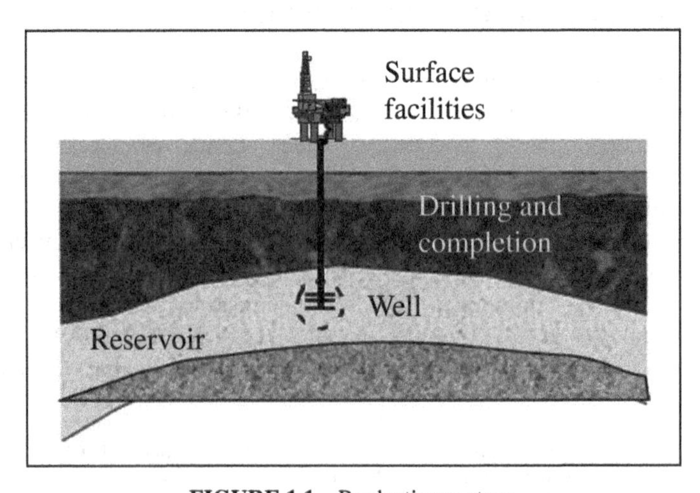

FIGURE 1.1 Production system.

arctic climates in Alaska and Siberia to deepwater environments in the Gulf of Mexico and off the coast of West Africa. They tend to specialize in one of three subdisciplines: drilling engineering, production engineering, and reservoir engineering. Drilling engineers are responsible for drilling and completing wells. Production engineers manage fluid flow between the reservoir and the well. Reservoir engineers seek to optimize hydrocarbon production using an understanding of fluid flow in the reservoir, well placement, well rates, and recovery techniques. The Society of Petroleum Engineers (SPE) is the largest professional society for petroleum engineers. A key function of the society is to disseminate information about the industry.

1.1.1 Alternative Energy Opportunities

Petroleum engineering principles can be applied to subsurface resources other than oil and gas (Fanchi, 2010). Examples include geothermal energy, geologic sequestration of gas, and compressed air energy storage (CAES). Geothermal energy can be obtained from temperature gradients between the shallow ground and surface, subsurface hot water, hot rock several kilometers below the Earth's surface, and magma. Geologic sequestration is the capture, separation, and long-term storage of greenhouse gases or other gas pollutants in a subsurface environment such as a reservoir, aquifer, or coal seam. CAES is an example of a large-scale energy storage technology that is designed to transfer off-peak energy from primary power plants to peak demand periods. The Huntorf CAES facility in Germany and the McIntosh CAES facility in Alabama store gas in salt caverns. Off-peak energy is used to pump air underground and compress it in a salt cavern. The compressed air is produced during periods of peak energy demand to drive a turbine and generate additional electrical power.

1.1.2 Oil and Gas Units

Two sets of units are commonly found in the petroleum literature: oil field units and metric units (SI units). Units used in the text are typically oil field units (Table 1.1). The process of converting from one set of units to another is simplified by providing frequently used factors for converting between oil field units and SI (metric) units in Appendix A. The ability to convert between oil field and SI units is an essential skill because both systems of units are frequently used.

TABLE 1.1 Examples of Common Unit Systems

Property	Oil Field	SI (Metric)	British
Length	ft	m	ft
Time	hr	sec	sec
Pressure	psia	Pa	lbf/ft^2
Volumetric flow rate	bbl/day	m^3/s	ft^3/s
Viscosity	cp	Pa·s	$lbf{\cdot}s/ft^2$

1.1.3 Production Performance Ratios

The ratio of one produced fluid phase to another provides useful information for understanding the dynamic behavior of a reservoir. Let q_o, q_w, q_g be oil, water, and gas production rates, respectively. These production rates are used to calculate the following produced fluid ratios:

Gas–oil ratio (GOR)

$$\mathrm{GOR} = \frac{q_g}{q_o} \tag{1.1}$$

Gas–water ratio (GWR)

$$\mathrm{GWR} = \frac{q_g}{q_w} \tag{1.2}$$

Water–oil ratio (WOR)

$$\mathrm{WOR} = \frac{q_w}{q_o} \tag{1.3}$$

One more produced fluid ratio is water cut, which is water production rate divided by the sum of oil and water production rates:

$$\mathrm{WCT} = \frac{q_w}{(q_o + q_w)} \tag{1.4}$$

Water cut (WCT) is a fraction, while WOR can be greater than 1.

Separator GOR is the ratio of gas rate to oil rate. It can be used to indicate fluid type. A separator is a piece of equipment that is used to separate fluid from the well into oil, water, and gas phases. Separator GOR is often expressed as MSCFG/STBO where MSCFG refers to one thousand standard cubic feet of gas and STBO refers to a stock tank barrel of oil. A stock tank is a tank that is used to store produced oil.

Example 1.1 Gas–oil Ratio

A well produces 500 MSCF gas/day and 400 STB oil/day. What is the GOR in MSCFG/STBO?

Answer

$$\mathrm{GOR} = \frac{500\ \mathrm{MSCFG/day}}{400\ \mathrm{STBO/day}} = 1.25\ \mathrm{MSCFG/STBO}$$

1.1.4 Classification of Oil and Gas

Surface temperature and pressure are usually less than reservoir temperature and pressure. Hydrocarbon fluids that exist in a single phase at reservoir temperature and pressure often transition to two phases when they are produced to the surface

TABLE 1.2 Rules of Thumb for Classifying Fluid Types

Fluid Type	Separator GOR (MSCF/STB)	Gravity (°API)	Behavior in Reservoir due to Pressure Decrease
Dry gas	No surface liquids		Remains gas
Wet gas	>50	40–60	Remains gas
Condensate	3.3–50	40–60	Gas with liquid dropout
Volatile oil	2.0–3.3	>40	Liquid with significant gas
Black oil	<2.0	<45	Liquid with some gas
Heavy oil	≈0		Negligible gas formation

Data from Raymond and Leffler (2006).

where the temperature and pressure are lower. There are a variety of terms for describing hydrocarbon fluids at surface conditions. Natural gas is a hydrocarbon mixture in the gaseous state at surface conditions. Crude oil is a hydrocarbon mixture in the liquid state at surface conditions. Heavy oils do not contain much gas in solution at reservoir conditions and have a relatively large molecular weight. By contrast, light oils typically contain a large amount of gas in solution at reservoir conditions and have a relatively small molecular weight.

A summary of hydrocarbon fluid types is given in Table 1.2. API gravity in the table is defined in terms of oil specific gravity as

$$\mathrm{API} = \left(\frac{141.5}{\gamma_o}\right) - 131.5 \tag{1.5}$$

The specific gravity of oil is the ratio of oil density ρ_o to freshwater density ρ_w:

$$\gamma_o = \frac{\rho_o}{\rho_w} \tag{1.6}$$

The API gravity of freshwater is 10°API, which is expressed as 10 degrees API. API denotes American Petroleum Institute.

Example 1.2 API Gravity

The specific gravity of an oil sample is 0.85. What is its API gravity?

Answer

$$\text{API gravity} = \frac{141.5}{\gamma_o} - 131.5 = \frac{141.5}{0.85} - 131.5 = 35°\text{API}$$

Another way to classify hydrocarbon liquids is to compare the properties of the hydrocarbon liquid to water. Two key properties are viscosity and density. Viscosity is a measure of the ability to flow, and density is the amount of material in a given volume.

TABLE 1.3 Classifying Hydrocarbon Liquid Types Using API Gravity and Viscosity

Liquid Type	API Gravity (°API)	Viscosity (cp)
Light oil	>31.1	
Medium oil	22.3–31.1	
Heavy oil	10–22.3	
Water	10	1 cp
Extra heavy oil	4–10	<10000 cp
Bitumen	4–10	>10000 cp

Water viscosity is 1 cp (centipoise) and water density is 1 g/cc (gram per cubic centimeter) at 60°F. A liquid with smaller viscosity than water flows more easily than water. Gas viscosity is much less than water viscosity. Tar, on the other hand, has very high viscosity relative to water.

Table 1.3 shows a hydrocarbon liquid classification scheme using API gravity and viscosity. Water properties are included in the table for comparison. Bitumen is a hydrocarbon mixture with large molecules and high viscosity. Light oil, medium oil, and heavy oil are different types of crude oil and are less dense than water. Extra heavy oil and bitumen are denser than water. In general, crude oil will float on water, while extra heavy oil and bitumen will sink in water.

1.2 LIFE CYCLE OF A RESERVOIR

The life cycle of a reservoir begins when the field becomes an exploration prospect and does not end until the field is properly abandoned. An exploration prospect is a geological structure that may contain hydrocarbons. The exploration stage of the project begins when resources are allocated to identify and assess a prospect for possible development. This stage may require the acquisition and analysis of more data before an exploration well is drilled. Exploratory wells are also referred to as wildcats. They can be used to test a trap that has never produced, test a new reservoir in a known field, and extend the known limits of a producing reservoir. Discovery occurs when an exploration well is drilled and hydrocarbons are encountered.

Figure 1.2 illustrates a typical production profile for an oil field beginning with the discovery well and proceeding to abandonment. Production can begin immediately after the discovery well is drilled or several years later after appraisal and delineation wells have been drilled. Appraisal wells are used to provide more information about reservoir properties and fluid flow. Delineation wells better define reservoir boundaries. In some cases, delineation wells are converted to development wells. Development wells are drilled in the known extent of the field and are used to optimize resource recovery. A buildup period ensues after first oil until a production plateau is reached. The production plateau is usually a consequence of facility limitations such as pipeline capacity. A production decline will eventually occur. Production continues until an economic limit is reached and the field is abandoned.

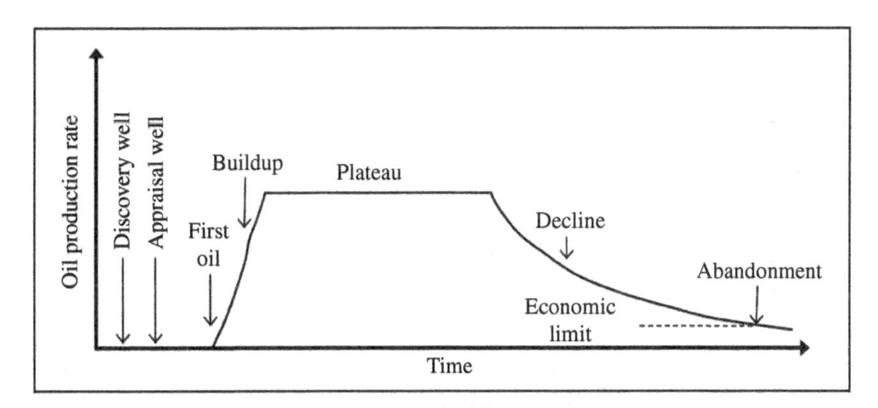

FIGURE 1.2 Typical production profile.

Petroleum engineers provide input to decision makers in management to help determine suitable optimization criteria. The optimization criteria are expected to abide by government regulations. Fields produced over a period of years or decades may be operated using optimization criteria that change during the life of the reservoir. Changes in optimization criteria occur for a variety of reason, including changes in technology, changes in economic factors, and the analysis of new information obtained during earlier stages of production.

Traditionally, production stages were identified by chronological order as primary, secondary, and tertiary production. Primary production is the first stage of production and relies entirely on natural energy sources to drive reservoir fluids to the production well. The reduction of pressure during primary production is often referred to as primary depletion. Oil recovery can be increased in many cases by slowing the decline in pressure. This can be achieved by supplementing natural reservoir energy. The supplemental energy is provided using an external energy source, such as water injection or gas injection. The injection of water or natural gas may be referred to as pressure maintenance or secondary production. Pressure maintenance is often introduced early in the production life of some modern reservoirs. In this case the reservoir is not subjected to a conventional primary production phase.

Historically, primary production was followed by secondary production and then tertiary production (Figure 1.3). Notice that the production plateau shown in Figure 1.2 does not have to appear if all of the production can be handled by surface facilities. Secondary production occurs after primary production and includes the injection of a fluid such as water or gas. The injection of water is referred to as water flooding, while the injection of a gas is called gas flooding. Typical injection gases include methane, carbon dioxide, or nitrogen. Gas flooding is considered a secondary production process if the gas is injected at a pressure that is too low to allow the injected gas to be miscible with the oil phase. A miscible process occurs when the gas injection pressure is high enough that the interface between gas and oil phases disappears. In the miscible case, injected gas mixes with oil and the process is considered an enhanced oil recovery (EOR) process.

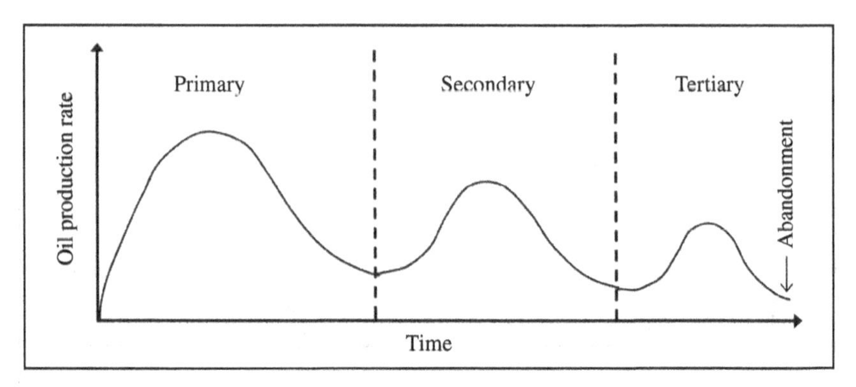

FIGURE 1.3 Sketch of production stages.

EOR processes include miscible, chemical, thermal, and microbial processes. Miscible processes inject gases that can mix with oil at sufficiently high pressures and temperatures. Chemical processes use the injection of chemicals such as polymers and surfactants to increase oil recovery. Thermal processes add heat to the reservoir. This is achieved by injecting heated fluids such as steam or hot water or by the injection of oxygen-containing air into the reservoir and then burning the oil as a combustion process. The additional heat reduces the viscosity of the oil and increases the mobility of the oil. Microbial processes use microbe injection to reduce the size of high molecular weight hydrocarbons and improve oil mobility. EOR processes were originally implemented as a third, or tertiary, production stage that followed secondary production.

EOR processes are designed to improve displacement efficiency by injecting fluids or heat. The analysis of results from laboratory experiments and field applications showed that some fields would perform better if the EOR process was implemented before the third stage in field life. In addition, it was found that EOR processes were often more expensive than just drilling more wells in a denser pattern. The process of increasing the density of wells in an area is known as infill drilling. The term improved oil recovery (IOR) includes EOR and infill drilling for improving the recovery of oil. The addition of wells to a field during infill drilling can also increase the rate of withdrawal of hydrocarbons in a process known as acceleration of production.

Several mechanisms can occur during the production process. For example, production mechanisms that occur during primary production depend on such factors as reservoir structure, pressure, temperature, and fluid type. Production of fluids without injecting other fluids will cause a reduction of reservoir pressure. The reduction in pressure can result in expansion of *in situ* fluids. In some cases, the reduction in pressure is ameliorated if water moves in to replace the produced hydrocarbons. Many reservoirs are in contact with water-bearing formations called aquifers. If the aquifer is much larger than the reservoir and is able to flow into the reservoir with relative ease, the reduction in pressure in the reservoir due to hydrocarbon production will be much less that hydrocarbon production from a reservoir that is not receiving support from an aquifer. The natural forces involved in primary production are called reservoir drives and are discussed in more detail in a later chapter.

Example 1.3 Gas Recovery

The original gas in place (OGIP) of a gas reservoir is 5 trillion ft^3 (TCF). How much gas can be recovered (in TCF) if recovery from analogous fields is between 70 and 90% of OGIP?

Answer

Two estimates are possible: a lower estimate and an upper estimate.
The lower estimate of gas recovery is 0.70×5 TCF = 3.5 TCF.
The upper estimate of gas recovery is 0.90×5 TCF = 4.5 TCF.

1.3 RESERVOIR MANAGEMENT

One definition of reservoir management says that the primary objective of reservoir management is to determine the optimum operating conditions needed to maximize the economic recovery of a subsurface resource. This is achieved by using available resources to accomplish two competing objectives: optimizing recovery from a reservoir while simultaneously minimizing capital investments and operating expenses. As an example, consider the development of an oil reservoir. It is possible to maximize recovery from the reservoir by drilling a large number of wells, but the cost would be excessive. On the other hand, drilling a single well would provide some of the oil but would make it very difficult to recover a significant fraction of the oil in a reasonable time frame. Reservoir management is a process for balancing competing objectives to achieve the key objective.

An alternate definition (Saleri, 2002) says that reservoir management is a continuous process designed to optimize the interaction between data and decision making. Both definitions describe a dynamic process that is intended to integrate information from multiple disciplines to optimize reservoir performance. The process should recognize uncertainty resulting from our inability to completely characterize the reservoir and fluid flow processes. The reservoir management definitions given earlier can be interpreted to cover the management of hydrocarbon reservoirs as well as other reservoir systems. For example, a geothermal reservoir is essentially operated by producing fluid from a geological formation. The management of the geothermal reservoir is a reservoir management task.

It may be necessary to modify a reservoir management plan based on new information obtained during the life of the reservoir. A plan should be flexible enough to accommodate changes in economic, technological, and environmental factors. Furthermore, the plan is expected to address all relevant operating issues, including governmental regulations. Reservoir management plans are developed using input from many disciplines, as we see in later chapters.

1.3.1 Recovery Efficiency

An important objective of reservoir management is to optimize recovery from a resource. The amount of resource recovered relative to the amount of resource originally in place is defined by comparing initial and final *in situ* fluid volumes.

The ratio of fluid volume remaining in the reservoir after production to the fluid volume originally in place is recovery efficiency. Recovery efficiency can be expressed as a fraction or a percentage. An estimate of recovery efficiency is obtained by considering the factors that contribute to the recovery of a subsurface fluid: displacement efficiency and volumetric sweep efficiency.

Displacement efficiency E_D is a measure of the amount of fluid in the system that can be mobilized by a displacement process. For example, water can displace oil in a core. Displacement efficiency is the difference between oil volume at initial conditions and oil volume at final (abandonment) conditions divided by the oil volume at initial conditions:

$$E_D = \frac{(S_{oi}/B_{oi}) - (S_{oa}/B_{oa})}{S_{oi}/B_{oi}} \tag{1.7}$$

where S_{oi} is initial oil saturation and S_{oa} is oil saturation at abandonment. Oil saturation is the fraction of oil occupying the volume in a pore space. Abandonment refers to the time when the process is completed. Formation volume factor (FVF) is the volume occupied by a fluid at reservoir conditions divided by the volume occupied by the fluid at standard conditions. The terms B_{oi} and B_{oa} refer to FVF initially and at abandonment, respectively.

Example 1.4 Formation Volume Factor

Suppose oil occupies 1 bbl at stock tank (surface) conditions and 1.4 bbl at reservoir conditions. The oil volume at reservoir conditions is larger because gas is dissolved in the liquid oil. What is the FVF of the oil?

Answer

$$\text{Oil FVF} = \frac{\text{vol at reservoir conditions}}{\text{vol at surface conditions}}$$

$$\text{Oil FVF} = \frac{1.4\,\text{RB}}{1.0\,\text{STB}} = 1.4\,\text{RB/STB}$$

Volumetric sweep efficiency E_{Vol} expresses the efficiency of fluid recovery from a reservoir volume. It can be written as the product of areal sweep efficiency and vertical sweep efficiency:

$$E_{Vol} = E_A \times E_V \tag{1.8}$$

Areal sweep efficiency E_A and vertical sweep efficiency E_V represent the efficiencies associated with the displacement of one fluid by another in the areal plane and vertical dimension. They represent the contact between *in situ* and injected fluids. Areal sweep efficiency is defined as

$$E_A = \frac{\text{swept area}}{\text{total area}} \tag{1.9}$$

and vertical sweep efficiency is defined as

$$E_V = \frac{\text{swept net thickness}}{\text{total net thickness}} \quad (1.10)$$

Recovery efficiency RE is the product of displacement efficiency and volumetric sweep efficiency:

$$\text{RE} = E_D \times E_{Vol} = E_D \times E_A \times E_V \quad (1.11)$$

Displacement efficiency, areal sweep efficiency, vertical sweep efficiency, and recovery efficiency are fractions that vary from 0 to 1. Each of the efficiencies that contribute to recovery efficiency can be relatively large and still yield a recovery efficiency that is relatively small. Reservoir management often focuses on finding the efficiency factor that can be improved by the application of technology.

Example 1.5 Recovery Efficiency

Calculate volumetric sweep efficiency E_{Vol} and recovery efficiency RE from the following data:

S_{oi}	0.75
S_{oa}	0.30
Area swept	750 acres
Total area	1000 acres
Thickness swept	10 ft
Total thickness	15 ft
Neglect FVF effects since $B_{oi} \approx B_{oa}$	

Answer

Displacement efficiency: $$E_D = \frac{(S_{oi}/B_{oi}) - (S_{oa}/B_{oa})}{S_{oi}/B_{oi}} \approx \frac{S_{oi} - S_{oa}}{S_{oi}} = 0.6$$

Areal sweep efficiency: $$E_A = \frac{\text{swept area}}{\text{total area}} = 0.75$$

Vertical sweep efficiency: $$E_V = \frac{\text{swept net thickness}}{\text{total net thickness}} = 0.667$$

Volumetric sweep efficiency: $$E_{vol} = E_A \times E_V = 0.5$$

Recovery efficiency: $$\text{RE} = E_D \times E_{Vol} = 0.3$$

1.4 PETROLEUM ECONOMICS

The decision to develop a petroleum reservoir is a business decision that requires an analysis of project economics. A prediction of cash flow from a project is obtained by combining a prediction of fluid production volume with a forecast of fluid price.

Production volume is predicted using engineering calculations, while fluid price estimates are obtained using economic models. The calculation of cash flow for different scenarios can be used to compare the economic value of competing reservoir development concepts.

Cash flow is an example of an economic measure of investment worth. Economic measures have several characteristics. An economic measure should be consistent with the goals of the organization. It should be easy to understand and apply so that it can be used for cost-effective decision making. Economic measures that can be quantified permit alternatives to be compared and ranked.

Net present value (NPV) is an economic measure that is typically used to evaluate cash flow associated with reservoir performance. NPV is the difference between the present value of revenue R and the present value of expenses E:

$$\text{NPV} = R - E \tag{1.12}$$

The time value of money is incorporated into NPV using discount rate r. The value of money is adjusted to the value associated with a base year using discount rate. Cash flow calculated using a discount rate is called discounted cash flow. As an example, NPV for an oil and/or gas reservoir may be calculated for a specified discount rate by taking the difference between revenue and expenses (Fanchi, 2010):

$$\begin{aligned} \text{NPV} &= \sum_{n=1}^{N} \frac{P_{on} q_{on} + P_{gn} q_{gn}}{(1+r)^n} - \sum_{n=1}^{N} \frac{\text{CAPEX}_n + \text{OPEX}_n + \text{TAX}_n}{(1+r)^n} \\ &= \sum_{n=1}^{N} \frac{P_{on} q_{on} + P_{gn} q_{gn} - \text{CAPEX}_n - \text{OPEX}_n - \text{TAX}_n}{(1+r)^n} \end{aligned} \tag{1.13}$$

where N is the number of years, P_{on} is oil price during year n, q_{on} is oil production during year n, P_{gn} is gas price during year n, q_{gn} is gas production during year n, CAPEX_n is capital expenses during year n, OPEX_n is operating expenses during year n, TAX_n is taxes during year n, and r is discount rate.

The NPV for a particular case is the value of the cash flow at a specified discount rate. The discount rate at which the maximum NPV is zero is called the discounted cash flow return on investment (DCFROI) or internal rate of return (IRR). DCFROI is useful for comparing different projects.

Figure 1.4 shows a typical plot of NPV as a function of time. The early time part of the figure shows a negative NPV and indicates that the project is operating at a loss. The loss is usually associated with initial capital investments and operating expenses that are incurred before the project begins to generate revenue. The reduction in loss and eventual growth in positive NPV are due to the generation of revenue in excess of expenses. The point in time on the graph where the NPV is zero after the project has begun is the discounted payout time. Discounted payout time on Figure 1.4 is approximately 2.5 years.

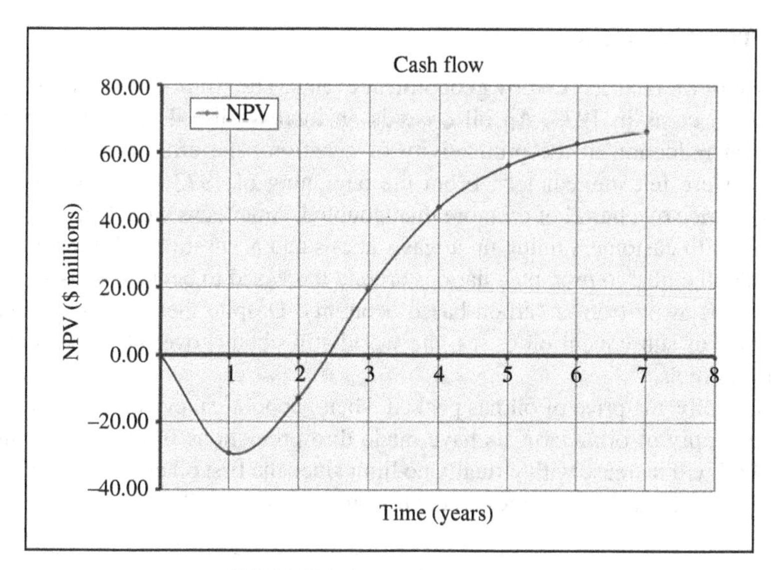

FIGURE 1.4 Typical cash flow.

TABLE 1.4 Definitions of Selected Economic Measures

Economic Measure	Definition
Discount rate	Factor to adjust the value of money to a base year
Net present value (NPV)	Value of cash flow at a specified discount rate
Discounted payout time	Time when NPV = 0
DCFROI or IRR	Discount rate at which maximum NPV = 0
Profit-to-investment (PI) ratio	Undiscounted cash flow without capital investment divided by total investment

Table 1.4 presents the definitions of several commonly used economic measures. DCFROI and discounted payout time are measures of the economic viability of a project. Another measure is the profit-to-investment (PI) ratio which is a measure of profitability. It is defined as the total undiscounted cash flow without capital investment divided by total investment. Unlike the DCFROI, the PI ratio does not take into account the time value of money. Useful plots include a plot of NPV versus time and a plot of NPV versus discount rate.

Production volumes and price forecasts are needed in the NPV calculation. The input data used to prepare forecasts includes data that is not well known. Other possible sources of error exist. For example, the forecast calculation may not adequately represent the behavior of the system throughout the duration of the forecast, or a geopolitical event could change global economics. It is possible to quantify uncertainty by making reasonable changes to input data used to calculate forecasts so that a range of NPV results is provided. This process is illustrated in the discussion of decline curve analysis in a later chapter.

1.4.1 The Price of Oil

The price of oil is influenced by geopolitical events. The Arab–Israeli war triggered the first oil crisis in 1973. An oil crisis is an increase in oil price that causes a significant reduction in the productivity of a nation. The effects of the Arab oil embargo were felt immediately. From the beginning of 1973 to the beginning of 1974, the price of a barrel of oil more than doubled. Americans were forced to ration gasoline, with customers lining up at gas stations and accusations of price gouging. The Arab oil embargo prompted nations around the world to begin seriously considering a shift away from a carbon-based economy. Despite these concerns and the occurrence of subsequent oil crises, the world still obtains over 80% of its energy from fossil fuels.

Historically, the price of oil has peaked when geopolitical events threaten or disrupt the supply of oil. Alarmists have made dire predictions in the media that the price of oil will increase with virtually no limit since the first oil crisis in 1973. These predictions neglect market forces that constrain the price of oil and other fossil fuels.

Example 1.6 Oil Security

A. If $100 billion is spent on the military in a year to protect the delivery of 20 million barrels of oil per day to the global market, how much does the military budget add to the cost of a barrel of oil?

Answer

$$\text{Total oil per year} = (20 \text{ million bbl/day}) \times (365 \text{ days/yr}) = 7.3 \text{ billion bbl/yr}$$

$$\text{Cost of military/bbl} = \frac{\$100 \text{ billion/yr}}{7.3 \text{ billion bbl/yr}} = \$13.70/\text{bbl}$$

B. How much is this cost per gallon?

Answer

$$\text{Cost/gal} = (\$13.70/\text{bbl}) \times (1 \text{ bbl}/42 \text{ gal}) = \$0.33/\text{gal}$$

1.4.2 How Does Oil Price Affect Oil Recovery?

Many experts believe we are running out of oil because it is becoming increasingly difficult to discover new reservoirs that contain large volumes of conventional oil and gas. Much of the exploration effort is focusing on less hospitable climates, such as arctic conditions in Siberia and deepwater offshore regions near West Africa. Yet we already know where large volumes of oil remain: in the reservoirs that have already been discovered and developed. Current development techniques have recovered approximately one third of the oil in known fields. That means roughly two thirds remains in the ground where it was originally found.

TABLE 1.5 Sensitivity of Oil Recovery Technology to Oil Price

	Oil Price Range	
Oil Recovery Technology	1997$/bbl	2016$/bbl 5% Inflation
Conventional	15–25	38–63
Enhanced oil recovery (EOR)	20–40	51–101
Extra heavy oil (e.g., tar sands)	25–45	63–114
Alternative energy sources	40–60	101–152

The efficiency of oil recovery depends on cost. Companies can produce much more oil from existing reservoirs if they are willing to pay for it and if the market will support that cost. Most oil-producing companies choose to seek and produce less expensive oil so they can compete in the international marketplace. Table 1.5 illustrates the sensitivity of oil-producing techniques to the price of oil. Oil prices in the table include prices in the original 1997 prices and inflation adjusted prices to 2016. The actual inflation rate for oil prices depends on a number of factors, such as size and availability of supply and demand.

Table 1.5 shows that more sophisticated technologies can be justified as the price of oil increases. It also includes a price estimate for alternative energy sources, such as wind and solar. Technological advances are helping wind and solar energy become economically competitive with oil and gas as energy sources for generating electricity. In some cases there is overlap between one technology and another. For example, steam flooding is an EOR process that can compete with conventional oil recovery techniques such as water flooding, while chemical flooding is one of the most expensive EOR processes.

1.4.3 How High Can Oil Prices Go?

In addition to relating recovery technology to oil price, Table 1.5 contains another important point: the price of oil will not rise without limit. For the data given in the table, we see that alternative energy sources become cost competitive when the price of oil rises above 2016$101 per barrel. If the price of oil stays at 2016$101 per barrel or higher for an extended period of time, energy consumers will begin to switch to less expensive energy sources. This switch is known as product substitution. The impact of price on consumer behavior is illustrated by consumers in European countries that pay much more for gasoline than consumers in the United States. Countries such as Denmark, Germany, and Holland are rapidly developing wind energy as a substitute to fossil fuels for generating electricity.

Historically, we have seen oil-exporting countries try to maximize their income and minimize competition from alternative energy and expensive oil recovery technologies by supplying just enough oil to keep the price below the price needed to justify product substitution. Saudi Arabia has used an increase in the supply of oil to drive down the cost of oil. This creates problems for organizations that are trying to develop more costly sources of oil, such as shale oil in the United States. It also creates problems for oil-exporting nations that are relying on a relatively high oil price to fund their government spending.

Oil-importing countries can attempt to minimize their dependence on imported oil by developing technologies that reduce the cost of alternative energy. If an oil-importing country contains mature oil reservoirs, the development of relatively inexpensive technologies for producing oil remaining in mature reservoirs or the imposition of economic incentives to encourage domestic oil production can be used to reduce the country's dependence on imported oil.

1.5 PETROLEUM AND THE ENVIRONMENT

Fossil fuels—coal, oil, and natural gas—can harm the environment when they are consumed. Surface mining of coal scars the environment until the land is reclaimed. Oil pollutes everything it touches when it is spilled on land or at sea. Pictures of wildlife covered in oil or natural gas appearing in drinking water have added to the public perception of oil and gas as "dirty" energy sources. The combustion of fossil fuels yields environmentally undesirable by-products. It is tempting to conclude that fossil fuels have always harmed the environment. However, if we look at the history of energy consumption, we see that fossil fuels have a history of helping protect the environment when they were first adopted by society as a major energy source.

Wood was the fuel of choice for most of human history and is still a significant contributor to the global energy portfolio. The growth in demand for wood energy associated with increasing population and technological advancements such as the development of the steam engine raised concerns about deforestation and led to a search for new source of fuel. The discovery of coal, a rock that burned, reduced the demand for wood and helped save the forests.

Coal combustion was used as the primary energy source in industrialized societies prior to 1850. Another fuel, whale oil, was used as an illuminant and joined coal as part of the nineteenth-century energy portfolio. Demand for whale oil motivated the harvesting of whales for their oil and was leading to the extinction of whales. The discovery that rock oil, what we now call crude oil, could also be used as an illuminant provided a product that could be substituted for whale oil if there was enough rock oil to meet growing demand. In 1861, the magazine *Vanity Fair* published a cartoon showing whales at a Grand Ball celebrating the production of oil in Pennsylvania. Improvements in drilling technology and the discovery of oil fields that could provide large volumes of oil at high flow rates made oil less expensive than coal and whale oil. From an environmental perspective, the substitution of rock oil for whale oil saved the whales in the latter half of the nineteenth century. Today, concern about the harmful environmental effects of fossil fuels, especially coal and oil, is motivating a transition to more beneficial sources of energy. The basis for this concern is considered next.

1.5.1 Anthropogenic Climate Change

One environmental concern facing society today is anthropogenic climate change. When a carbon-based fuel burns in air, carbon reacts with oxygen and nitrogen in the air to produce carbon dioxide (CO_2), carbon monoxide, and nitrogen oxides

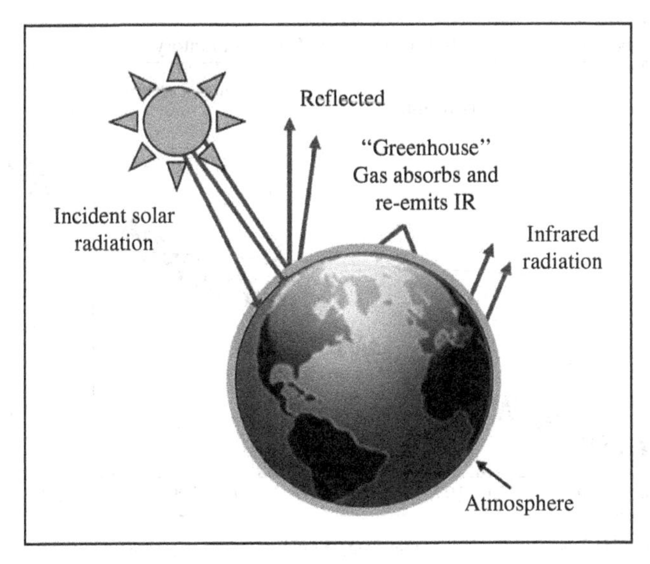

FIGURE 1.5 The greenhouse effect. (Source: Fanchi (2004). Reproduced with permission of Elsevier Academic Press.)

(often abbreviated as NOx). The by-products of unconfined combustion, including water vapor, are emitted into the atmosphere in gaseous form.

Some gaseous combustion by-products are called greenhouse gases because they absorb heat energy. Greenhouse gases include water vapor, carbon dioxide, methane, and nitrous oxide. Greenhouse gas molecules can absorb infrared light. When a greenhouse gas molecule in the atmosphere absorbs infrared light, the energy of the absorbed photon of light is transformed into the kinetic energy of the gas molecule. The associated increase in atmospheric temperature is the greenhouse effect illustrated in Figure 1.5.

Much of the solar energy arriving at the top of the atmosphere does not pass through the atmosphere to the surface of the Earth. A study of the distribution of light energy arriving at the surface of the Earth shows that energy from the sun at certain frequencies (or, equivalently, wavelengths) is absorbed in the atmosphere. Several of the gaps are associated with light absorption by a greenhouse gas molecule.

One way to measure the concentration of greenhouse gases is to measure the concentration of a particular greenhouse gas. Charles David Keeling began measuring atmospheric carbon dioxide concentration at the Mauna Loa Observatory on the Big Island of Hawaii in 1958. Keeling observed a steady increase in carbon dioxide concentration since he began his measurements. His curve, which is now known as the Keeling curve, is shown in Figure 1.6. It exhibits an annual cycle in carbon dioxide concentration overlaying an increasing average. The initial carbon dioxide concentration was measured at a little over 310 parts per million. Today it is approximately 400 parts per million. These measurements show that carbon dioxide concentration in the atmosphere has been increasing since the middle of the twentieth century.

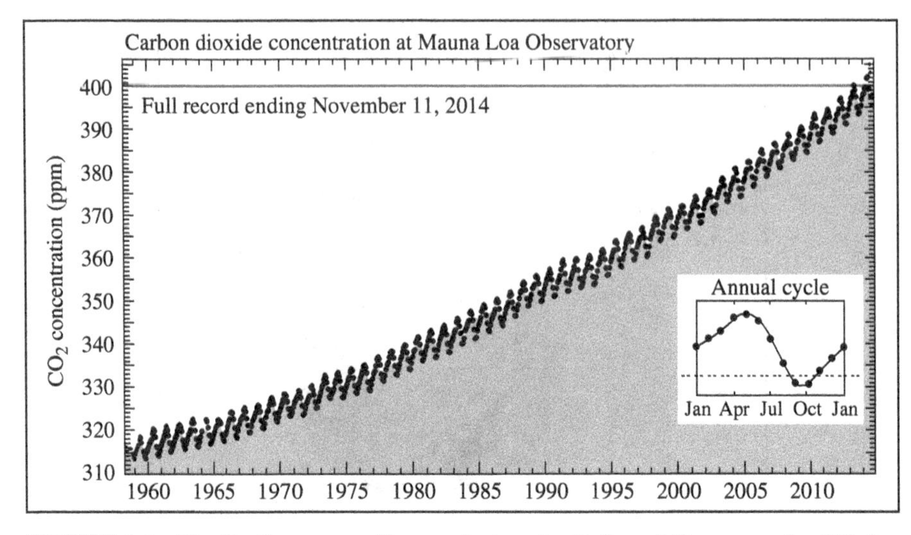

FIGURE 1.6 The Keeling curve. (Source: Scripps Institution of Oceanography, UC San Diego, https://scripps.ucsd.edu/programs/keelingcurve/wp-content/plugins/sio-bluemoon/graphs/mlo_full_record.png)

Samples of air bubbles captured in ice cores extracted from glacial ice in Vostok, Antarctica, are used to measure the concentration of gases in the past. Measurements show that CO_2 concentration has varied from 150 to 300 ppm for the past 400 000 years. Measurements of atmospheric CO_2 concentration during the past two centuries show that CO_2 concentration is greater than 300 ppm and continuing to increase. Ice core measurements show a correlation between changes in atmospheric temperature and CO_2 concentration.

Wigley et al. (1996) projected ambient CO_2 concentration through the twenty-first century. They argued that society would have to reduce the rate that greenhouse gases are being emitted into the atmosphere to keep atmospheric concentration beneath 550 ppm, which is the concentration of CO_2 that would establish an acceptable energy balance. Some scientists have argued that optimum CO_2 concentration is debatable since higher concentrations of carbon dioxide can facilitate plant growth.

People who believe that climate change is due to human activity argue that combustion of fossil fuels is a major source of CO_2 in the atmosphere. Skeptics point out that the impact of human activity on climate is not well established. For example, they point out that global climate model forecasts are not reliable because they do not adequately model all of the mechanisms that affect climate behavior. Everyone agrees that climate does change over the short term. Examples of short-term climate change are seasonal weather variations and storms. We refer to long-term climate change associated with human activity as anthropogenic climate change to distinguish it from short-term climate change.

Evidence that human activity is causing climate to change more than it would naturally change has motivated international attempts by proponents of anthropogenic climate change to regulate greenhouse gas emissions and transition as quickly as possible from fossil fuels to energy sources such as wind and solar. Skeptics typically argue that reducing our dependence on fossil fuels is important, but they believe that the transition should occur over a period of time that does not significantly harm the global economy. One method for reducing the emission of CO_2 into the atmosphere is to collect and store carbon dioxide in geologic formations in a process known as CO_2 sequestration. Recent research has suggested that large-scale sequestration of greenhouse gases could alter subsurface stress to cause fault slippage and seismic activity at the surface.

1.5.2 Environmental Issues

Fossil fuel producers should be good stewards of the Earth. From a personal perspective, they share the environment with everyone else. From a business perspective, failure to protect the environment can lead to lawsuits, fines, and additional regulation. There are many examples of society imposing penalties on operators for behavior that could harm the environment or already harmed the environment. A few examples are discussed here.

Shell UK reached an agreement with the British government in 1995 to dispose an oil storage platform called the Brent Spar in the deep waters of the Atlantic. The environmental protection group Greenpeace and its allies were concerned that oil left in the platform would leak into the Atlantic. Greenpeace challenged the Shell UK plan by occupying the platform and supporting demonstrations that, in some cases, became violent. Shell UK abandoned the plan to sink the Brent Spar in the Atlantic and instead used the structure as a ferry quay. As a consequence of this incident, governments throughout Europe changed their rules regulating disposal of offshore facilities (Wilkinson, 1997; Offshore Staff, 1998).

Another example is shale oil and gas development in populated areas. Shale oil and gas development requires implementation of a technique known as hydraulic fracturing. The only way to obtain economic flow rates of oil and gas from shale is to fracture the rock. The fractures provide flow paths from the shale to the well. Hydraulic fracturing requires the injection of large volumes of water at pressures that are large enough to break the shale. The injected water carries chemicals and small solid objects called proppants that are used to prop open fractures when the fracturing process is completed, and the well is converted from an injection well operating at high pressure to a production well operating at much lower pressure.

Some environmental issues associated with hydraulic fracturing include meeting the demand for water to conduct hydraulic fracture treatments and disposing produced water containing pollutants. One solution is to recycle the water. Another solution is to inject the produced water in disposal wells. Both the fracture process and the water disposal process can result in vibrations in the Earth that can be measured as seismic events. The fracture process takes place near the depth of the shale and is typically a very low magnitude seismic event known as a microseismic event. Water injection into disposal wells can lead to seismic events, and possibly earthquakes that can be felt at the surface, in a process known as injection-induced seismicity (Rubinstein and Mahani, 2015; Weingarten et al., 2015). King (2012) has provided an extensive review of hydraulic fracturing issues associated with oil and gas production from shale. Concern about environmental effects has led some city, county, and state governments in the United States to more closely regulate shale drilling and production.

Oil spills in marine environments can require expensive cleanup operations. Two such oil spills were the grounding of the 1989 Exxon Valdez oil tanker in Alaska and the 2010 explosion and sinking of the BP Deepwater Horizon offshore platform in the Gulf of Mexico. Both incidents led to significant financial penalties, including remediation costs, for the companies involved. In the case of the BP Deepwater Horizon incident, 11 people lost their lives. The Exxon Valdez spill helped motivate the passage of US government regulations requiring the use of double-hulled tankers.

Example 1.7 Environmental Cost

A. A project is expected to recover 500 million STB of oil. The project will require installing an infrastructure (e.g., platforms, pipelines, etc.) that costs $1.8 billion and another $2 billion in expenses (e.g., royalties, taxes, operating costs). Breakeven occurs when revenue = expenses. Neglecting the time value of money, what price of oil (in $/STB) is needed to achieve breakeven? STB refers to stock tank barrel.

Answer

Total expenses = $3.8 billion

Oil price = $3.8 billion/0.5 billion STB = $7.6/STB

B. Suppose an unexpected environmental disaster occurs that adds another $20 billion to project cost. Neglecting the time value of money, what price of oil (in $/STB) is needed to achieve breakeven?

Answer

Total expenses = $23.8 billion

Oil price = $23.8 billion/0.5 billion STB = $47.6/STB

1.6 ACTIVITIES

1.6.1 Further Reading

For more information about petroleum in society, see Fanchi and Fanchi (2016), Hyne (2012), Satter et al. (2008), Raymond and Leffler (2006), and Yergin (1992). For more information about reservoir management and petroleum economics, see Hyne (2012), Fanchi (2010), Satter et al. (2008), and Raymond and Leffler (2006).

1.6.2 True/False

1.1 A hydrocarbon reservoir must be able to trap and retain fluids.

1.2 API gravity is the weight of a hydrocarbon mixture.

1.3 Separator GOR is the ratio of gas rate to oil rate.

1.4 The first stage in the life of an oil or gas reservoir is exploration.

1.5 Volumetric sweep efficiency is the product of areal sweep efficiency and displacement efficiency.

1.6 Net present value is usually negative at the beginning of a project.

1.7 DCFROI is discounted cash flow return on interest.

1.8 Nitrogen is a greenhouse gas.

1.9 Water flooding is an EOR process.

1.10 Geological sequestration of carbon dioxide in an aquifer is an EOR process.

1.6.3 Exercises

1.1 Suppose the density of oil is 48 lb/ft^3 and the density of water is 62.4 lb/ft^3. Calculate the specific gravity of oil γ_o and its API gravity.

1.2 Estimate recovery efficiency when displacement efficiency is 30%, areal sweep efficiency is 65%, and vertical sweep efficiency is 70%.

1.3 Calculate volumetric sweep efficiency E_{Vol} and recovery efficiency RE from the following data where displacement efficiency can be estimated as $E_D = (S_{oi} - S_{or})/S_{oi}$.

Initial oil saturation S_{oi}	0.75
Residual oil saturation S_{or}	0.30
Area swept	480 acres
Total area	640 acres
Thickness swept	80 ft
Total thickness	100 ft

1.4 **A.** If the initial oil saturation of an oil reservoir is $S_{oi} = 0.70$ and the residual oil saturation from water flooding a core sample in the laboratory is $S_{or} = 0.30$, calculate the displacement efficiency E_D assuming displacement efficiency can be estimated as $E_D = (S_{oi} - S_{or})/S_{oi}$.

B. In actual floods, the residual oil saturation measured in the laboratory is seldom achieved. Suppose $S_{or} = 0.35$ in the field, and recalculate displacement efficiency. Compare displacement efficiencies.

1.5 **A.** A project is expected to recover 200 million STB of oil. The project will require installing an infrastructure (e.g., platforms, pipelines, etc.) that costs \$1.2 billion and another \$0.8 billion in expenses (e.g., royalties, taxes, operating costs). Breakeven occurs when revenue = expenses. Neglecting the time value of money, what price of oil (in \$/STB) is needed to achieve breakeven?

B. Suppose a fire on the platform adds another \$0.5 billion to project cost. Neglecting the time value of money, what price of oil (in \$/STB) is needed to achieve breakeven?

1.6 **A.** The water cut of an oil well that produces 1000 STB oil per day is 25%. What is the water production rate for the well? Express your answer in STB water per day.

B. What is the WOR?

1.7 **A.** Fluid production from a well passes through a separator at the rate of 1200 MSCF gas per day and 1000 STB oil per day. What is the separator GOR in MSCF/STB?

B. Based on this information, would you classify the fluid as black oil or volatile oil?

1.8 **A.** How many acres are in 0.5 mi^2?

B. If one gas well can drain 160 acres, how many gas wells are needed to drain 1 mi^2?

1.9 **A.** A wellbore has a total depth of 10000 ft. If it is full of water with a pressure gradient of 0.433 psia/ft, what is the pressure at the bottom of the wellbore?

B. The pressure in a column of water is 1000 psia at a depth of 2300 ft. What is the pressure at a shallower depth of 2200 ft.? Assume the pressure gradient of water is 0.433 psia/ft. Express your answer in psia.

1.10 **A.** Primary recovery from an oil reservoir was 100 MMSTBO where 1 MMSTBO = 1 million STB of oil. A water flood was implemented following primary recovery. Incremental recovery from the water flood was 25% of original oil in place (OOIP). Total recovery (primary recovery plus recovery from water flooding) was 50% of OOIP. How much oil (in MMSTBO) was recovered by the water flood?

B. What was the OOIP (in MMSTBO)?

1.11 **A.** A core contains 25% water saturation and 75% oil saturation before it is flooded. Core floods show that the injection of water into the core leaves a residual oil saturation of 25%. If the same core is resaturated with oil and then flooded with carbon dioxide, the residual oil saturation is 10%. What is the displacement efficiency of the water flood? Assume displacement efficiency can be estimated as $E_D = (S_{oi} - S_{or})/S_{oi}$.

B. What is the displacement efficiency of the carbon dioxide flood?

1.12 The revenue from gas produced by a well is \$6 million per year. The gas drains an area of 640 acres. Suppose you have 1 acre in the drainage area and are entitled to 25% of the revenue for your fraction of the drainage area, which is 1 acre/640 acres. How much revenue from the gas well is yours? Express your answer in \$/yr.

2

THE FUTURE OF ENERGY

The global energy mix is in a period of transition from fossil fuels to more sustainable energy sources. Recognition that oil and gas are nonrenewable resources, increasing demand for energy, concerns about the security of oil and gas supply, and possibility of anthropogenic climate change are among the factors that are motivating changes to the global energy mix. In this chapter, we describe the global distribution of oil and gas production and consumption, introduce M. King Hubbert's concept of peak oil, and discuss the role oil and gas will play in the future energy mix.

2.1 GLOBAL OIL AND GAS PRODUCTION AND CONSUMPTION

The global distribution of oil and gas production and consumption is illustrated by presenting the leading nations in production and consumption categories. Lists of top producing and consuming nations change from year to year. For example, Figure 2.1 shows the five countries with the largest production of oil in 2014. The United States was the top producer in the 1980s, while production in Saudi Arabia was relatively low. By the 1990s, Saudi Arabia replaced the United States as the top producing country. The development of techniques for economically producing hydrocarbons from shale, which is rock with very low permeability, made it possible for the United States to become the top producing country in the 2010s.

Introduction to Petroleum Engineering, First Edition. John R. Fanchi and Richard L. Christiansen.
© 2017 John Wiley & Sons, Inc. Published 2017 by John Wiley & Sons, Inc.
Companion website: www.wiley.com/go/Fanchi/IntroPetroleumEngineering

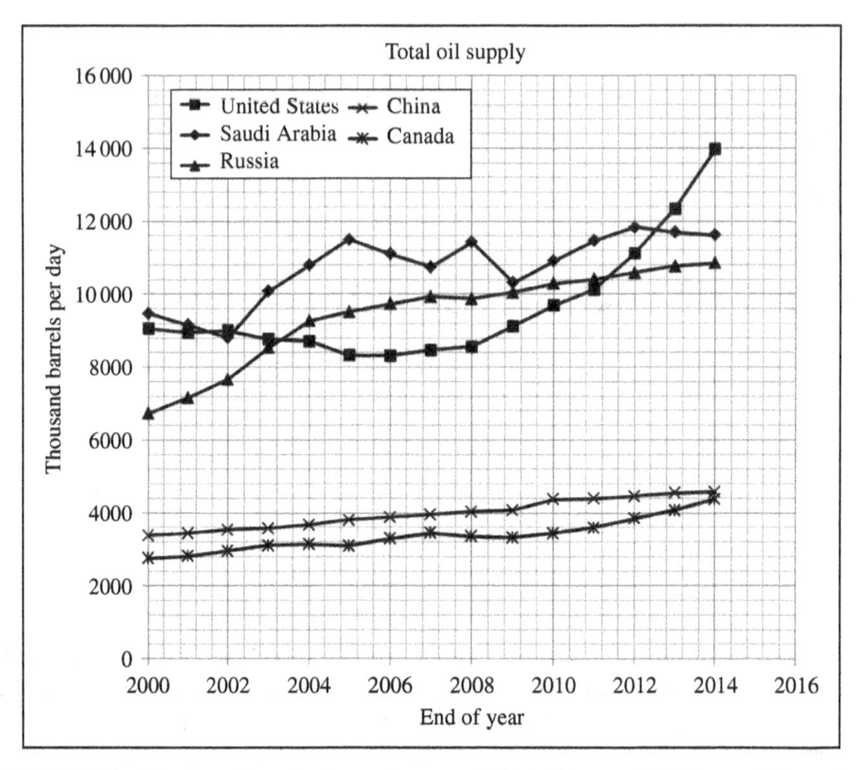

FIGURE 2.1 Top five oil-producing nations as of 2014. (Source: U.S. Energy Information Administration Petroleum (2015).)

Figure 2.2 presents the five countries with the largest consumption of petroleum in 2014. The United States is the top consuming nation, followed by China, Japan, India, and Russia. We can see if a country is a net importer or exporter of oil and gas by comparing production and consumption in a particular country. The United States is a net oil-importing nation, while Saudi Arabia is a net oil-exporting nation.

Figure 2.3 shows the five countries with the largest production of dry natural gas in 2014. The discovery of drilling and completion methods capable of producing natural gas from very low-permeability rock such as tight sandstone and shale has helped the United States increase its production of natural gas.

Figure 2.4 presents the five countries with the largest consumption of dry natural gas in 2014. The United States is the leading consumer of natural gas. The global demand for natural gas is expected to increase as countries like the United States replace coal-fired power plants with power plants that burn cleaner, dry natural gas.

2.2 RESOURCES AND RESERVES

The distribution of a resource can be displayed using the resource triangle illustrated in Figure 2.5 (Masters, 1979). Masters suggested that the distribution of a natural resource can be represented by a triangle with high-quality deposits at the top and

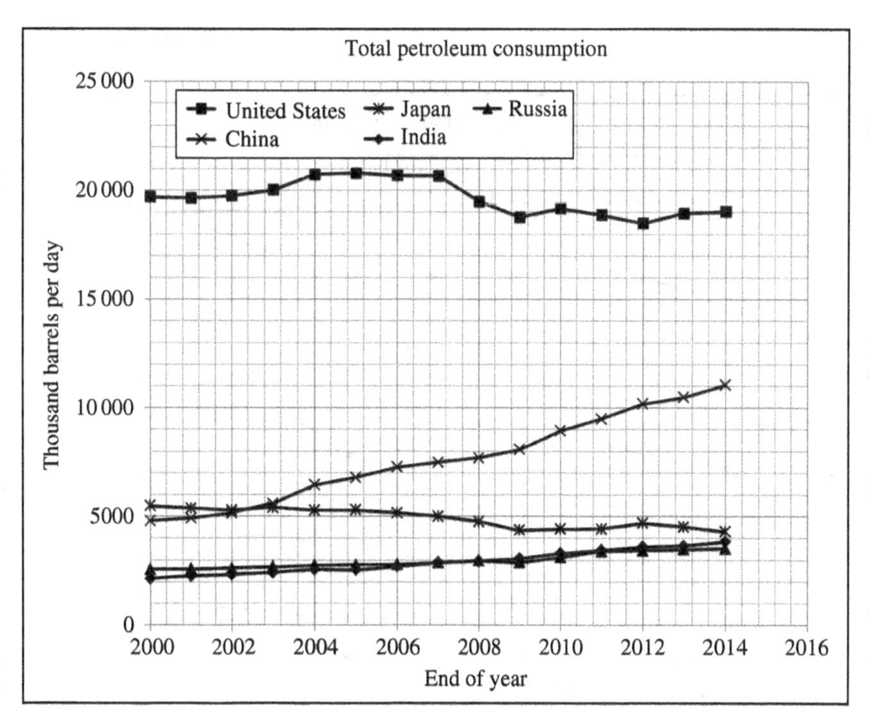

FIGURE 2.2 Top five oil-consuming nations as of 2014. (Source: U.S. Energy Information Administration Petroleum (2015).)

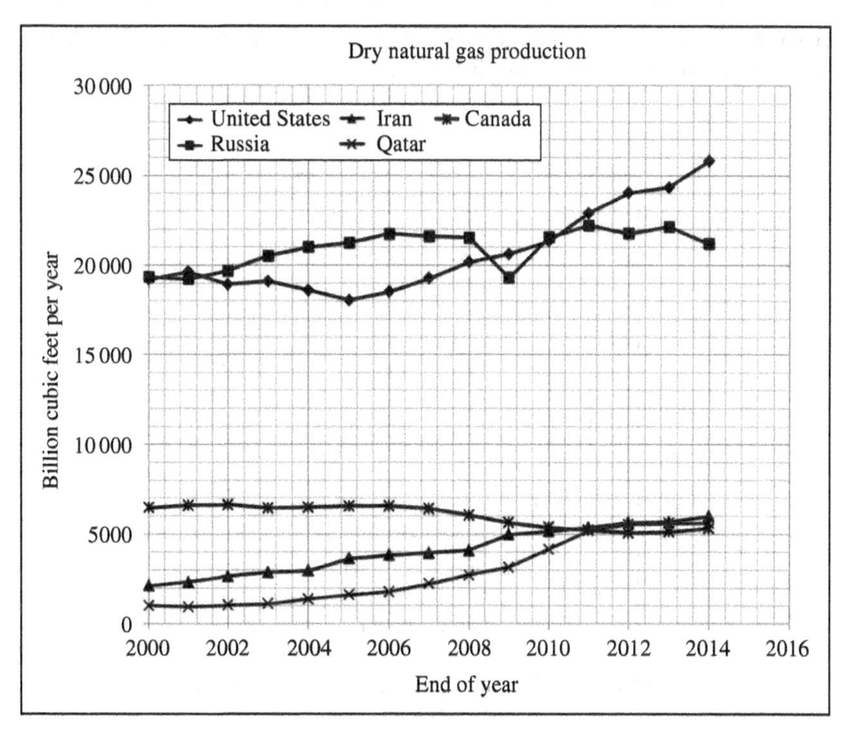

FIGURE 2.3 Top five dry natural gas-producing nations as of 2014. (Source: U.S. Energy Information Administration Petroleum (2015); BP, 2015.)

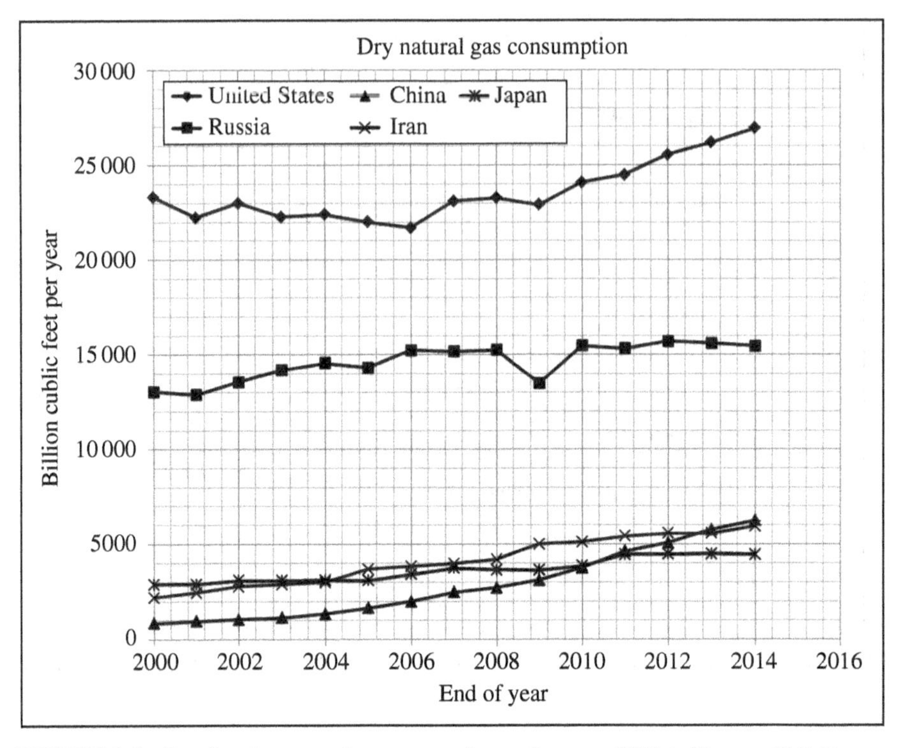

FIGURE 2.4 Top five dry natural gas-consuming nations as of 2014. (Source: U.S. Energy Information Administration Petroleum (2015); BP, 2015.)

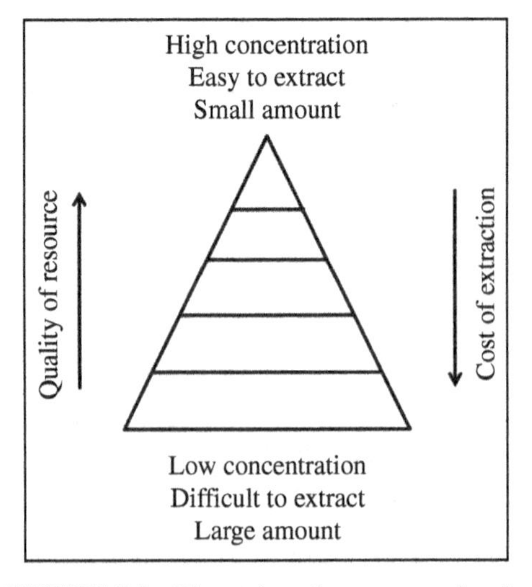

FIGURE 2.5 Illustration of a resource triangle.

lower-quality deposits at the lower part of the triangle. A resource deposit at the top of the triangle has a relatively large concentration of resource that is relatively inexpensive to extract. A resource deposit at the lower part of the triangle has a lower concentration of resource, so the extraction is more difficult or expensive. For example, a large, shallow, light oil reservoir that is onshore would be at the top of the triangle, while a small, deep, heavy oil reservoir in an offshore environment would be at the base of the triangle. The amount of resource that is both easy to extract and has a high concentration is expected to be small, while the amount of resource in a low-concentration deposit is difficult or expensive to extract and is expected to be large.

2.2.1 Reserves

Resource size tells us how much of a resource is present in a deposit. The amount of the resource that can be extracted is discussed in terms of reserves. The definition of reserves is presented in the Petroleum Reservoir Management System maintained by the Society of Petroleum Engineers (SPE-PRMS, 2011). Reserves classifications are summarized in Table 2.1.

The probability distribution associated with the SPE-PRMS reserves definitions recognizes that there is a statistical distribution of resource deposits in nature. For example, oil reservoirs vary in size from reservoirs with relatively small volumes of oil to reservoirs containing relatively large volumes of oil. If we plot the volume of oil in the reservoir versus the number of reservoirs with that volume, we can develop a distribution of reservoirs as a function of size. This distribution can be represented by a frequency distribution and can be interpreted as a probability distribution of reservoir size. If we combine reservoir size and recovery factor, we can obtain a probability distribution of reserves. As an illustration, suppose we assume the probability distribution of reserves is a normal distribution. Normal distributions

TABLE 2.1 SPE-PRMS Reserves Definitions

Proved reserves	Those quantities of petroleum, which by analysis of geoscience and engineering data, can be estimated with reasonable certainty to be commercially recoverable, from a given date forward, from known reservoirs, and under defined economic conditions, operating methods, and government regulations There should be at least a 90% probability (P_{90}) that the quantities actually recovered will equal or exceed the low estimate
Probable reserves	Those additional reserves which analysis of geoscience and engineering data indicate are less likely to be recovered than proved reserves but more certain to be recovered than possible reserves There should be at least a 50% probability (P_{50}) that the quantities actually recovered will equal or exceed the best estimate
Possible reserves	Those additional reserves which analysis of geoscience and engineering data suggests are less likely to be recoverable than probable reserves There should be at least a 10% probability (P_{10}) that the quantities actually recovered will equal or exceed the high estimate

are characterized by the mean μ and standard deviation σ of the distribution. Based on the SPE-PRMS reserves definitions for proved, probable, and possible reserves, we have

$$\begin{aligned} \text{Proved reserves} &= P_{90} = \mu - 1.28\sigma \\ \text{Probable reserves} &= P_{50} = \mu \\ \text{Possible reserves} &= P_{10} = \mu + 1.28\sigma \end{aligned} \tag{2.1}$$

for a normal distribution with mean μ and standard deviation σ.

Example 2.1 Reserves

Figure 2.6 shows a distribution of reserves for a normal distribution with mean of 200 MMSTB oil and three different standard deviations: 20 MMSTB, 40 MMSTB, and 60 MMSTB. What are the proved, probable, and possible reserves assuming a normal distribution with standard deviation of 20 MMSTB?

Answer

We can read the values from Figure 2.6 or calculate the values using Equation 2.1:

$$\begin{aligned} \text{Proved reserves} &= P_{90} \approx 175\ \text{MMSTB} \\ \text{Probable reserves} &= P_{50} = 200\ \text{MMSTB} \\ \text{Possible reserves} &= P_{10} \approx 225\ \text{MMSTB} \end{aligned}$$

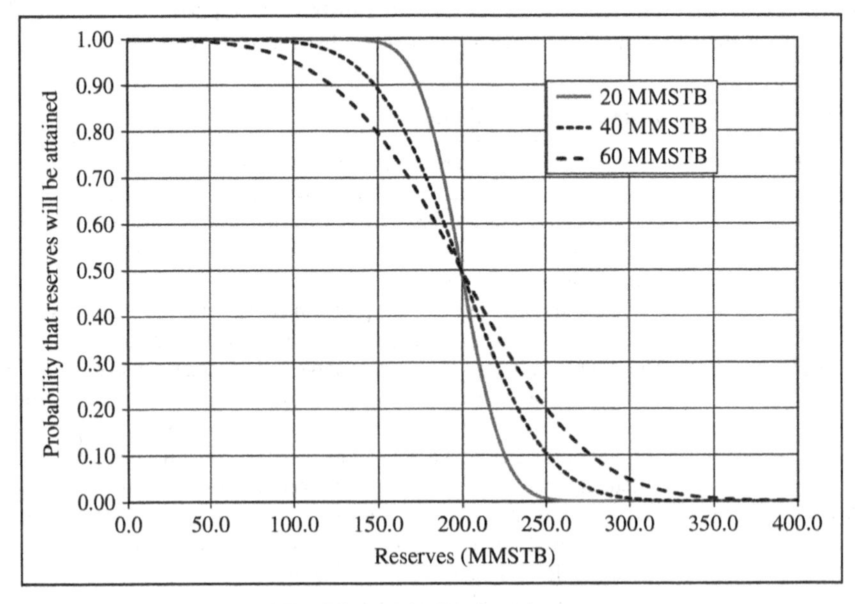

FIGURE 2.6 Distribution of reserves.

2.3 OIL AND GAS RESOURCES

Oil and gas resources may be characterized as conventional and unconventional resources. Snyder and Seale (2011) defined conventional oil and gas resources as formations that can be produced at economic flow rates or that produce economic volumes of oil and gas without stimulation treatments or special recovery processes and technologies. Unconventional oil and gas resources refer to formations that cannot be produced at economic flow rates or do not produce economic volumes of oil and gas without stimulation treatments or special recovery processes and technologies. Figure 2.7 presents a classification of oil and gas resources that is consistent with these definitions. Following Fanchi and Fanchi (2016), more information about several unconventional oil and gas resources near the base of the resource triangle in Figure 2.7 is provided in the following text.

Large oil and gas fields can be characterized as giant or supergiant fields. A giant oil field contains from 500 million barrels to 5 billion barrels of recoverable oil. Oil fields with more than five billion barrels of recoverable oil are supergiant oil fields. A giant gas field contains from 3 to 30 trillion ft^3 of recoverable gas. Gas fields with more than 30 trillion ft^3 of recoverable gas are supergiant gas fields.

2.3.1 Coal Gas

Gas recovered from coalbeds is known as coal gas (Jenkins et al., 2007). The gas can be present as liberated gas in the fracture system or as a monomolecular layer on the internal surface of the coal matrix. The composition of coal gas is predominately methane but can also include constituents such as ethane, carbon dioxide, nitrogen, and hydrogen (Mavor et al., 1999). Gas content in coal can range from approximately 20 standard cubic feet (SCF) gas per ton of coal in the Powder River

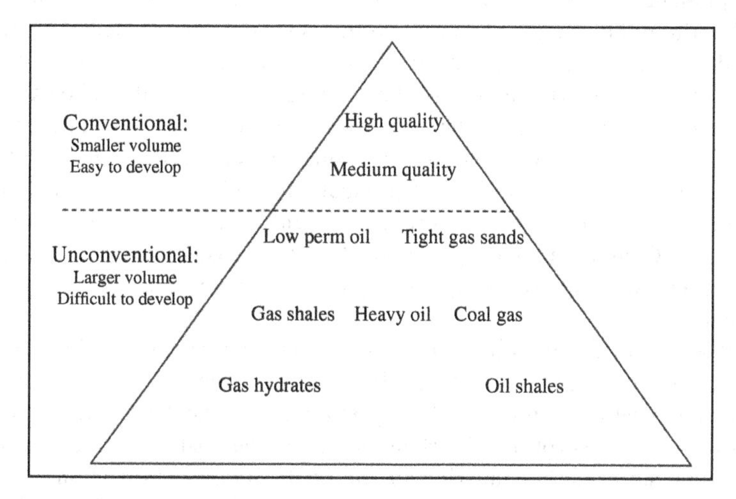

FIGURE 2.7 Resource triangle. (Source: Adapted from Snyder and Seale (2011) and Holditch, 2007.)

Basin of Wyoming (Mavor et al., 1999) to 600 SCF per ton in the Appalachian Basin (Gaddy, 1999).

Coal gas was historically known as coalbed methane. The term coal gas is used to better convey that gas from coalbeds is usually a mixture. Other terms for gas from coal include coal seam methane, coal mine methane, and abandoned mine methane. The practice of degasifying, or removing gas, from a coal seam was originally used to improve coal miner safety. Today, people recognize that coal gas has commercial value as a fuel.

Coal gas bound in the micropore structure of the coalbed can diffuse into the natural fracture network when a pressure gradient exists between the matrix and the fracture network. Fractures in coalbeds are called "cleats." Flow in the fractures is typically Darcy flow which implies that flow rate between two points A and B is proportional to the change in pressure between the points.

The ability to flow between two points in a porous medium is characterized by a property called permeability, and a unit of permeability is the darcy. It is named after Henry Darcy, a nineteenth-century French engineer. Permeability typically ranges from 1 millidarcy = 1 md (or $1.0 \times 10^{-15}\,m^2$) to 1 darcy = 1 D = 1000 md (or $1.0 \times 10^{-12}\,m^2$) for conventional oil and gas fields. Permeability in the cleat system typically ranges from 0.1 to 50 md.

Recovery of coal gas depends on three processes (Kuuskraa and Brandenburg, 1989). Gas recovery begins with desorption of gas from the internal surface to the coal matrix and micropores. The gas then diffuses through the coal matrix and micropores into the cleats. Finally, gas flows through the cleats to the production well. Gas flow rate through the cleats depends on such factors as the pressure gradient in the cleats, the density of cleats, and the distribution of cleats. The flow rate in cleats obeys Darcy's Law in many systems but may also depend on stress-dependent permeability or gas slippage (the Klinkenberg effect).

The production performance of a well producing gas from a coalbed will typically exhibit three stages. The well produces water from the cleat system in the first production stage. The withdrawal of water reduces pressure in the cleat system relative to the coal matrix and establishes a pressure gradient that allows coal gas to flow into the cleat system. The gas production rate increases during the first stage of cleat system dewatering and pressure depletion. The amount of water produced during the second stage of production is relatively small compared to gas production because there is more gas present in the cleat system relative to mobile water. Consequently, the gas production rate peaks during the second stage of production and gradually declines during the third stage of production as coalbed pressure declines.

The injection of carbon dioxide into a coal seam can increase coal gas recovery because carbon dioxide preferentially displaces methane in the coal matrix. The displaced methane flows into the cleat system where it can be extracted by a production well. The adsorption of carbon dioxide in the coal matrix can be used to sequester, or store, carbon dioxide in the coal seam. Sequestration of carbon dioxide in a coal seam is a way to reduce the amount of carbon dioxide emitted into the atmosphere.

2.3.2 Gas Hydrates

The entrapment of natural gas molecules in an ice-like crystalline form of water at very low temperatures forms an ice-like solid called a gas hydrate. Gas hydrates are also called clathrates, which is a chemical complex that is formed when one type of molecule completely encloses another type of molecule in a lattice. In the case of gas hydrates, hydrogen-bonded water molecules form a cage-like structure around low molecular weight gas molecules such as methane, ethane, and carbon dioxide. For more discussion of hydrate properties and technology, see Sloan (2006, 2007) and references therein.

Gas hydrates have historically been a problem for oil and gas field operations. For example, the existence of hydrates on the ocean floor can affect drilling operations in deep water. The simultaneous flow of natural gas and water in tubing and pipelines can result in the formation of gas hydrates that can impede or completely block the flow of fluids through pipeline networks. The formation of hydrates can be inhibited by heating the gas or treating the gas–water system with chemical inhibitors, but these inhibition techniques increase operating costs.

Today, the energy industry recognizes that gas hydrates may have commercial value as a clean energy resource or as a means of sequestering greenhouse gases. The potential of gas hydrates as a source of methane or ethane is due to the relatively large volume of gas contained in the gas hydrate complex. In particular, Makogon et al. (1997) reported that 1 m^3 of gas hydrate contains 164.6 m^3 of methane. This is equivalent to one barrel of gas hydrate containing 924 ft^3 of methane and is approximately six times as much gas as the gas contained in an unimpeded gas-filled pore system (Selley, 1998, page 25). The gas in gas hydrates occupies approximately 20% of the volume of the gas hydrate complex. Water occupies the remaining 80% of the gas hydrate complex volume.

Gas hydrates are naturally present in arctic sands, marine sands, and nonsandstone marine reservoirs. They are common in marine sediments on continental margins and below about 600 ft in permafrost regions. Ruppel (2011) reported that approximately 99% of gas hydrates occurs in the sediments of marine continental margins. Methane hydrates form when both methane and water are present at appropriate pressure and temperature. The size of the hydrate resource is not well known. Boswell (2009) said that gas hydrates may contain approximately 680 000 trillion ft^3 of methane. Development of technology for commercially producing the hydrate resource is ongoing.

2.3.3 Tight Gas Sands, Shale Gas, and Shale Oil

Low-permeability hydrocarbon resources include tight gas sands (Holditch, 2007) and shale (Kuuskraa and Bank, 2003; King, 2012). Both tight gas sands and shale are characterized by very low permeability. The permeability of tight gas sand is on the order of microdarcies (1 microdarcy is 1 thousandth of a millidarcy), while the permeability of shale is on the order of nanodarcies (1 nanodarcy is 1 millionth of a millidarcy).

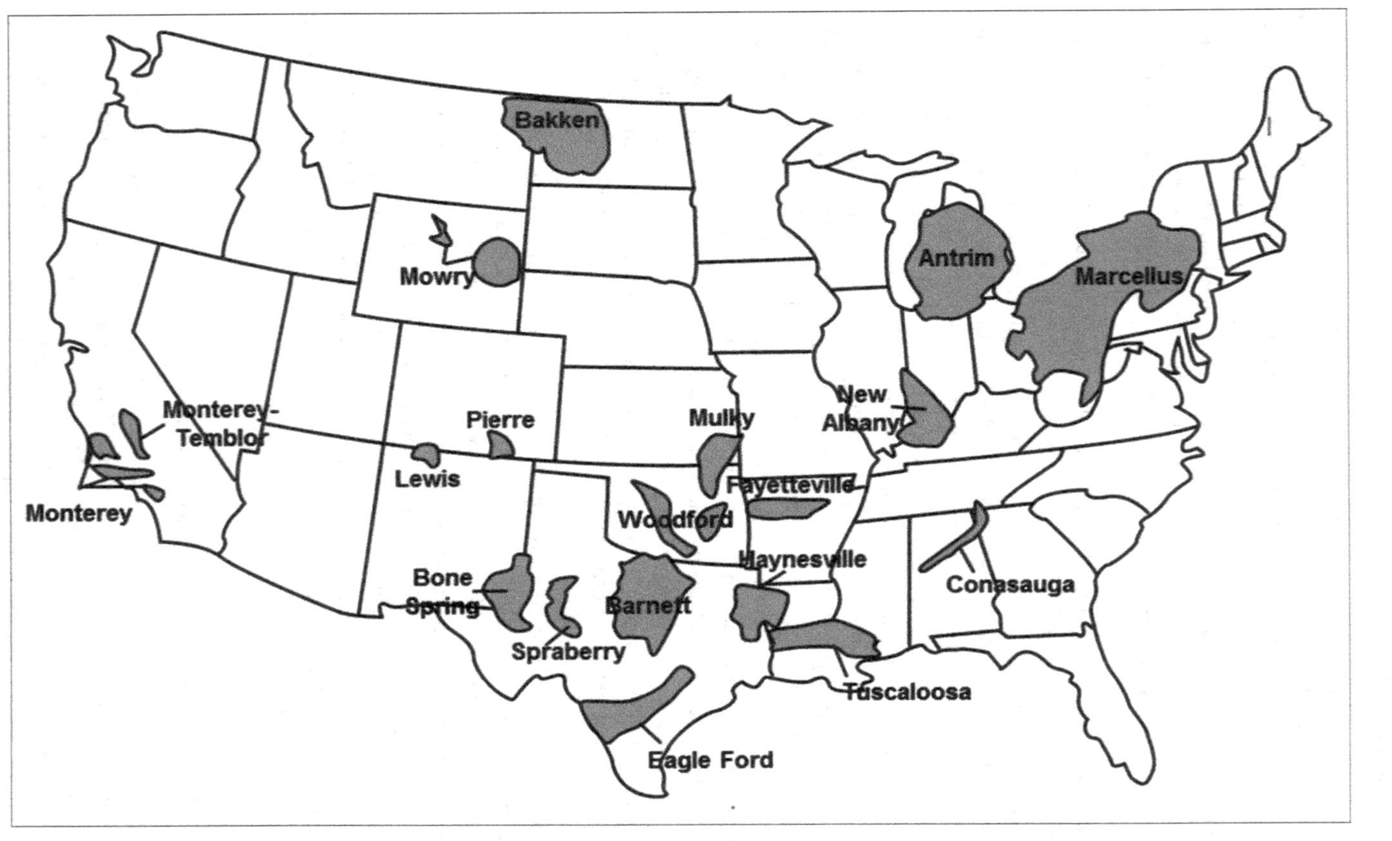

FIGURE 2.8 Selection of shale plays in the contiguous United States. (Source: Adapted from U.S. Energy Information Administration (August 18, 2015).)

Economic production of hydrocarbons from shale or tight sand became possible with the development of directional drilling and hydraulic fracturing technology. Directional drilling is the ability to drill wells at angles that are not vertically downward. Hydraulic fracturing is the creation of fractures in rock by injecting a water-based mixture into a formation at a pressure that exceeds the fracture pressure of the formation. The orientation and length of the induced fracture depends on formation characteristics such as thickness and stress. Once fractures have been created in the formation, a proppant such as manmade pellets or coarse grain sand is injected into the fracture to prevent it from closing, or healing, when injection pressure is removed. The proppant keeps the fractures open enough to provide a higher permeability flow path for fluid to flow to the production well.

Shales are typically rich in organic materials and often serve as source rock for conventional oil and gas fields. Production of oil and gas from shale is considered unconventional because shale functions as both the source rock and the reservoir. Shale deposits can be found throughout the world. Figure 2.8 shows shale plays in the contiguous United States. Shale gas deposits include the Barnett Shale in North Texas and the Marcellus Shale in Pennsylvania. Shale oil deposits include the Bakken Shale in North Dakota and the Eagle Ford Shale in South Texas.

Holditch (2013) and McGlade et al. (2013) provided estimates of the global volume of unconventional gas resources. Table 2.2 summarizes their estimates of technically recoverable reserves for coal gas, tight gas, and shale gas. The differences in tight gas and shale gas estimates illustrate the range of uncertainty.

Example 2.2 The Darcy Unit

Express the following permeabilities in darcies: 1 md, 1 μd (microdarcy), and 1 nd (nanodarcy).

Answer

$$1\,\text{md} = 1\,\text{md} \times \left(\frac{1\,\text{D}}{1000\,\text{md}}\right) = 1 \times 10^{-3}\,\text{D}$$

$$1\,\mu\text{d} = 1\,\mu\text{d} \times \left(10^{-3}\,\text{md}/\mu\text{d}\right) \times \left(\frac{1\,\text{D}}{1000\,\text{md}}\right) = 1 \times 10^{-6}\,\text{D}$$

$$1\,\text{nd} = 1\,\text{nd} \times \left(10^{-6}\,\text{md}/\text{nd}\right) \times \left(\frac{1\,\text{D}}{1000\,\text{md}}\right) = 1 \times 10^{-9}\,\text{D}$$

2.3.4 Tar Sands

Sand grains that are cemented together by tar or asphalt are called tar sands. Tar and asphalt are highly viscous plastic or solid hydrocarbons. Extensive tar sand deposits are found throughout the Rocky Mountain region of North America, as well as in

TABLE 2.2 Global Estimate of Technically Recoverable Reserves of Unconventional Gas Trillion Standard Cubic Feet

Source	Coal Gas	Tight Gas	Shale Gas	Total
Holditch (2013)	1453	43 551	12 637	57 641
McGlade et al. (2013)	1384	1914	6823	10 121

other parts of the world. Although difficult to produce, the volume of hydrocarbon in tar sands has stimulated efforts to develop production techniques.

The hydrocarbon in tar sands can be extracted by mining when they are close enough to the surface. Near-surface tar sands have been found in many locations around the world. In locations where oil shale and tar sands are too deep to mine, it is necessary to increase the mobility of the hydrocarbon.

An increase in permeability or a decrease in viscosity can increase mobility. Increasing the temperature of high API gravity oil, tar, or asphalt can significantly reduce viscosity. If there is enough permeability to allow injection, steam or hot water can be used to increase formation temperature and reduce hydrocarbon viscosity. In many cases, however, permeability is too low to allow significant injection of a heated fluid. An alternative to fluid injection is electromagnetic heating. Radio frequency heating has been used in Canada, and electromagnetic heating techniques are being developed for other parts of the world.

2.4 GLOBAL DISTRIBUTION OF OIL AND GAS RESERVES

The global distribution of oil and gas is illustrated by presenting the size of a nation's reserves. Table 2.3 lists 15 countries with the largest proved oil reserves and 15 countries with the largest proved gas reserves. National reserves are found by summing the reserves for all of the reservoirs in the nation.

Table 2.4 shows the regional distribution of oil and natural gas reserves. It is notable that the Middle East is at the top of both lists, and Europe is at the bottom of both lists. Political instability in the Middle East, which is home to many oil and gas exporting nations, has raised concerns about the stability of the supply. The lack of oil and natural gas reserves means that Europeans cannot rely on oil and natural gas as primary energy sources. Concerns about the security of their energy supply and the impact of fossil fuel combustion on the environment have helped motivate the European Union to become a leader in the development and installation of renewable energy facilities such as wind farms and solar plants. France has adopted nuclear fission energy as its primary energy source for electricity generation and has enough installed nuclear fission capacity to export electricity to other European nations.

TABLE 2.3 Nations with Largest Crude Oil and Natural Gas Proved Reserves in 2014

Crude Oil Reserves (Billion Barrels)		Natural Gas Reserves (Trillion Cubic Feet)	
World	1656	World	6972
Venezuela	298	Russia	1688
Saudi Arabia	268	Iran	1193
Canada	173	Qatar	885
Iran	157	United States	338
Iraq	140	Saudi Arabia	291
Kuwait	104	Turkmenistan	265
United Arab Emirates	98	United Arab Emirates	215
Russia	80	Venezuela	196
Libya	48	Nigeria	181
Nigeria	37	Algeria	159
United States	36	China	155
Kazakhstan	30	Iraq	112
Qatar	25	Indonesia	105
China	24	Mozambique	100
Brazil	15	Kazakhstan	85

Source: U.S. Energy Information Administration Petroleum (2015).

TABLE 2.4 Regional Distribution of Crude Oil and Natural Gas Proved Reserves in 2014

Crude Oil Reserves (Billion Barrels)		Natural Gas Reserves (Trillion Cubic Feet)	
World	1656	World	6972
Middle East	804	Middle East	2813
Central and South America	328	Eurasia	2178
North America	220	Africa	606
Africa	127	Asia and Oceania	540
Eurasia	119	North America	422
Asia and Oceania	42	Central and South America	277
Europe	12	Europe	136

Source: U.S. Energy Information Administration Petroleum (2015).

Figure 2.9 shows that world proved crude oil and natural gas reserves have increased from 2000 to 2014. Much of the recent increase in reserves is associated with advances in technology that made the development of unconventional resources economically viable.

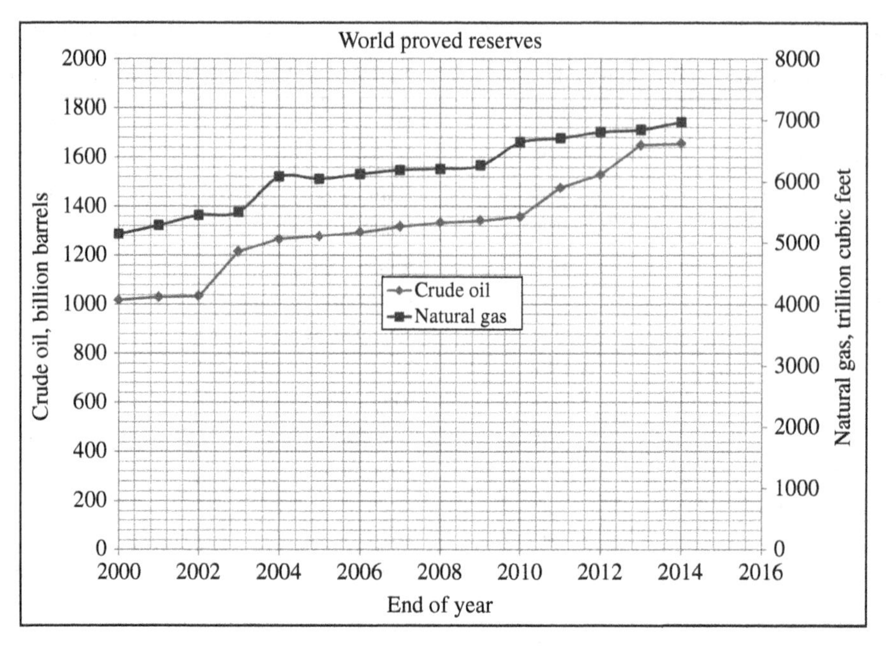

FIGURE 2.9 World proved reserves from 2000 to 2014. (Source: U.S. Energy Information Administration Petroleum (2015).)

2.5 PEAK OIL

Efforts to change from an energy mix that depends on fossil fuels to a more sustainable energy mix are motivated by environmental concerns and by the concern that oil production is finite and may soon be coming to an end. M. King Hubbert studied the production of oil in the contiguous United States (excluding Alaska and Hawaii) as a nonrenewable resource. Hubbert (1956) found that oil production in this limited geographic region could be modeled as a function of time. The annual production of oil increased steadily until a maximum was reached and then began to decline as it became more difficult to find and produce. The maximum oil production is considered a peak. Hubbert used his method to predict peak oil production in the contiguous United States, which excludes Alaska and Hawaii. Hubbert predicted that the peak would occur between 1965 and 1970. Hubbert then used the methodology that he developed for the contiguous United States to predict the peak of global oil production. He predicted that global oil production would peak around 2000 at a peak rate of 12–13 billion barrels per year or approximately 33–36 million barrels per day.

Crude oil production in the contiguous United States peaked at 9.4 million barrels per day in 1970. A second peak for the United States occurred in 1988 when Alaskan oil production peaked at 2.0 million barrels per day. The second peak is not considered the correct peak to compare to Hubbert's prediction because Hubbert restricted his analysis to the production from the contiguous United States. Many modern experts consider the 1970 oil peak to be a validation of Hubbert's methodology and have tried to apply the methodology to global oil production. Analyses of historical

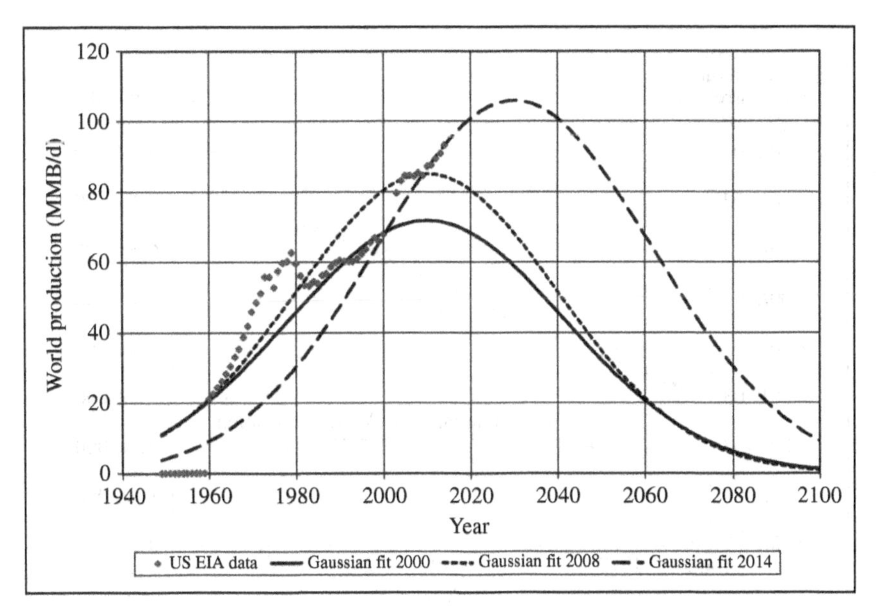

FIGURE 2.10 World oil production rate forecast using Gaussian curves.

data using Hubbert's methodology typically predict that world oil production will peak in the first quarter of the twenty-first century.

2.5.1 World Oil Production Rate Peak

Forecasts based on an analytical fit to historical data can be readily prepared using publicly available data. Figure 2.10 shows a fit of world oil production rate (in millions of barrels per day) from the US EIA database. Three fits are displayed. The fit labeled "Gaussian Fit 2000" matches peak data at year 2000, the fit labeled "Gaussian Fit 2008" matches peak data at year 2008, and the fit labeled "Gaussian Fit 2014" matches peak data at year 2014. Each fit was designed to match the most recent part of the production curve most accurately. The fits give oil production rate peaks between 2010 and 2030.

The increase in actual world oil production rate between 2000 and 2010 is due to a change in infrastructure capacity in Saudi Arabia. This period coincided with a significant increase in oil price per barrel, which justified an increase in facilities needed to produce, collect, and transport an additional one to two million barrels per day of oil. The increase in actual world oil production rate since 2000 is largely due to an increase in oil production from shale oil in the United States. Technological advances have added hydrocarbon resources to the global energy mix and shifted the date when peak oil rate occurs.

2.5.2 World Per Capita Oil Production Rate Peak

The evidence for a peak in world oil production rate is inconclusive. On the other hand, suppose we consider world per capita oil production rate, which is annual world oil production rate divided by world population for that year. Figure 2.11

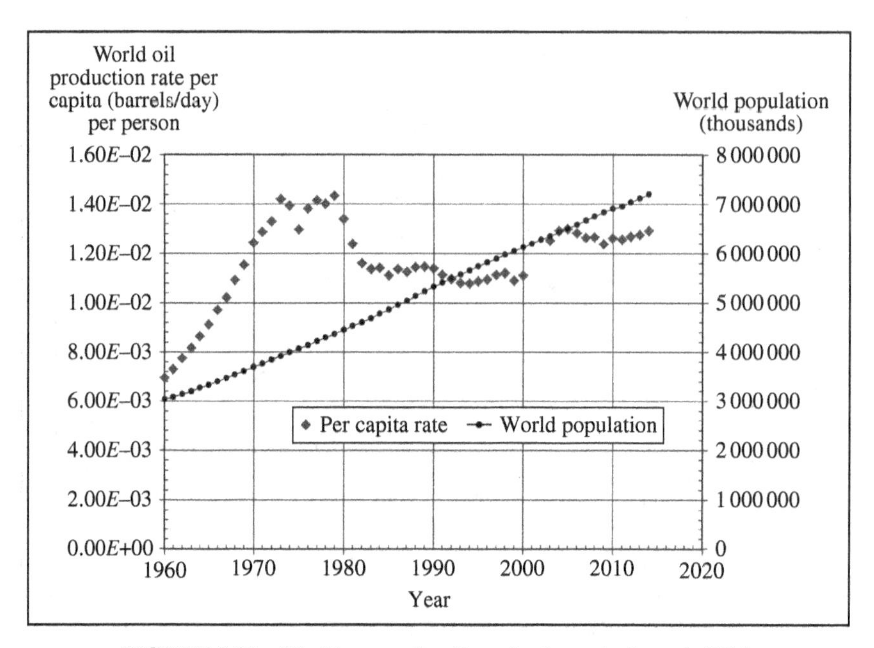

FIGURE 2.11 World per capita oil production rate through 2014.

shows world per capita oil production rate (in barrels of oil produced per day per person) for the period from 1960 through 2014.

Figure 2.11 shows two peaks in the 1970s. The first peak was at the time of the first oil crisis, and the second peak occurred when Prudhoe Bay, Alaska, oil production came online. World per capita oil production rate has been significantly below the peak since the end of the 1970s. The increase in 2000 is largely due to the increase in production in Saudi Arabia and the United States. It appears that world per capita oil production rate peaked in the 1970s. We do not know if another higher peak is possible given the continuing growth in world population.

Example 2.3 Future Power Demand

Assume 10 billion people will consume 200 000 MJ energy per person per year in 2100. How many power plants will be needed to provide the energy consumed each year? Assume an average power plant provides 1000 MW power.

Answer

$200\,000\ \text{MJ/person/yr} \times (10 \times 10^9\ \text{people}) = 2.0 \times 10^{15}\ \text{MJ/yr}$

Power in MW: $(2.0 \times 10^{15}\ \text{MJ/yr}) \times (1\ \text{yr}/3.1536 \times 10^7\ \text{s}) \approx 6.34 \times 10^7\ \text{MW}$

Number of powerplants needed: $\dfrac{6.34 \times 10^7\ \text{MW}}{1000\ \text{MW/plant}} \approx 63\,400$ plants

2.6 FUTURE ENERGY OPTIONS

Future energy demand is expected to grow substantially as global population increases and developing nations seek a higher quality of life (Fanchi and Fanchi, 2016). Social concern about nuclear waste and proliferation of nuclear weapons is a significant deterrent to reliance on nuclear fission power. These concerns are alleviated to some extent by the safety record of the modern nuclear power industry. Society's inability to resolve the issues associated with nuclear fusion makes fusion an unlikely contributor to the energy mix until at least the middle of the twenty-first century. If nuclear fusion power is allowed to develop and eventually becomes commercially viable, it could become the primary energy source. Until then, the feasible sources of energy for use in the future energy mix are fossil fuels, nuclear fission, and renewable energy.

Concerns about the environmental impact of combustible fuels and security of the energy supply are encouraging a movement away from fossil fuels. Political instability in countries that export oil and gas, the finite size of known oil and gas supplies, increases in fossil fuel prices, and decreases in renewable energy costs, such as wind energy costs, are motivating the adoption of renewable energy. On the other hand, the development of technology that makes unconventional sources of fossil fuels economically competitive is encouraging continued use of fossil fuels, especially as the natural gas infrastructure is improved. These conflicting factors have an impact on the rate of transition from fossil fuels to a sustainable energy mix. A key decision facing society is to determine the rate of transition.

2.6.1 Goldilocks Policy for Energy Transition

Fanchi and Fanchi (2015) introduced a Goldilocks policy for determining the rate of transition from one energy source to another. An appropriate rate can be estimated using the historical energy consumption data from the United States shown in Figure 2.12 (US EIA Annual Energy Review, 2001). The United States is a developed country with a history of energy transitions over the past few centuries.

The data in Figure 2.12 are presented in Figure 2.13 as the percent of total US energy consumption by source. Wood was the principal energy source when the United States was founded in the eighteenth century. Coal began to take over the energy market in the first half of the nineteenth century and peaked in the early twentieth century. Oil began to compete with coal during the latter half of the nineteenth century and became the largest component of the energy mix by the middle of the twentieth century. The period of time from the appearance of an energy source in the market place to its peak can be seen in Figure 2.13 for coal and oil. Historically, energy transition periods in the United States last approximately 60–70 years.

Our future energy mix depends on choices we make, which depends, in turn, on energy policy. Several criteria need to be considered when establishing energy policy. We need to consider the capacity of the energy mix, its cost, safety, reliability, and effect on the environment. We need to know that the energy mix can meet our needs (capacity) and be available when it is needed (reliability). The energy mix should

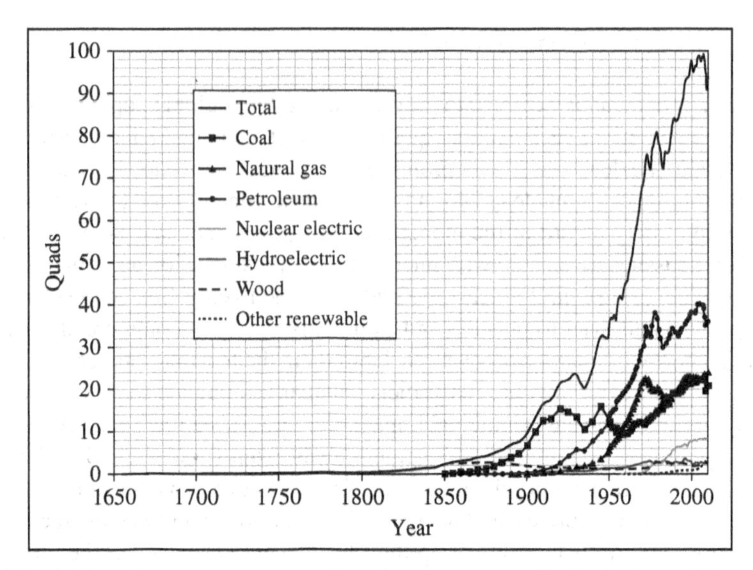

FIGURE 2.12 US energy consumption by source, 1650–2010 (quadrillion BTU). (Source: U.S. Energy Information Administration (2001).)

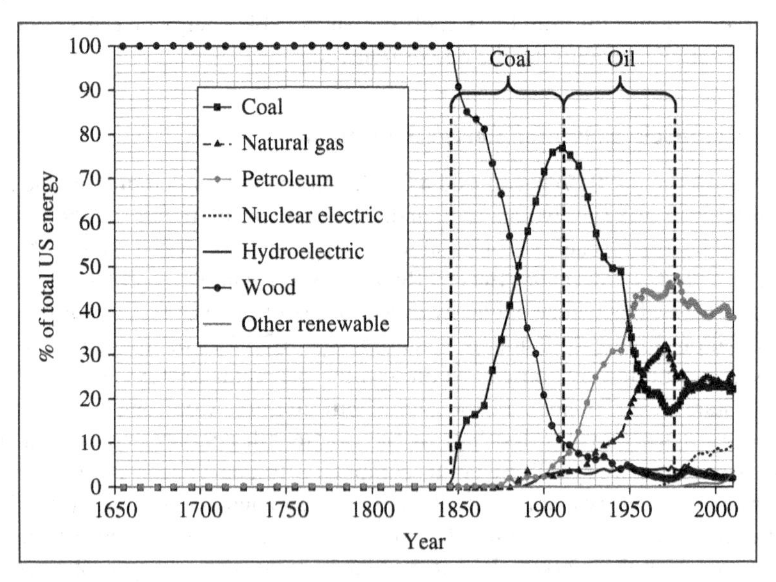

FIGURE 2.13 Coal and oil transition periods based on US energy consumption by source, 1650–2010 (%). (Source: Fanchi and Fanchi, 2015.)

have a negligible or positive effect on the environment, and it should be safe. When we consider cost, we need to consider both tangible and intangible costs associated with each component of the energy mix.

The transition from wood to coal and from coal to oil has been interpreted as a trend toward decarbonization or the reduction in the relative amount of carbon in

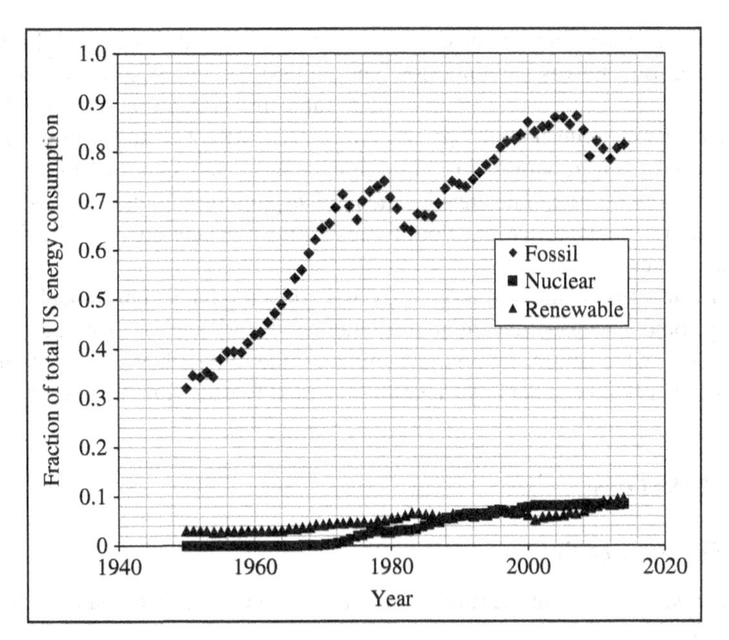

FIGURE 2.14 Fraction of US annual energy consumption by source, 1950–2014. (Source: U.S. Energy Information Administration (2015).)

combustible fuels. Figure 2.14 shows the fraction of annual energy consumption in the United States since 1950. Fossil energy includes coal, oil, and natural gas; nuclear energy refers to fission energy; and renewable energy includes hydroelectric, wind, and solar. The next step in decarbonization would be a transition to natural gas. A greater reliance on natural gas rather than wood, coal, or oil would reduce the emission of greenhouse gases into the atmosphere.

The twenty-first-century energy mix will depend on technological advances, including some advances that cannot be anticipated, and on choices made by society. There are competing visions for reaching a sustainable energy mix. Some people see an urgent need to replace fossil fuels with sustainable/renewable energy sources because human activity is driving climate change. Others believe that it is necessary to replace fossil fuels with sustainable/renewable energy sources, but the need is not urgent. They argue that the economic health of society outweighs possible climate effects. If the energy transition is too fast, it could significantly damage the global economy. If the energy transition is too slow, damage to the environment could be irreversible.

The "Goldilocks Policy for Energy Transition" is designed to establish a middle ground between these competing visions. We need the duration of the energy transition to be just right; that is, we need to adopt a reasonable plan of action that reduces uncertainty with predictable public policy and reduces environmental impact.

Based on historical data, we could plan an energy transition to a sustainable energy mix by the middle of the twenty-first century. The European Union is operating on this timetable with the EU Supergrid. In addition, natural gas could serve as a transition fuel because it is relatively abundant, continues the trend to decarbonization, halves greenhouse gas emissions relative to oil and coal combustion, and requires reasonably affordable infrastructure changes that take advantage of available technology. A natural gas infrastructure would be a step toward a hydrogen economy infrastructure in the event that hydrogen becomes a viable energy carrier. The development of a game-changing technology, such as commercial nuclear fusion, would substantially accelerate the transition to a sustainable energy mix.

The twenty-first-century energy mix will depend on technological advances, including some advances that cannot be anticipated, and on choices made by society. For the foreseeable future, oil and gas will continue to be a key source of energy in the global energy mix as society makes a transition to a sustainable energy mix.

2.7 ACTIVITIES

2.7.1 Further Reading

For more discussion about the future of petroleum, see Fanchi and Fanchi (2016) and Yergin (2011).

2.7.2 True/False

2.1 Gas hydrates are clathrates.

2.2 Shale gas is produced from reservoirs with high permeability.

2.3 Coal gas production requires gas desorption from the coal matrix.

2.4 Coal gas is primarily propane.

2.5 The first oil crisis began in 1973.

2.6 M. King Hubbert predicted that global oil production would peak between 1965 and 1970.

2.7 Per capita global oil production peaked by 1980.

2.8 Recovery factor is the fraction of original fluid in place that can be produced from a reservoir.

2.9 Shale oil and gas are unconventional resources.

2.10 Probable reserves are more likely to be recoverable than possible reserves.

2.7.3 Exercises

2.1 Complete the table that follows and estimate proved, probable, and possible reserves. Assume the reserves are normally distributed.
Hint: Reserves = OOIP times recovery factor.

Model	OOIP (MMSTB)	Recovery Factor	Reserves (MMSTB)
1	400	0.30	
2	650	0.43	
3	550	0.48	
4	850	0.35	
5	700	0.38	

2.2 **A.** A geothermal power plant was able to provide 2000 MWe (megawatts electric) power when it began production. Twenty years later the plant is only able to provide 1000 MWe from the geothermal source. Assuming the decline in power production is approximately linear, estimate the average annual decline in power output (in MWe/yr).

B. Suppose the plant operator has decided to close the plant when the electric power output declines to 10 MWe. How many more years will the plant operate if the decline in power output calculated in Part A continues?

2.3 **A.** A coal seam is 600 ft wide, 1 mile long, and 15 ft thick. The volume occupied by the fracture network is 1%. The volume of coal is the bulk volume of the coal seam minus the fracture volume. What is the volume of coal in the coal seam? Express your answer in ft^3.

B. If the density of coal is 1.7 lbm/ft^3, how many tons of coal are in the coal seam?

C. The gas content G_{coal} of gas in the coal matrix is 500 SCF methane per ton of coal. What is the total volume of methane contained in the coal seam? Neglect any gas that may be in the fracture.

2.4 Porosity (fraction) is related to bulk density by $\rho_b = (1-\phi)\times(\rho_{ma}) + (\phi \times \rho_f)$ where bulk density ρ_b is 2.40 g/cc from the density log, density of rock matrix ρ_{ma} is 2.70 g/cc, and fluid density ρ_f is 1.03 g/cc for brine. Estimate porosity.

2.5 **A.** A news article reported that 535 MMSCF gas could supply gas to 8150 homes for a year. Calculate the value of gas. Note: 1 MMSCF = 1 million standard cubic feet.

Gas Volume (MSCF)	Gas Price/Volume ($/MSCF)	Value of Gas ($)
535 000	2.50	
535 000	5.00	

B. If the price of gas to the consumer is $5.00 per MSCF, how much does the average home have to pay for gas each year?

Hint: First estimate the volume of gas in MSCF that is used by each home each year, where 1 MSCF = 1000 SCF.

2.6 Typical energy densities for coal, oil, and methane are shown in the table. The relative value of the energy in each material can be estimated by using energy density to calculate the cost per unit of energy. Use the information in the table and unit conversion factors to complete the table.

Fuel	Price	Energy Density	$ per MJ
Coal	$50 per tonne	42 MJ/kg	
Oil	$60 per barrel	42000 MJ/m^3	
Methane	$3 per MSCF	38 MJ/m^3	

2.7 Suppose Country A imports 55% of its oil. Of this amount, 24% is imported from Region A. What is the percent of oil imported into Country A from Region A?

2.8 **A.** A national oil company (NOC) reports that it has 700 billion bbl OOIP. The resource is classified in the table that follows. Fill in the % OOIP column.

Classification Category	Volume (billion bbl)	% OOIP
Produced	99	
Remaining proved	260	
Probable	32	
Possible	71	
Contingent	238	

B. How long can the NOC produce at a production capacity of 10 million bbl/day using remaining proved reserves? Express your answer in years.

C. How long can the NOC produce at a production capacity of 15 million bbl/day using remaining proved reserves? Express your answer in years.

2.9 Estimate the year when peak oil occurs using Gaussian fits to data for years 2000, 2008, and 2014 shown in Figure 2.10.

2.10 **A.** Use Figure 2.13 to estimate the length of time (in years) it took for the transition from wood to coal.

B. Use Figure 2.13 to estimate the length of time (in years) it took for the transition from coal to oil.

C. What will be the next transition and how long will it last?

3

PROPERTIES OF RESERVOIR FLUIDS

A key question to ask upon discovering oil and gas is, "How much is there?" In addition to oil or gas, there will be water and probably some gas dissolved in the liquids. Gas also may appear as a separate phase along with oil. Properties of these fluids are used to determine their amounts in a formation and their fluid flow characteristics. This chapter describes the origin and the common methods for classifying oil and gas resources, introduces keywords and definitions of fluid properties, provides methods for estimating fluid properties, and shows how these properties are measured and used in petroleum engineering.

3.1 ORIGIN

The details of the formation of oil and gas are largely unknown. From about 1860 to 1960, some debate centered on whether these fluids derived from biological or nonbiological sources. Geochemical analysis has uncovered many molecules in oils that share structure with chemicals in living organisms. These molecules, or biomarkers, are fossil remains of life from millions of years ago. Hence, it is generally accepted that oil and gas have biological origin. Many geochemists around the world continue to research the processes of oil and gas formation. Although the details are still being determined, we can describe the process in broad-brush terms.

Introduction to Petroleum Engineering, First Edition. John R. Fanchi and Richard L. Christiansen.
© 2017 John Wiley & Sons, Inc. Published 2017 by John Wiley & Sons, Inc.
Companion website: www.wiley.com/go/Fanchi/IntroPetroleumEngineering

According to the "biogenic" theory, hydrocarbon gases and liquids found in formations today are the product of short- and long-term processes that acted on remnants of organisms such as algae and plankton that lived millions of years ago in aqueous environments. Upon death, the organisms formed organic-rich sediments. Oil and gas were formed by bacterial action in the anaerobic conditions of the sediment and subsequent thermal processes that occurred after the sediment was buried under many additional layers of sediment. These thermal processes break larger organic molecules into smaller molecules. Although some oil and gas may remain in the original sediment, which is also known as source rock, much of it has migrated upward as a result of buoyancy effects. Some of the upward migrating fluid was trapped by impermeable formations; the remainder continued migrating to the surface where it dissipated. Accumulation of oil and gas in a trap forms a reservoir, which is the target of modern exploration activities.

Much of the oil and gas that is produced from shales today is thought to be in the original source rock. If the temperature of a shale formation is at least 200°F, the thermal process of producing oil and gas, or "maturation," from organic material in the shale is ongoing to some extent. The rate of maturation increases with formation temperature, which increases with depth at about 0.01–0.02°F per ft. Hence, a formation's temperature could exceed 200°F even at 8000ft below the surface. As a result of increasing maturation rate with depth, the size of the hydrocarbon molecules in oil decreases with increasing depth. Thus, deep formations are more likely to contain gas and not oil.

The compositions of hydrocarbon resources are quite variable from location to location around the world. Here are some interesting and unusual examples:

1. In the Athabasca oil sands of northern Alberta, the oil is usually called "tar." The density oil in the oil sands is greater than water density and its viscosity is very high. Much of the oil produced from these sands is obtained by mining and then transporting the oily sand to a central location for separation.
2. Gilsonite, which is mined from formations in northeastern Utah, is a solid with a composition similar to the very largest molecules found in oils.
3. Natural gas hydrates are solids formed when water and gas mix at low temperatures and high pressures, such as in the oceans off the continental shelves where natural gas that has seeped from subsurface formations is trapped on the ocean floor.
4. A smelly wax mined from fractured formations, ozokerite is thought to be produced by natural processes that have stripped away all but the waxy portion of oils.
5. Gas associated with most coal formations was generated by thermal maturation of the coal, which was formed from compressed plant matter. As a result, the composition of "coalbed" gas differs from gases that originated from the accumulation of algae and plankton in sediment.
6. Natural gases from formations in western Kansas and Oklahoma and northern Texas contain up to 2% helium, which is extracted from the natural gas. The helium is produced by alpha decay of radioactive minerals in the formations of that area.

There are many reasons for compositional variations such as those mentioned earlier. First, the compositions of the fluids in the source rock depend on the organisms that were present in the original sediment. Second, the processes of biological and thermal maturation vary, depending on the type of organisms and temperature history. Third, trapping processes vary, depending on geologic structures, properties of those structures, and flux of water and other fluids through the structures. Given all these variables, even oil samples taken from a single formation can vary with depth and horizontal location. For example, exposure to an actively flowing aquifer below an oil accumulation can produce compositional variations. A compositional gradient, or variation of composition with spatial location, can also be induced by thermal diffusion and gravity segregation.

3.2 CLASSIFICATION

Oils and gases can be classified in many ways. The most common of these are summarized in this section.

Oils and gases are mixtures of predominantly hydrocarbon molecules with some inorganic molecules, such as nitrogen, carbon dioxide, and hydrogen sulfide. Oils and gases are called "sweet" if they contain only negligible amounts of sulfur compounds such as hydrogen sulfide (H_2S) or mercaptans—organic molecules with a sulfur–hydrogen functional group. If these fluids contain sulfur compounds such as H_2S or a mercaptan, they are called "sour."

Oils and gases are also classified by their specific gravities. For an oil, specific gravity equals oil density divided by density of water at standard conditions. Heavy oils have densities close to and even exceeding water density. Light oils are much less dense than water. For gases, the reference density is that of air at standard conditions. Gases can be rich or lean, depending on the amount of hydrocarbon molecules larger than methane.

Oil is also classified by the relative amounts of different kinds of hydrocarbon molecules: paraffins, naphthenes, and aromatics. Paraffin molecules such as methane, ethane, and propane have a single bond between carbon atoms and are considered saturated hydrocarbons. Paraffins have the general chemical formula C_nH_{2n+2}. Naphthenes have the general chemical formula C_nH_{2n} and are saturated hydrocarbons with a ringed structure, as in cyclopentane. The term "saturated" indicates that all carbon–carbon bonds are single bonds. By contrast, unsaturated hydrocarbons have at least one double or triple bond. Aromatics are unsaturated hydrocarbons with a ringed structure. Aromatics are relatively stable and unreactive because of their unique ring structure. Benzene is a well-known example of an aromatic.

Petroleum can exist in a reservoir as a gas, liquid, or solid, depending on fluid composition, temperature, and pressure. Natural gas is typically methane with lesser amounts of heavier hydrocarbon molecules like ethane, propane, and butanes. Oil is a liquid mixture of some natural gas dissolved in greater amounts of heavier hydrocarbons with carbon numbers up to about 60. The carbon number designates the number of carbon atoms in the molecule. Asphaltenes and waxes, the largest hydrocarbon molecules in oil, can precipitate as solids under some conditions.

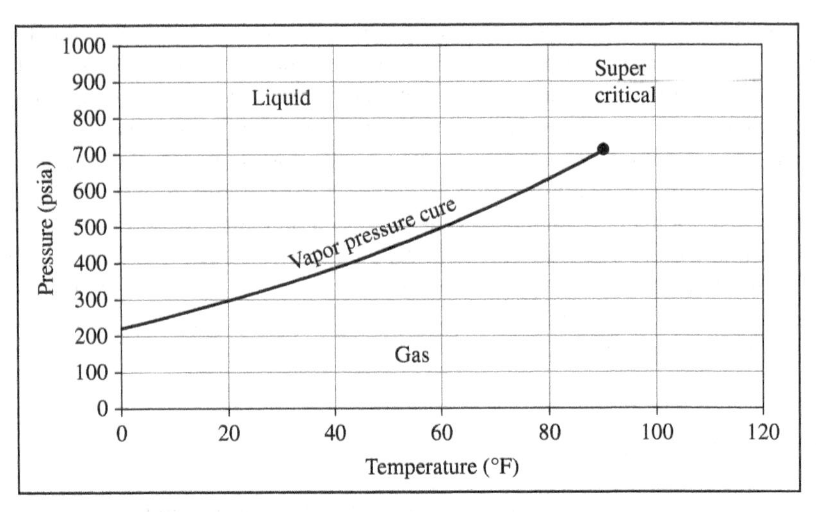

FIGURE 3.1 *P–T* diagram for ethane.

The gas–liquid–solid phase behavior of a petroleum sample is often shown in a pressure–temperature (*P–T*) diagram. *P–T* diagrams are another tool for classifying oil and gas. The *P–T* diagram is relatively simple for a single-component system, as shown in Figure 3.1 for ethane. The curve in Figure 3.1 is known as the vapor pressure curve, and it ends at the right at the critical point for ethane. Above the vapor pressure curve, ethane exists as a liquid; below the vapor pressure curve, it exists as a gas. At 80°F and 800 psi, ethane is a liquid, while at 80°F and 300 psi, it is a gas. If ethane exists at a temperature and pressure above the critical point, there is no way to distinguish between liquid and gas, and it is called a supercritical fluid.

The *P–T* diagram for a two-component system is more complex, as shown in Figure 3.2 for a mixture of 59 mol% ethane and 41 mol% *n*-heptane (which may also be written as 59/41 mol% ethane/*n*-heptane). For comparison, the vapor pressure curves for pure ethane and pure *n*-heptane are included in the figure. The curve consisting of both the bubble-point (BP) and dew-point (DP) curves is termed the phase envelope. The BP curve and the DP curve meet at the critical point. Anywhere inside the phase envelope, there are two phases: gas and liquid. Only one phase exists outside the phase envelope. Liquid exists in the *P–T* region above the BP curve. Elsewhere outside the phase envelope, the fluid is a gas. For example, at 350°F and 200 psia, the fluid is gaseous, while at 350°F and 1200 psia, the fluid is liquid. If the temperature is 400°F, the fluid is gaseous at both 200 psia and also at 1200 psia. How do we know it is a gas at 400°F and 1200 psia? If pressure on that fluid is reduced to the phase envelope pressure, a droplet of dew forms. By contrast, at 300°F and 1200 psia, the fluid is a liquid. If pressure on that fluid is reduced to the phase envelope, a bubble of gas forms at the phase envelope pressure.

The *P–T* diagram for 59/41 mol% ethane/*n*-heptane is shown in Figure 3.3 without the vapor pressure curves for ethane and *n*-heptane. The highest temperature on the phase envelope is called the cricondentherm. It appears in later discussions.

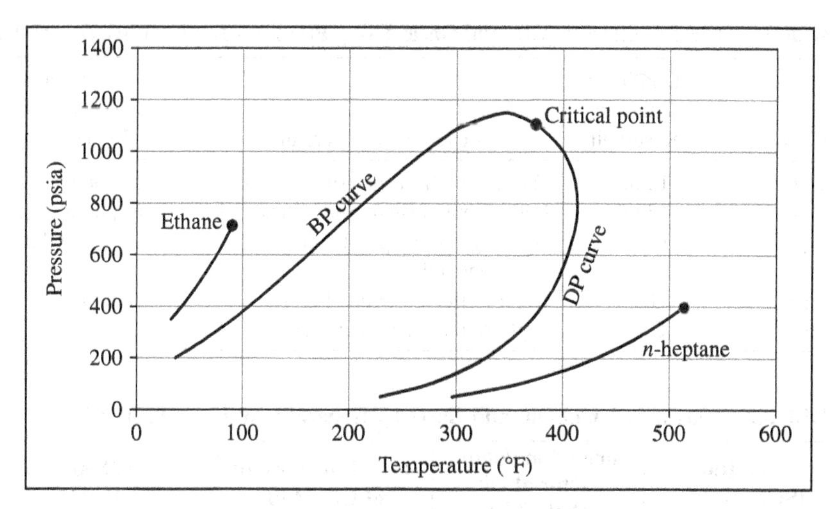

FIGURE 3.2 Comparison of vapor pressure curves for ethane and *n*-heptane to the phase envelope for a mixture of 59 mol% ethane and 41 mol% *n*-heptane.

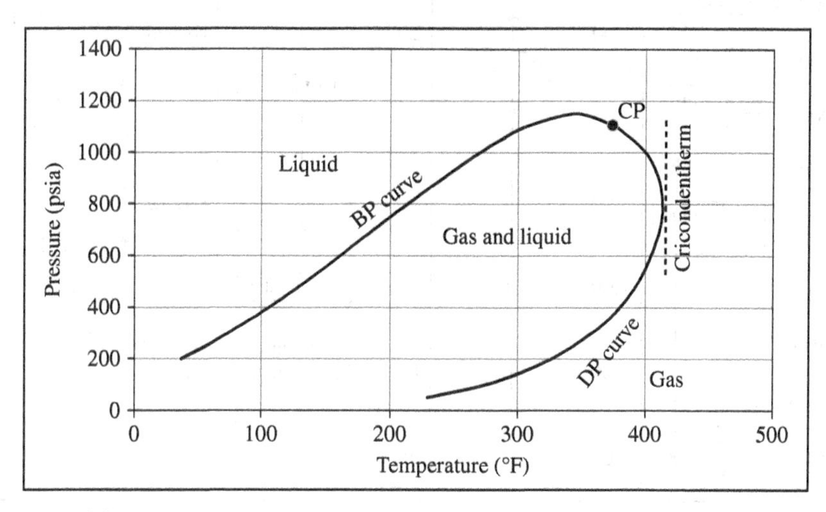

FIGURE 3.3 Additional nomenclature for *P–T* diagrams using data for 59/41 mol% ethane/*n*-heptane.

P–T diagrams for oils and gases have shape and other features similar to the diagram of Figure 3.3. Based on the *P–T* diagrams, five patterns of behavior are often identified for these fluids. They are listed in Table 3.1. The separator is a piece of equipment that is used to separate the fluid produced from the reservoir into distinct fluid phases. In most of this text, we focus on black oils and dry gases because these fluids are relatively easy to model and understand.

The *P–T* diagram can be used to anticipate phase behavior changes when pressure changes. Consider a reservoir containing black oil. Reservoir pressure is initially greater

TABLE 3.1 Classifications of Oils and Gases Using Pressure–Temperature Diagrams

Fluid Type	Dominant Phase in Reservoir	Reservoir Temperature	Phases at Separator Pressure and Temperature
Black oil	Liquid	Far to the left of the critical point	Liquid and gas
Volatile oil	Liquid	Left of, but close to, the critical point	Liquid and gas
Retrograde gas	Gas	Between critical point and cricondentherm	Liquid and gas
Wet gas	Gas	Right of the cricondentherm	Liquid and gas
Dry gas	Gas	Right of the cricondentherm	Gas

TABLE 3.2 Classification of Oils and Gases by Generally Available Properties

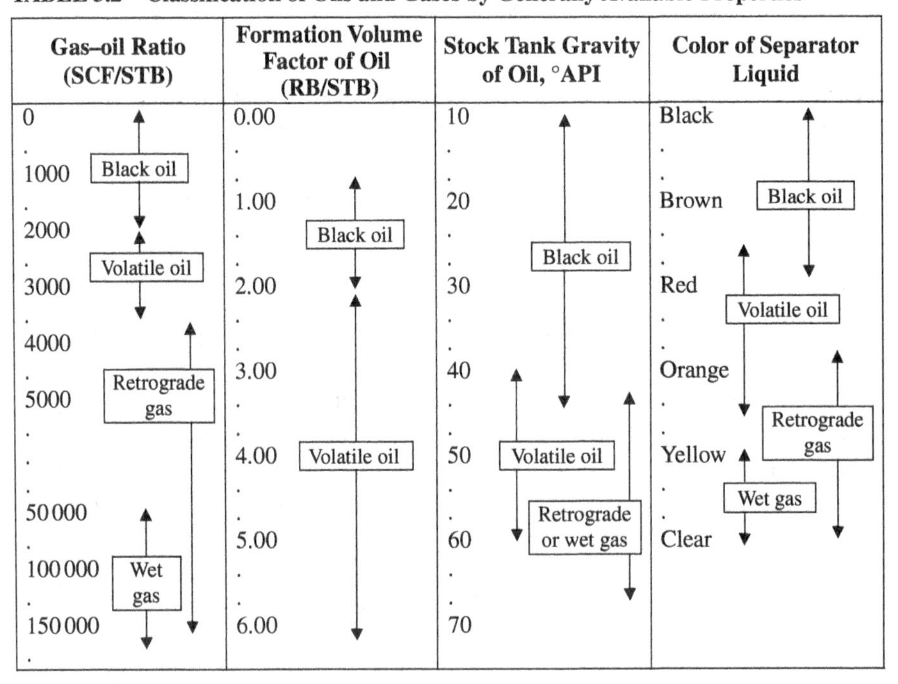

Gas–oil Ratio (SCF/STB)	Formation Volume Factor of Oil (RB/STB)	Stock Tank Gravity of Oil, °API	Color of Separator Liquid
0	0.00	10	Black
1000 Black oil			
	1.00	20	Brown Black oil
2000	Black oil		
Volatile oil		Black oil	
3000	2.00	30	Red
			Volatile oil
4000			
Retrograde gas	3.00	40	Orange
5000			Retrograde gas
	4.00 Volatile oil	50 Volatile oil	Yellow
			Wet gas
50 000		Retrograde or wet gas	
	5.00	60	Clear
100 000 Wet gas			
150 000	6.00	70	

than BP pressure, which is the pressure on the BP pressure curve at reservoir temperature. Production of oil will result in the decline of reservoir pressure. Reservoir temperature does not change significantly, if at all, and can be considered constant. When reservoir pressure declines below BP pressure, the remaining hydrocarbon mixture enters the two-phase gas–oil region where both an oil phase and a gas phase coexist.

P–T diagrams provide a useful perspective for classifying hydrocarbon fluids, but just small portions of the *P–T* diagram are measured for any particular oil or gas. Another perspective for classification focuses on more easily observed properties: gas–oil ratio, oil formation volume factor (FVF), stock tank gravity of separator oil, and the color of separator liquid. Such a classification is shown in Table 3.2.

The dissolved gas–oil ratio is the volume of dissolved gas measured in standard cubic feet (SCF) to the volume of oil measured in stock tank barrels (STB) as obtained from surface separation equipment. The oil FVF is the corresponding volume of oil at reservoir conditions (RB for reservoir barrels) of temperature and pressure divided by the stock tank volume of oil (STB). As fluids flow from the reservoir to the surface separator and stock tank, pressure and temperature decrease. In response to decreasing pressure, gases bubble out of the oil phase, and the oil volume decreases. Stock tank gravity is a measure of specific gravity. In Table 3.2, it is reported in °API. The relation of °API to specific gravity is described in another section of this chapter.

3.3 DEFINITIONS

In this section, we provide brief definitions and units for many of the fluid properties encountered in the oil and gas industry. Unit conversion factors are found in Appendix A. In subsequent sections, we include examples of correlations for the fluid properties most often needed by petroleum engineers.

Pressure. Fluid properties depend on pressure, temperature, and composition. Pressure is defined as normal force divided by the area to which it is applied. Some common units of pressure include pounds/in^2 or psi, pascals (Newtons/m^2), atmospheres, and bars.

Temperature. Temperature is a measure of the average kinetic energy of a system. The most commonly used temperature scales are the Fahrenheit and Celsius scales. The relationship between these scales is

$$T_C = \frac{5}{9}(T_F - 32) \tag{3.1}$$

where T_C and T_F are temperatures in degrees Celsius and degrees Fahrenheit, respectively.

Some applications, such as equations of state, require the use of absolute temperature expressed in Kelvin or Rankine degrees. The absolute temperature scale in degrees Kelvin is related to the Celsius scale by

$$T_K = T_C + 273 \tag{3.2}$$

where T_K is temperature in degrees Kelvin. The absolute temperature scale in degrees Rankine is related to the Fahrenheit scale by

$$T_R = T_F + 460 \tag{3.3}$$

where T_R is temperature in degrees Rankine.

Example 3.1 Temperature of the Earth's Crust

The temperature in some parts of the Earth's crust increases by about 1°F for every 100 ft of depth. Estimate the temperature of the Earth at a depth of 8000 ft. Assume the temperature at the surface is 60°F. Express your answer in °F and °C.

Answer

$$\text{Temp} = 60°\text{F} + (8000\,\text{ft})\left(\frac{1°\text{F}}{100\,\text{ft}}\right) = 140°\text{F}$$

$$T_\text{C} = \frac{5}{9}(140 - 32) = 60°\text{C}$$

Composition. The composition of a fluid refers to the types and amounts of molecules that comprise the fluid. Oil is a mixture of hydrocarbon compounds and minor amounts of inorganic molecules such as carbon dioxide and nitrogen. *In situ* water usually contains dissolved solids and dissolved gases.

The relative amount of each component in a mixture may be expressed in such units as volume fraction, weight fraction, or molar fraction. The unit of concentration should be clearly expressed to avoid errors. The elemental composition of oil is primarily carbon (84–87% by mass) and hydrogen (11–14% by mass). Oil can contain other elements as well, including sulfur, nitrogen, oxygen, and various metals.

The symbols x_i and y_i are often used to denote the mole fraction of component i in the liquid and gas phases, respectively. The mole fraction of component i in a gas mixture is the number of moles $n_{i\text{V}}$ of the component in the gas (vapor phase) divided by the total number of moles in the gas phase:

$$y_i = \frac{n_{i\text{V}}}{\sum_{j=1}^{N_\text{c}} n_{j\text{V}}} \tag{3.4}$$

where N_c is the number of components in the mixture. The mole fraction of component i in an oil mixture x_i is similarly defined as the number of moles $n_{i\text{s}}$ of the component in the liquid phase divided by the total number of moles in the liquid phase:

$$x_i = \frac{n_{i\text{L}}}{\sum_{j=1}^{N_\text{c}} n_{j\text{L}}} \tag{3.5}$$

Specific Gravity. Specific gravity is the ratio of the density of a fluid divided by a reference density. Gas specific gravity is calculated at standard conditions using air density as the reference density. Taking the ratio of gas density to a reference gas density and canceling common terms, lets us write the specific gravity of gas as

$$\gamma_\text{g} = \frac{M_\text{a,gas}}{M_\text{a,air}} \cong \frac{M_\text{a,gas}}{29} \tag{3.6}$$

where M_a is apparent molecular weight. Apparent molecular weight is calculated as

$$M_a = \sum_{i=1}^{N_c} y_i M_i \tag{3.7}$$

where N_c is the number of components, y_i is the mole fraction of component i, and M_i is the molecular weight of component i. Gas density is calculated from the ideal gas equation of state as

$$\rho_g = \frac{pM_{a,gas}}{RT} = \frac{p\gamma_g M_{a,air}}{RT} \tag{3.8}$$

where p is the pressure, T is the absolute temperature, R is the gas constant, and mole fraction n is the mass of gas m_g divided by the apparent molecular weight (or molar mass) of gas.

Oil specific gravity is calculated at standard conditions using the density of freshwater as the reference density. The American Petroleum Institute characterizes oil in terms of API gravity. API gravity is calculated from oil specify gravity γ_o at standard temperature and pressure by the equation

$$°\text{API} = \frac{141.5}{\gamma_o} - 131.5 \tag{3.9}$$

If specific gravity γ_o greater than 1, the oil is denser than water and API less than 10. If specific gravity γ_o less than 1, the oil is less dense than water and API greater than 10. Heavy oils with API less than 20 do not contain much gas in solution and have a relatively large molecular weight and specific gravity γ_o. By contrast, light oils with API greater than 30 typically contain a large amount of volatile hydrocarbons in solution and have a relatively small molecular weight and specific gravity γ_o. The equation for API gravity shows that heavy oil has a relatively low API gravity because it has a large γ_o, while light oils have a relatively high API gravity.

Gas–Liquid Ratio. The gas–liquid ratio (GLR) is the ratio of a volume of gas divided by a volume of liquid at the same temperature and pressure. The choice of GLR depends on the fluids in the reservoir. Two commonly used GLR are gas–oil ratio (GOR) and gas–water ratio (GWR). The GWR is the ratio of gas volume to water volume at the same temperature and pressure. GOR is the ratio of gas volume to oil volume at the same temperature and pressure. The ratios can be calculated using volume ratios or ratios of flow rates.

Viscosity. Viscosity is a measure of resistance of a fluid to shearing. Fluids like honey and heavy oil have a very high viscosity, while fluids like water have a relatively low viscosity. In the oil industry, viscosity is often expressed in centipoise, which equals 0.01 poise. One centipoise (1 cp) equals 1 millipascal second (1 mPa s = 0.001 Pa s), which is the metric unit for viscosity.

Compressibility. Compressibility is a measure of the change in volume resulting from the change in pressure applied to the system. The fractional volume change of a system is the ratio of the change in volume ΔV to the initial volume V. The fractional volume change $\Delta V/V$ may be estimated from

$$\frac{\Delta V}{V} = -c\Delta p \tag{3.10}$$

where c is the average compressibility of the system and Δp is the pressure change. The minus sign is applied so that an increase in pressure ($\Delta p > 0$) will result in a decrease in the volume of the system. Similarly, a decrease in pressure ($\Delta p < 0$) will result in an increase in the volume of the system.

Formation Volume Factor. The volume of oil swells when gas is dissolved in the oil. The FVF for oil, B_o, expresses this swelling as a ratio of the swollen volume to the volume of the oil phase at a reference condition, usually the stock tank pressure and temperature. This ratio is expressed as reservoir volume divided by stock tank volume. In this sense "reservoir" refers to pressure, temperature, and composition that exist in a reservoir. An example of a unit for oil FVF in oil field units is RB/STB where "RB" refers to reservoir barrels and STB refers to stock tank barrels or it could be rm^3/sm^3 for reservoir meters cubed per stock tank meters cubed in metric units. For example, an FVF of 1.5 RB/STB means that for every barrel of oil produced to the stock tank, 1.5 barrels were taken from the reservoir. The 0.5 barrel volume difference represents the volume of oil phase lost as volatile species escaped from the liquid phase during the reduction in pressure from the reservoir up through the well to the separator and stock tank.

Usually, most of the change in volume from stock tank to reservoir results from the volume of gas dissolved in the oil. But pressure and temperature also play a role. The increase in pressure from stock tank to reservoir compresses the oil a small amount, while the increase in temperature from stock tank to reservoir thermally expands the oil.

FVF for oil usually ranges from 1 to 2 RB/STB. FVF for water is usually about 1 RB/STB because gas is much less soluble in water than in oil. Gas FVF varies over a wider range than oil FVF because gas volume is more sensitive to changes in pressure.

3.4 GAS PROPERTIES

Formation Volume Factor. For this text, we use the ideal gas law to estimate the FVF B_g for gas:

$$B_g\left(\mathrm{RB/MCF}\right) = 5.03\frac{T\left(^\circ R\right)}{p\left(\mathrm{psia}\right)} \tag{3.11}$$

where the units of each variable B_g, T, p are given in parentheses and MCF denotes $1000\,ft^3$. The coefficient on the right-hand side of the equation includes conversion factors. For this correlation, the temperature in degrees Rankine is required. To improve the estimate of B_g, the real gas law should be used in place of the ideal gas law.

The real gas equation of state can be written in the form

$$Z = \frac{pV}{nRT} \tag{3.12}$$

where Z is the dimensionless gas compressibility factor, R is the gas constant, and n is the number of moles of gas in volume V at pressure p and temperature T. The gas is an ideal gas if $Z=1$ and a real gas if $Z \neq 1$. Gas FVF for a given temperature and pressure is calculated from the real gas equation of state as

$$B_g = \frac{p_{sc}}{Z_{sc}T_{sc}}\frac{ZT}{p} = \frac{\text{reservoir volume}}{\text{standard volume}} \tag{3.13}$$

The subscript *sc* denotes standard conditions (typically 60°F and 14.7 psia).

Viscosity. The viscosity of gases at reservoir conditions usually ranges from 0.02 to 0.04 cp. Correlations are available for more precise estimates. Viscosities of gases are rarely measured for oil and gas applications—they are normally estimated with correlations.

Heating Value. The heating value of a gas can be estimated from the composition of the gas and heating values associated with each component of the gas. The heating value of the mixture H_m is defined as

$$H_m = \sum_{i=1}^{N_c} y_i H_i \tag{3.14}$$

where N_c is the number of components, y_i is the mole fraction of component i, and H_i is the heating value of component i. Heating values of individual components are tabulated in reference handbooks. The heating value of a natural gas is often between 1000 and 1200 BTU/SCF where BTU refers to energy in British thermal units and SCF refers to standard cubic feet of gas.

3.5 OIL PROPERTIES

Examples of correlations for estimating three properties of oils are provided in this section: BP pressure, FVF, and viscosity. Many correlations have been published. They often represent a particular geographic region or selection oil. The selection of a correlation should take into account the source of the data that was used to prepare the correlation. Correlations based on McCain (1990) are used here.

Bubble-Point Pressure. If a container is filled partly with oil and the remainder with gas, the amount of gas dissolved in the oil increases as pressure in the container increases. As long as some gas phase remains in the container, the applied pressure is the saturation pressure, and it is often called a BP pressure even though there may be more than a tiny bubble of gas in the container. At pressures higher than that needed to dissolve all the available gas in the container, the oil is considered undersaturated. The BP pressure, or P_{bp} in psi, can be related to the amount of gas in solution (R_s in SCF/STB), gas gravity (γ_g), temperature (T) in°F, and API gravity (°API) with the following correlation:

$$p_b = 18.2(A - 1.4) \tag{3.15}$$

with

$$A = \left(\frac{R_s}{\gamma_g}\right)^{0.83} 10^{0.00091T - 0.0125\times°API} \tag{3.16}$$

Example 3.2 Bubble-point Pressure

Calculate bubble-point pressure for reservoir temperature of 220°F, oil gravity of 35°API, and gas gravity of 0.68. The amount of gas dissolved in the oil is 350 SCF/STB.

Answer

Use Equations 3.15 and 3.16 with the values provided:

$$A = \left(\frac{350\ \text{SCF/STB}}{0.68}\right)^{0.83} 10^{0.00091(220°F) - 0.0125(35°API)} = 103.11$$

$$P_b = 18.2(103.11 - 1.4) = 1851\ \text{psi}$$

The plot in Figure 3.4 was created using the correlation of Equations 3.15 and 3.16 with the properties of Example 3.2 with R_s varying from near 0 up to 350 SCF/STB. Above the BP pressure of 1851 psi, R_s is constant at 350 SCF/STB.

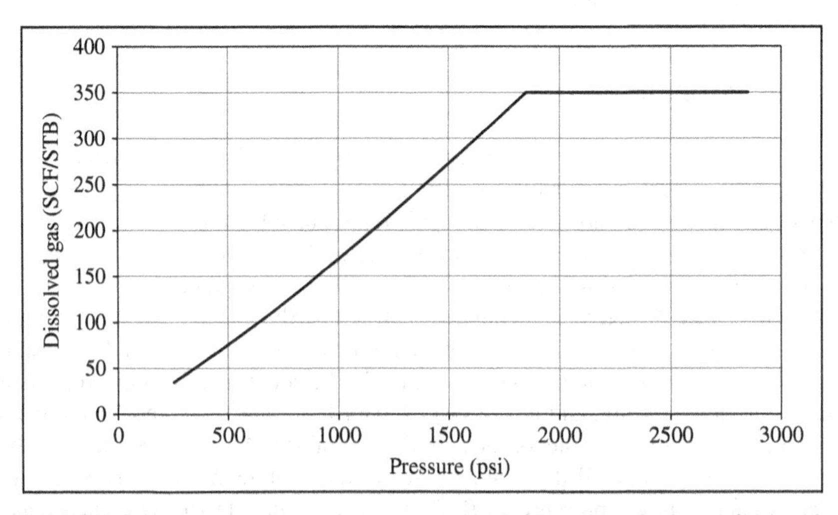

FIGURE 3.4 Demonstration of the correlation in Equations 3.15 and 3.16 with properties from Example 3.2.

Oil Formation Volume Factor. The FVF of oil at the BP pressure (B_{ob} in RB/STB) can be estimated with the following correlation in terms of solution GOR (R_s in SCF/STB), gas gravity (γ_g), oil specific gravity (not API gravity), and temperature (T in °F):

$$B_{ob} = 0.98 + 0.00012\,\mathrm{A}^{1.2} \tag{3.17}$$

with

$$\mathrm{A} = R_s\left(\frac{\gamma_g}{\gamma_o}\right)^{0.5} + 1.25\mathrm{T} \tag{3.18}$$

Example 3.3 Oil Formation Volume Factor

Calculate formation volume factor for the same conditions as Example 3.2, that is, reservoir temperature of 220°F, oil gravity of 35°API, gas gravity of 0.68, and 350 SCF/STB of dissolved gas.

Answer

First, convert 35°API to oil specific gravity, and then use Equations 3.17 and 3.18 with the appropriate values:

$$\gamma_o = \frac{141.5}{°\mathrm{API} + 131.5} = \frac{141.5}{35 + 131.5} = 0.85$$

$$\mathrm{A} = (350\ \mathrm{SCF/STB})\left(\frac{0.68}{0.85}\right)^{0.5} + 1.25(220) = 588$$

$$B_{ob} = 0.98 + 0.00012(588)^{1.2} = 1.23\ \mathrm{RB/STB}$$

The plot in Figure 3.5 shows results for the correlation of Equations 3.17 and 3.18 with the properties of the previous example. We consider R_s varying from near 0 up to 350 SCF/STB.

Oil FVF decreases because of compression of the oil above BP pressure. Oil FVF B_o at pressure p above BP pressure p_b is calculated as

$$B_o = B_{ob} + \delta_p B_o (p - p_b) \tag{3.19}$$

where B_{ob} is oil FVF at BP pressure and $\delta_p B_o$ is the change in oil FVF above BP pressure due to increasing pressure. The value of $\delta_p B_o$ for the oil FVF shown in Figure 3.6 is approximately -1.4×10^{-5} RB/STB/psi for pressures greater than BP pressure. The slope $\delta_p B_o$ is negative since oil FVF decreases as pressure increases for pressures greater than BP pressure.

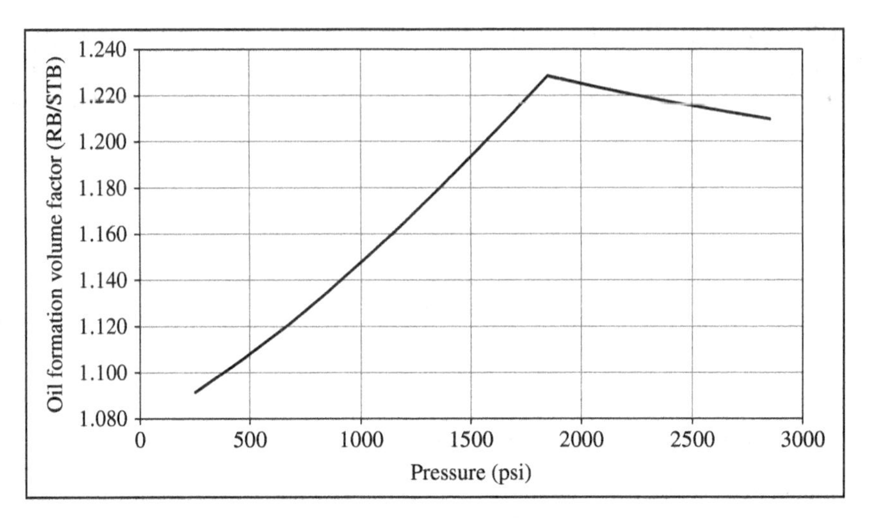

FIGURE 3.5 Demonstration of the correlation in Equations 3.17 and 3.19 with properties from Examples 3.2 and 3.3.

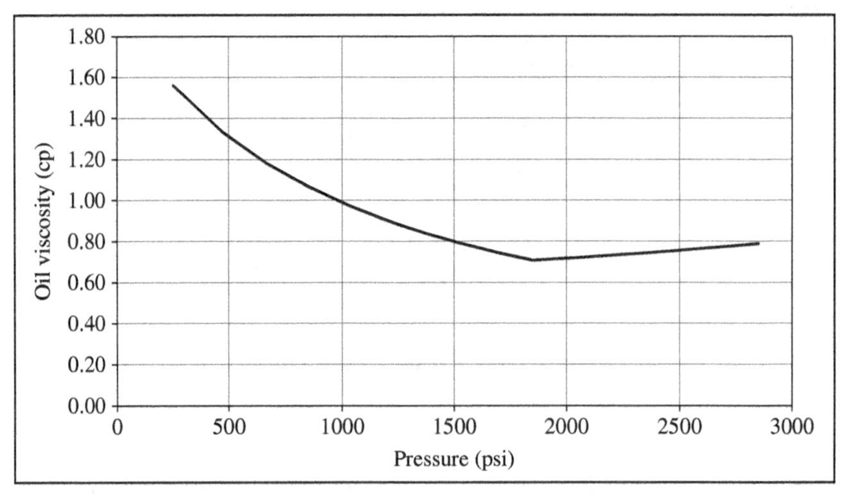

FIGURE 3.6 Demonstration of the correlation in Equations 3.20 through 3.23 with properties from Examples 3.4 and 3.5.

Viscosity. The following correlation for oil viscosity requires two steps. First, the viscosity (cp) of "dead" oil is estimated from API gravity and temperature (°F):

$$\log_{10}\left(\mu_{oD}+1\right)=73.3\frac{10^{-0.0251\times°API}}{T^{0.564}} \tag{3.20}$$

Dead oil refers to oil that has little dissolved gas; it is equivalent to stock tank oil. The second step accounts for the decrease in oil viscosity that occurs as gas dissolves into it:

$$\mu_o = A\mu_{oD}^B \tag{3.21}$$

with

$$A = 10.7(R_s + 100)^{-0.515} \tag{3.22}$$

$$B = 5.44(R_s + 150)^{-0.338} \tag{3.23}$$

where μ_{oD} is dead oil viscosity calculated in the first step and R_s is solution GOR (SCF/STB). Oil with dissolved gas is often called "live" oil. Live oil and dead oil have analogs in the world of carbonated beverages.

Example 3.4 Dead Oil Viscosity

Calculate dead oil viscosity for a 35°API oil at 220°F.

Answer
Substitute values into Equation 3.20:

$$\log_{10}(\mu_{oD} + 1) = 73.3\frac{10^{-0.0251(35)}}{(220)^{0.564}} = 0.46$$

$$\mu_{oD} = 10^{0.46} - 1 = 1.90\,\text{cp}$$

Example 3.5 Live Oil Viscosity

Calculate live oil viscosity for a 35°API oil at 220°F with 350 SCF/STB of dissolved gas.

Answer
Combine the dead oil viscosity from the previous example with the above values and Equations 3.21 through 3.23 to find live oil viscosity:

$$A = 10.7(350 + 100)^{-0.515} = 0.46$$

$$B = 5.44(350 + 150)^{-0.338} = 0.67$$

$$\mu_o = A\mu_{oD}^B = (0.44)(1.90\,\text{cp})^{0.64} = 0.71\,\text{cp}$$

The plot in Figure 3.6 shows results for the correlation of Equations 3.20 through 3.23 with the properties of Examples 3.4 and 3.5. We consider R_s varying from near 0 up to 350 SCF/STB.

Oil viscosity increases because of compression of the oil above BP pressure. Oil viscosity μ_o at pressure p above BP pressure p_b can be estimated using

$$\mu_o = \mu_{ob} + \delta_p \mu_o (p - p_b) \quad (3.24)$$

where μ_{ob} is live oil viscosity at BP pressure and $\delta_p \mu_o$ is the change in oil viscosity above BP pressure due to increasing pressure. The value of $\delta_p \mu_o$ for the oil viscosity shown in Figure 3.6 is approximately 8×10^{-5} cp/psi for pressures greater than BP pressure. The value of the slope $\delta_p \mu_o$ is positive since oil viscosity increases as pressure increases for pressures greater than BP pressure.

3.6 WATER PROPERTIES

The presence of water in geologic formations means that the properties of water must be considered. Water properties are discussed in this section.

Formation Volume Factor. The effects of pressure and temperature on the volume of water very nearly cancel, so the FVF of water is approximately 1.0 RB/STB for most reservoirs.

Viscosity. The viscosity of water depends on pressure, temperature, and composition. At reservoir conditions, water will contain dissolved solids (mostly salts) as well as some dissolved hydrocarbon gases and small amounts (<1000 ppm) of other hydrocarbons. The effects of temperature and composition on water viscosity at 14.7 psi (1 atm) of pressure are shown in Figure 3.7. Viscosity of pure water is near 1 cp at room temperature, but it is much less at typical reservoir temperatures.

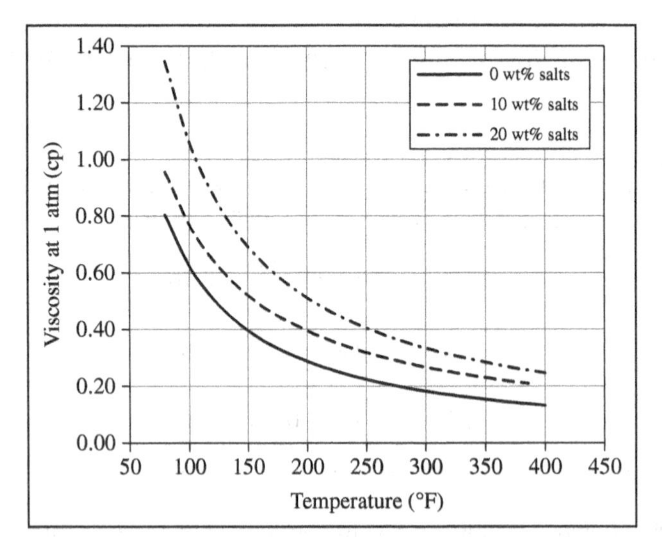

FIGURE 3.7 Effect of temperature and dissolved salts on viscosity of water at 14.7 psi (1 atm).

Viscosity of water increases by about 50% from 14.7 psi (1 atm) to 8000 psi. The effect of pressure on water viscosity can be estimated using the following correlation with pressure in units of psi:

$$\frac{\mu_w}{\mu_{w,1atm}} = 1.0 + 0.00004\,P + 3.1\times10^{-9}\,P^2 \qquad (3.25)$$

3.7 SOURCES OF FLUID DATA

The best information about oil, gas, and water in a formation is obtained from fluid samples that are representative of the original *in situ* fluids. A well should be conditioned before sampling by producing sufficient fluid to flush any contaminating drilling or completion fluids from the well.

Surface sampling at the separator is easier and less expensive than subsurface sampling. For surface samples, the original *in situ* fluid is obtained by combining separator gas and separator oil samples at the appropriate GOR. The recombination step assumes accurate measurements of flow data at the surface. Subsurface sampling from a properly conditioned well avoids the recombination step, but is more difficult and costly than surface sampling, and usually provides a smaller volume of sample fluid.

Once a sample has been acquired, it is necessary to verify the quality of the sample. This can be done by compositional analysis and measurement of such physical properties as density and molecular weight. Gas chromatography is the most typical instrument for compositional analysis.

After sample integrity is verified, several procedures may be performed to measure fluid properties that are suitable for reservoir engineering studies. The most common procedures include the following tests: constant composition expansion (CCE), differential liberation (DL), and separator tests.

3.7.1 Constant Composition Expansion

A CCE test provides information about pressure–volume behavior of a fluid without changes in fluid composition. The CCE test begins with a sample of reservoir fluid in a high-pressure cell at reservoir temperature and at a pressure in excess of the reservoir pressure. Traditionally, the cell contained oil and mercury. Pressure was altered by changing the volume of mercury in the cell. Modern systems are designed to be mercury-free by replacing mercury with a piston. The piston is used to alter pressure in the cell, as illustrated in Figure 3.8. The cell pressure is lowered in small increments, and the change in volume at each pressure is recorded. The procedure is repeated until the cell pressure is reduced to a pressure that is considerably lower than the saturation pressure. The original composition of the fluid in the cell does not change at any time during the test because no material is removed from the cell. The fluid may be either oil or a gas with condensate. If the fluid is oil, the saturation pressure is the BP pressure. If the fluid is a gas with condensate, the saturation pressure is the DP pressure.

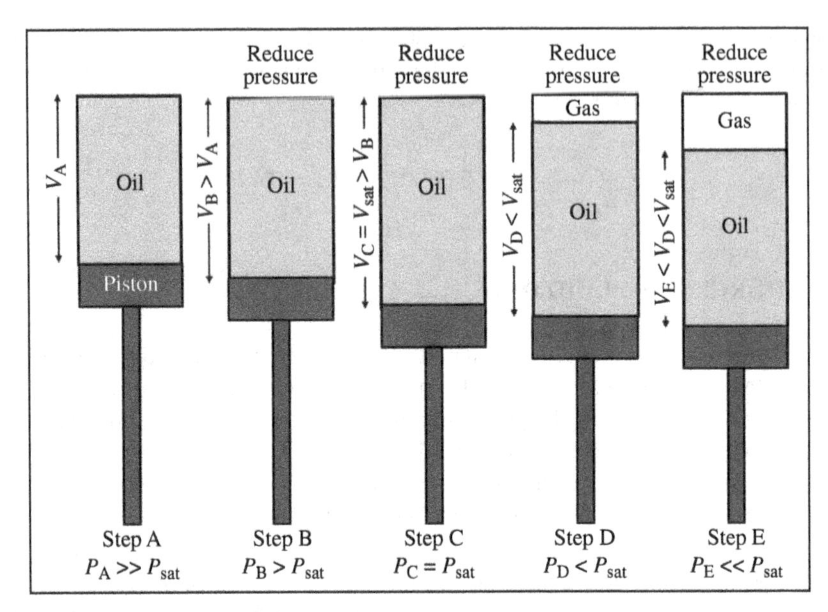

FIGURE 3.8 Constant composition expansion.

3.7.2 Differential Liberation

The DL test is used to determine the liberation of gas from live oil, that is, oil containing dissolved gas. A live oil sample is placed in a PVT cell at reservoir temperature and BP pressure as in Figure 3.9. The pressure is lowered in small increments and the evolved gas is removed at each stage. The volume of the evolved gas and the volume of oil remaining in the cell are recorded. Oil viscosity is typically measured in a DL test.

3.7.3 Separator Test

The separator test is used to study the behavior of a fluid as it flashes from reservoir to surface conditions. A flash is the one-step change from a relatively high-pressure and high-temperature environment to a relatively low-pressure and low-temperature environment. The primary difference between a flash and a differential process is the magnitude of the pressure differential between stages. The pressure differential is generally much smaller in the differential process than in the flash process. Figure 3.10 illustrates a multistage flash where the input stream goes through multiple flashes. Each flash has a different temperature and pressure relative to the preceding flash.

The PVT cell in a separator test is charged with a carefully measured volume of reservoir fluid at reservoir temperature and saturation pressure. The cell pressure and temperature are then changed. Each change in pressure and temperature corresponds to a separator stage, and one or more stages may be used in the test. The volume of gas from each separation stage and the volume of the liquid remaining in the last stage are measured.

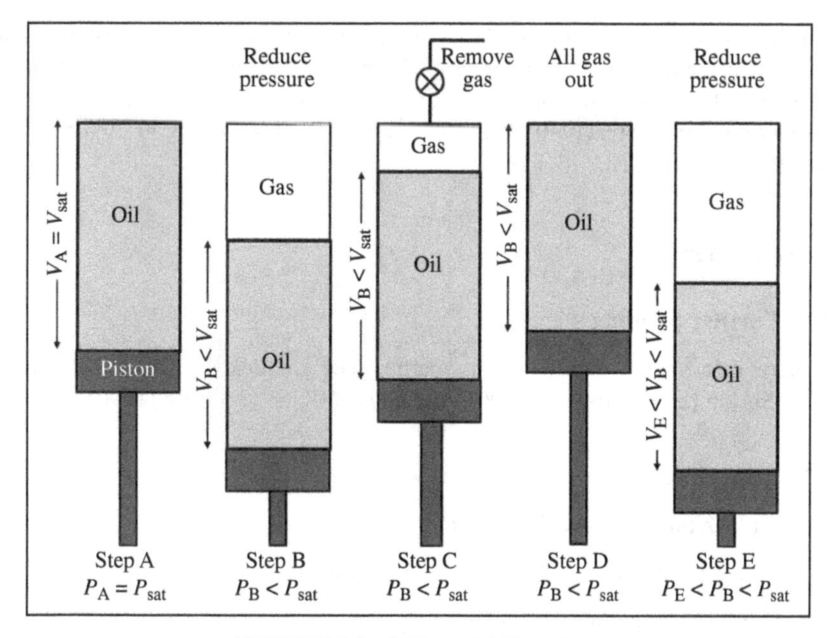

FIGURE 3.9 Differential liberation.

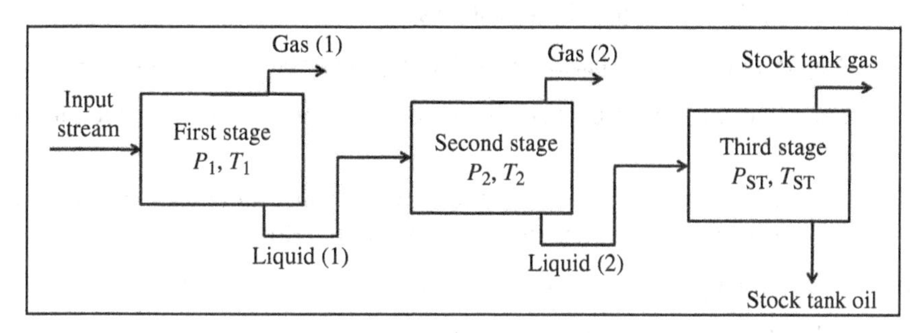

FIGURE 3.10 Multistage flash.

3.8 APPLICATIONS OF FLUID PROPERTIES

Petroleum engineers use fluid properties in a variety of applications, including volumetric estimation of oil and gas in place, material balances, well testing, and reservoir modeling. For volumetric estimates, they use FVF. For material balances, they use a combination of FVF, GOR, and fluid compressibilities. For well testing and reservoir modeling, they use fluid viscosities in addition to the fluid properties needed for the other applications.

Over the past 40 or 50 years, engineers have developed software for modeling of reservoirs with the five types of hydrocarbon fluids described previously. These models predict changes in reservoir pressures as fluids are produced from or injected to the wells. Model predictions are compared to field observations, yielding insight to guide future operations.

Of these models, those for black oils and dry gas are the simplest. The pressure-dependent properties previously described for gas, oil, and water, including the properties in Figures 3.4 through 3.6, are sufficient. If reliable fluid data are missing for one or more of the reservoir fluids, fluid properties from analogous fields or from correlations can be used.

3.9 ACTIVITIES

3.9.1 Further Reading

For more information about fluid properties, see Dandekar (2013), Satter et al. (2008), Sutton (2006), Towler (2006), Ahmed (2000), and McCain (1990).

3.9.2 True/False

3.1 A black oil is always black in color.

3.2 Unlike a volatile oil, a black oil contains no gas.

3.3 Centipoise is a unit of viscosity.

3.4 Condensate exists in the liquid phase if reservoir pressure in a gas condensate reservoir is greater than dew-point pressure.

3.5 A hydrocarbon gas can have a dew-point temperature but not a dew-point pressure.

3.6 The mole fraction of methane is usually larger in oil reservoirs than in gas reservoirs.

3.7 Most scientists believe that fossil fuels were formed from various biological organisms that lived on the surface of the Earth millions of years ago.

3.8 The cricondentherm is the maximum pressure in the *P–T* diagram.

3.9 Oil and gas are carbon-based materials.

3.10 Single-phase oil becomes two-phase gas/oil when pressure drops below the bubble-point pressure.

3.9.3 Exercises

3.1 **A.** The density of water is 1 g/cc or 62.4 lbm/cu ft. The density of air is 0.0765 lbm/cu ft. What is the density of air in g/cc?

B. What is the density of gas in g/cc if the specific gravity of the gas is 0.7?

Hint: The specific gravity of gas is the density of gas divided by the density of air, or $\gamma_g = \rho_g / \rho_{air}$.

3.2 Fluid density for a volume with oil and water phases can be estimated using $\rho_f = S_o \times \rho_o + S_w \times \rho_w$ where ρ_o, ρ_w refer to oil density and water density, S_o, S_w refer to oil saturation and water saturation, and $S_o + S_w = 1$. Estimate fluid density when oil density is 0.9 g/cc, water density is 1.03 g/cc, and water saturation is 30%.

Hint: Convert water saturation to a fraction and calculate oil saturation from $S_o + S_w = 1$.

3.3 **A.** Use the real gas law $pV = ZnRT$ to find a general expression for gas formation volume factor B_g. Use subscripts "s" and "r" to denote surface conditions and reservoir conditions, respectively.

B. Calculate B_g using $\{p_s = 14.7\,\text{psia}, T_s = 60°\text{F}, Z_s = 1\}$ and $\{p_r = 2175\,\text{psia}, T_r = 140°\text{F}, Z_r = 0.9\}$. Express B_g as reservoir cubic feet per standard cubic feet (RCF/SCF).

C. Calculate B_g using $\{p_s = 1\,\text{atm}, T_s = 20°\text{C}, Z_s = 1\}$ and $\{p_r = 15\,\text{MPa}, T_r = 60°\text{C}, Z_r = 0.9\}$. Express B_g as reservoir cubic meters per standard cubic meter (Rm^3/Sm^3).

D. What is the difference between the calculation in Part B and the calculation in Part C?

3.4 **A.** A well produces 1000 MSCF gas and 400 STB oil. What is the GOR in MSCFG/STBO? Note: 1 MSCFG = 1 MSCF gas and 1 STBO = 1 STB oil.

B. Assuming all of the gas is solution gas (no free gas), is the oil a black oil or a volatile oil?

3.5 Suppose bubble-point pressure of oil is 2400 psia in a reservoir with 70% oil saturation and 30% irreducible water saturation. Is the reservoir saturated or undersaturated at an initial reservoir pressure of 2515 psia?

3.6 A set of saturated oil fluid property data is given below as a function of pressure P. Bubble-point pressure PBO is 2014.7 psia. The undersaturated oil viscosity MUO is approximated as a line with slope MUOSLP from bubble-point pressure (PBO) to reservoir pressure above PBO. The undersaturated oil formation volume factor BO is approximated as a line with slope BOSLP from bubble-point pressure (PBO) to reservoir pressure above PBO.

A. Calculate oil viscosity MUO at 3000 psia pressure.

B. Calculate oil formation volume factor BO at 3000 psia pressure.

C. Calculate solution gas–oil ratio RSO at 3000 psia pressure.

PBO (psia)	MUOSLP (cp/psia)	BOSLP (RB/STB/psia)	
2014.7	0.000046	-2.3×10^{-5}	
P (psia)	MUO (cp)	BO (RB/STB)	RSO (SCF/STB)
14.7	1.04	1.062	1
514.7	0.91	1.111	89
1014.7	0.83	1.192	208
1514.7	0.765	1.256	309
2014.7	0.695	1.32	392
2514.7	0.641	1.38	457
3014.7	0.594	1.426	521
4014.7	0.51	1.472	586

3.7 Plot the saturated values of MUO, BO, and RSO in Exercise 3.6 as functions of P.

3.8 **A.** A gas sample has the composition shown in the following table. Estimate the apparent molecular weight of the gas sample by completing the table. The number of moles of each component is in weight/molecular weight. The mole fraction of each component is the number of moles of the component divided by the sum of the number of moles of each component. The apparent molecular weight of each component is the mole fraction × molecular weight of each component. The sum of the apparent molecular weights of the components gives the apparent molecular weight of the mixture.

Component	Weight (lbs)	Molecular Weight (lbs/lb mol)	Number of Moles	Mole Fraction	Apparent Molecular Weight (lbs/lb mol)
Methane (C_1)	20	16			
Ethane (C_2)	6	30			
Propane (C_3)	3	44			
Sum					

B. Calculate the specific gravity of the gas sample when the apparent molecular weight of air is 29.

3.9 **A.** How much energy is in 1 MSCF gas? Assume the energy density is 1000 BTU/SCF. Express your answer in BTU, J, and kWh.

B. Suppose the efficiency of converting the energy in gas to electrical energy is 50%. How much gas is needed to produce 1 kWh of electrical energy?

C. If gas costs $7/MSCF, how much does it cost to produce 1 kWh of electrical energy?

3.10 **A.** Suppose the gas compressibility factor of a gas in a reservoir is 0.9 at a temperature of 160°F and pressure of 3500 psia. Calculate gas formation volume factor at reservoir conditions assuming standard pressure is 14.7 psia and standard temperature is 60°F.

B. Estimate original gas in place given

Net acre feet = 1600 acre-ft (maps)
Initial water saturation = 0.20 (log data, cores)
Porosity = 0.25 (log data, cores)

and

$$\text{OGIP} = 7758 \frac{\text{NAF}(1 - S_{\text{wi}})\phi}{B_{\text{gi}}}$$

where

OGIP = original gas in place, scf
7758 = conversion factor, res bbl/acre-ft
NAF = net acre feet, acre-ft
S_{wi} = initial water saturation, fraction
B_{gi} = gas FVF at initial pressure and temperature, res bbl/scf
ϕ = porosity, fraction

4

PROPERTIES OF RESERVOIR ROCK

Porosity and permeability are fundamental rock properties. Porosity is a measure of the storage capacity of the rock and permeability is a measure of rock flow capacity. Storage capacity tells us how much resource can be contained in the rock, and flow capacity tells us how fast we can produce the resource. Porosity, permeability, and associated topics are the subject of this chapter.

4.1 POROSITY

Consider a handful of sand that contains sand grains and space between grains. The volume of the sand is the sum of the volume of the sand grains plus the volume of space between grains. Figure 4.1 presents a sketch of a block of rock with grains of sand filling the block and an example of Berea sandstone from Berea, Ohio. The Berea sandstone image is a microscopic view of sandstone surface that has been expanded 25 times. Rock grains and the space between grains, called pore space, are visible in the image.

What is the difference between a handful of sand and a piece of sandstone? The handful of sand consists of loose grains of sand. By contrast, grains in the sandstone are cemented together by minerals that have precipitated from the mineral-rich water that has occupied the pore space over tens of thousands, perhaps millions, of years. Furthermore, sandstone is typically composed of a variety of grain sizes. Small grains can fill pore space between larger grains and reduce the porosity

Introduction to Petroleum Engineering, First Edition. John R. Fanchi and Richard L. Christiansen.
© 2017 John Wiley & Sons, Inc. Published 2017 by John Wiley & Sons, Inc.
Companion website: www.wiley.com/go/Fanchi/IntroPetroleumEngineering

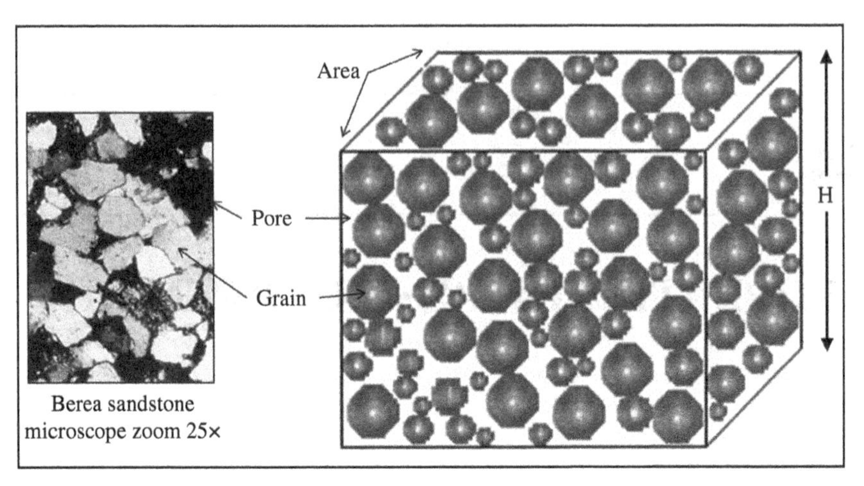

FIGURE 4.1 Porous medium.

of sandstone. Some rocks have pore spaces that are not connected. Our interest here is confined to connected pores that contribute to commercial storage and flow capacity.

Bulk volume is the volume of the block shown in Figure 4.1. It includes both grain volume and the volume of space between grains in the block. The bulk volume V_B of the block is the product of area A in the horizontal plane and gross thickness H:

$$V_B = AH \tag{4.1}$$

The volume of space between grains is called pore volume. Porosity ϕ is the ratio of pore volume V_p to bulk volume V_B:

$$\phi = \frac{V_p}{V_B} \tag{4.2}$$

Typically, 30–40% of the bulk volume of sand is open volume. Porosities of commercially viable reservoir systems range from a few percent for shales and coals to about 50% for diatomaceous formations in California. The porosities of most conventional oil and gas formations range from 15 to 25%. Porosities for a variety of media are listed in Table 4.1.

Consider another example. A beaker contains 500 ml of marbles. In terms of bulk volume, the marbles occupy a bulk volume of 500 ml. How much water must be added to bring the water level to the top of the marbles? If 200 ml must be added, then the pore volume is 200 ml and the porosity is 40%. What is the actual volume of the marbles? That volume, typically referred to as grain volume for rocks, would be 300 ml. In summary, bulk volume V_B is the sum of grain volume V_G and pore volume V_p:

$$V_B = V_G + V_p \tag{4.3}$$

TABLE 4.1 Porosities of Media of Geologic Origin

Media	Porosity (%)
Coal	2–3
Shale	5–10
Sandstones	<25
Limestones	<25
Dolomites	<30
Loose sand (well sorted)	32–42
Diatomites	40–60

Example 4.1 Core Porosity

A sandstone core sample is cleanly cut and carefully measured in a laboratory. The cylindrical core has a length of 3 in. and a diameter of 0.75 in. The core is dried and weighed. The dried core weighs 125 g. The core is then saturated with freshwater. The water-saturated core weighs 127.95 g. Determine the porosity of the sandstone core. Neglect the weight of air in the dried core and assume the density of water is 1 g/cc.

Answer
Bulk volume:

$$V_B = \pi r^2 h = 3.14159 \times \left(\frac{0.75\text{ in.}}{2}\right)^2 (3\text{ in.}) = 1.325\text{ in.}^3 \times \left(\frac{2.54\text{ cm}}{1\text{ in.}}\right)^3 = 21.7\text{ cc}$$

Pore volume = mass of water in core divided by density of freshwater:

$$V_P = \frac{m_w}{\rho_w} = \frac{(127.95\text{ g} - 125\text{ g})}{1\text{ g/cc}} = 2.95\text{ cc}$$

Porosity = pore volume divided by bulk volume:

$$\phi = \frac{V_P}{V_B} = \frac{2.95}{21.7} = 0.136$$

4.1.1 Compressibility of Pore Volume

The city of Long Beach, California, sank in the 1940s, 1950s, and 1960s as oil and water were pumped from the underlying formations, most significantly the Wilmington Oil Field. The extent of sinking, or subsidence, ranged from inches up to 29 ft. Considerable problems arose with damage to roads, buildings, and wells. Extensive flooding occurred in the area. The cause of the subsidence was compression of the pore volume in the subsurface formations. Pore volume compression coincided with extensive oil and water production and corresponding pressure reduction in producing formations. Starting in the mid-1960s, oil producers were required to inject about 1.05 barrels of water into the Wilmington Field for every barrel of oil and water produced. The water injection stopped surface subsidence. Most of the lost surface elevation has since been recovered.

Subsidence is a common problem in oil and gas operations, but it is usually much less than the subsidence seen in the Long Beach area. The amount of subsidence depends on rock strength and thickness of the producing formations. The strength of the rock can be measured in terms of pore volume compressibility

$$c_{\mathrm{f}} = \frac{1}{V_p}\frac{\Delta V_{\mathrm{p}}}{\Delta p} \tag{4.4}$$

where V_{p} is the pore volume and p is the pressure in the pore volume. Equation 4.4 can be rewritten as

$$\Delta V_{\mathrm{p}} = c_{\mathrm{f}} V_{\mathrm{p}} \Delta p \tag{4.5}$$

Pore volume compressibility is positive because the pore volume increases as pressure in the pore volume increases. Alternatively, as pressure in the pore volume decreases, the pore volume also decreases. We see from Equation 4.3 that a decrease in pore volume leads to a corresponding decrease in bulk volume. Typically, pore volume compressibility c_{f}, which is also referred to as formation compressibility, ranges from about 10×10^{-6} psi^{-1} to 60×10^{-6} psi^{-1}.

4.1.2 Saturation

Pore spaces in reservoir rock are occupied by fluid phases, including oil, water, and gas phases. The fraction of the pore volume that is occupied by any phase is called the saturation of that phase. For example, S_{o}, S_{w}, and S_{g} are the saturations of oil, water, and gas phases, respectively. Saturation can be expressed as a percentage or as a fraction. For a reservoir with oil, water, and gas phases, the sum of the saturations must satisfy the constraint

$$1 = S_{\mathrm{o}} + S_{\mathrm{w}} + S_{\mathrm{g}} \tag{4.6}$$

since the sum of the pore space occupied by each phase must equal the total pore space. If just oil and water are present in a reservoir, then gas saturation $S_{\mathrm{g}} = 0$ and

$$1 = S_{\mathrm{o}} + S_{\mathrm{w}} \tag{4.7}$$

Example 4.2 Core Saturation

A sandstone core is completely saturated with 2.95 ml water. Oil is injected into the core and 2.10 ml of water is collected from the core. What are the oil and water saturations in the core?

Answer

Pore volume = initial volume of water = 2.95 ml
Oil saturation = 2.10 ml/2.95 ml = 0.712
Water saturation = (2.95 ml − 2.10 ml)/2.95 ml = 0.288
Alternatively $S_w = 1 - S_o = 0.288$

4.1.3 Volumetric Analysis

The volume of oil in place V_o in a reservoir with bulk volume V_B can be expressed with quantities introduced in this chapter:

$$V_o = V_B \phi S_o \tag{4.8}$$

If A is the areal extent of the reservoir and h its thickness, then

$$V_o = Ah\phi S_o \tag{4.9}$$

As written, Equations 4.8 and 4.9 give the volume of oil at reservoir conditions. To obtain oil volume at stock tank conditions, the formation volume factor for oil B_o must be included:

$$V_o = \frac{Ah\phi S_o}{B_o} \tag{4.10}$$

Example 4.3 OIP in a Square Mile

In the Sanish field of North Dakota, the porosity is 0.06 and the oil saturation is 75%. The formation is 30 ft thick. What is the oil in place in STB in a square mile? The oil formation volume factor is 1.5 RB/STB.

Answer

$$V_o = \frac{Ah\phi S_o}{B_o} = \frac{(5280\,\text{ft})^2(30\,\text{ft})(0.06)(0.75)}{1.5\,\text{RB/STB}} \frac{1\,\text{Bbl}}{5.6\,\text{ft}^3} = 4500000\,\text{STB}$$

4.2 PERMEABILITY

The concept of permeability was developed by Henry Darcy in the early 1800s while studying the flow of water through sand filters for water purification. Darcy found that flow rate is proportional to the pressure change between the inlet and outlet of the porous medium. Darcy's relationship between flow rate and pressure change, now known as Darcy's law, states that flow rate q of a fluid of viscosity μ is related to pressure drop Δp by the equation

$$q = -\frac{kA}{\mu L}\Delta p \tag{4.11}$$

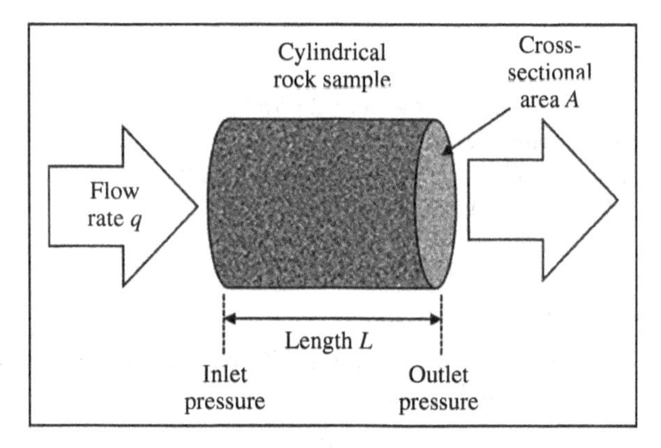

FIGURE 4.2 Definition of terms for Darcy's law.

where the terms of the physical system are defined in Figure 4.2. As defined by Darcy, the terms in Equation 4.11 have the following units:

Term	Description	Units
q	Flow rate	cm^3/s
k	Permeability	darcies
A	Cross-sectional area	cm^2
$-\Delta p$	Pressure drop	atmospheres
μ	Viscosity	centipoise
L	Length	cm

Permeability units are called darcies. Most conventional reservoir applications with sandstone and limestone use millidarcies (md) where 1 darcy = 1000 md. In oil field units, Equation 4.11 has the form

$$q = -0.001127 \frac{kA}{\mu L} \Delta p \tag{4.12}$$

where the constant 0.001127 is the appropriate conversion factor and the units for Equation 4.12 are the following:

Term	Description	Units
q	Flow rate	bbl/day
k	Permeability	millidarcies (md)
A	Cross-sectional area	ft^2
$-\Delta p$	Pressure drop from inlet to outlet	psi
μ	Viscosity	centipoise
L	Length	ft

Permeability is qualitatively proportional to the average cross-sectional area of pores in a porous material, and its dimensions are those of area. For example, 1 darcy $= 9.87 \times 10^{-13}\,m^2$. Coarse-grain sandstone might have a permeability of

TABLE 4.2 Examples of Permeability

Porous Medium	Permeability
Coal	0.1–200 md
Shale	<0.005 md
Loose sand (well sorted)	1–500 md
Partially consolidated sandstone	0.2–2 d
Consolidated sandstone	0.1–200 md
Tight gas sandstone	<0.01 md
Limestone	0.1–200 md
Diatomite	1–10 md

500 md, while fine-grain sandstone might have a permeability of just a few millidarcies. Rough estimates of permeabilities for a variety of media are listed in Table 4.2.

Example 4.4 Flow Rate From Darcy's Law

Assume the permeability of a cylindrical rock sample (often called a core) is 150 md, the length of the core is 6 in., the diameter of the core is 1 in., the pressure drop across the core is 20 psi, and the viscosity of brine passing through the core is 1.03 cp. Use Darcy's law to calculate the magnitude of volumetric flow rate in bbl/day.

Answer

Cross-sectional area of the core: $A = \frac{1}{4}\pi d^2 = \frac{1}{4}\pi\left(\frac{1}{12}\text{ft}\right)^2 = 0.0054\,\text{ft}^2$

Volumetric flow rate through the core:

$$q = 0.001127\frac{(150\,\text{md})(0.0054\,\text{ft}^2)}{(1.03\,\text{cp})(0.5\,\text{ft})}(20\,\text{psi}) = 0.036\,\text{bbl/day}$$

4.2.1 Pressure Dependence of Permeability

Rock above a formation is known as the overburden. Pore pressure decreases in the formation as fluid is withdrawn from the formation during production. The weight of the overburden compresses formation rock as pore pressure decreases. The decrease in pore space due to compression leads to reduction in formation porosity and permeability. The rate of permeability decrease can be quite variable, depending on the strength of the rock and the structure of pores. Figure 4.3 shows the change in permeability for a reservoir with an initial pore pressure of 2500 psi. As pressure declines, the permeability decreases from the initial value of about 33 md to about 26 md at pore pressure of 1000 psi.

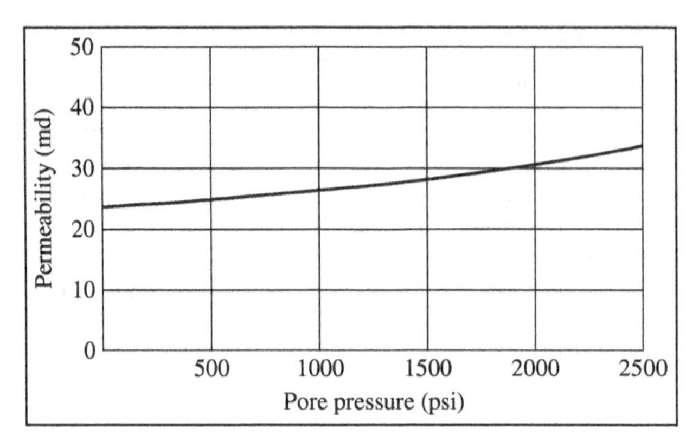

FIGURE 4.3 Pressure dependence of permeability.

4.2.2 Superficial Velocity and Interstitial Velocity

Two types of velocities often appear in reservoir engineering discussions. Sometimes called Darcy velocity, superficial velocity is the quotient of flow rate and cross-sectional area:

$$u = \frac{q}{A} \tag{4.13}$$

The term "superficial" is a good descriptor for this velocity because it ignores the effect of porosity. On the other hand, interstitial velocity does account for porosity:

$$v = \frac{u}{\phi} = \frac{q}{\phi A} \tag{4.14}$$

Interstitial velocity is the average velocity of fluid moving through a porous rock. Superficial velocity is much less than interstitial velocity.

4.2.3 Radial Flow of Liquids

Equations 4.11 and 4.12 describe flow rate for a linear geometry. For flow of a liquid, such as oil, in radial geometry, Darcy's law becomes

$$q_o = 0.00708 \frac{kh}{\mu_o B_o \ln(r_e / r_w)} (p_e - p_w) \tag{4.15}$$

The units for Equation 4.15 are as follows:

Term	Description	Units
q_o	Flow rate	STB/D
k	Permeability	Millidarcies (md)
h	Thickness	ft
p_e	Pressure at external boundary	psi
p_w	Pressure in well (inside boundary)	psi
μ_o	Viscosity	Centipoise
B_o	Formation volume factor	RB/STB
r_w	Well radius (inside boundary)	ft
r_e	Radius of external boundary	ft

Switching from linear flow in Equation 4.12 to radial flow produced the constant 0.00708, which equals $2\pi \times 0.001127$. When pressure at the external boundary is greater than well pressure, the flow rate is positive, corresponding to a producing well. When p_w is greater than p_e, the flow rate is negative, corresponding to an injection well.

The external radius r_e is frequently called the drainage radius of the well. The flow rate in Equation 4.15 is less sensitive to an error in the estimate of r_e than a similar error in a parameter like permeability because the radial flow calculation depends on the logarithm of r_e. It is therefore possible to tolerate larger errors in r_e than other flow parameters and still obtain a reasonable value for radial flow rate. Usually, the drainage radius is equated to the radius of a circle that has the same area as that based on well spacing. For example, if there is one well centered on an area of 40 acres, the total area drained by that well in square feet is $(40\,\text{acres})(43560\,\text{ft}^2/\text{acre}) = 1.74 \times 10^6\,\text{ft}^2$ and the drainage radius is $r_e = (1.74 \times 10^6\,\text{ft}^2/\pi)^{1/2} = 745\,\text{ft}$. This estimate assumes that the well is centered in the 40-acre area. Methods for noncentered wells are beyond the scope of this text.

4.2.4 Radial Flow of Gases

Darcy's law for radial flow of gases is complicated by changes in gas viscosity and gas formation volume factor with pressure. To manage these complications, a new property was developed called real gas pseudopressure, *m*(*p*). The details of real gas pseudopressure are beyond the scope of this text, but in short, *m*(*p*) is a pressure-weighted average of gas viscosity and gas compressibility factor *z*, which is defined for real gases as follows:

$$z = \frac{pV}{nRT} \tag{4.16}$$

The compressibility factor *z* indicates departure from ideal gas behavior. For an ideal gas, *z* equals unity. For real gases, *z* can be as low as about 0.3 and as high as 1.6. As a result, Darcy's law for radial flow of gases is as follows:

$$q_s = 0.703 \frac{kh}{T_r \ln(r_e/r_w)} \left[m(p_e) - m(p_w) \right] \tag{4.17}$$

The units for Equation 4.17 are as follows:

Term	Description	Units
q_s	Gas flow rate at standard conditions (60 °F, 14.7 psia)	SCF/D
k	Permeability	Millidarcies (md)
h	Thickness	ft
$m(p_e)$	Real gas pseudopressure at external boundary	psi^2/cp
$m(p_w)$	Pressure in well (inside boundary)	psi^2/cp
T_r	Reservoir temperature	°R (Rankine temperature)
r_w	Well radius (inside boundary)	ft
r_e	Radius of external boundary	ft

4.3 RESERVOIR HETEROGENEITY AND PERMEABILITY

Rock formations are quite heterogeneous. In pieces of sandstone, bedding planes are usually visible at a scale of 1–5 mm, looking somewhat like the lines in a piece of wood. These bedding planes are visible because of variations in grain size and composition. These variations lead to variations in porosity and permeability. In addition to these small-scale variations, variations at larger scale also occur and are often seen in road cuts through sedimentary formations. In this section, features of permeability heterogeneity are considered.

4.3.1 Parallel Configuration

Consider linear flow through two parallel layers of porous material with permeabilities, thicknesses, and length as shown in Figure 4.4. The inlet pressure is the same for both layers. Similarly, the outlet pressure is the same for both layers. The total flow rate is the sum of the flow rates for the two layers. Using Darcy's law, we calculate the average permeability for the two layers as

$$k_{\text{ave}} = \frac{k_1 h_1 + k_2 h_2}{h_1 + h_2} \tag{4.18}$$

This result can be generalized to multiple layers as follows:

$$k_{\text{ave}} = \frac{\sum_i k_i h_i}{\sum_i h_i} \tag{4.19}$$

Equation 4.19 holds for linear and radial flow.

4.3.2 Series Configuration

Consider linear flow through two porous materials in series with permeabilities, lengths, and thickness as shown in Figure 4.5. In this case, the flow rate is the same in each serial segment. The total pressure drop is the sum of the pressure

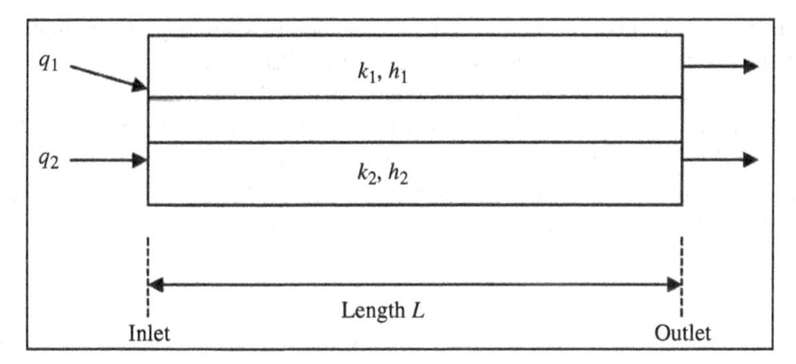

FIGURE 4.4 Flow through two layers of porous material.

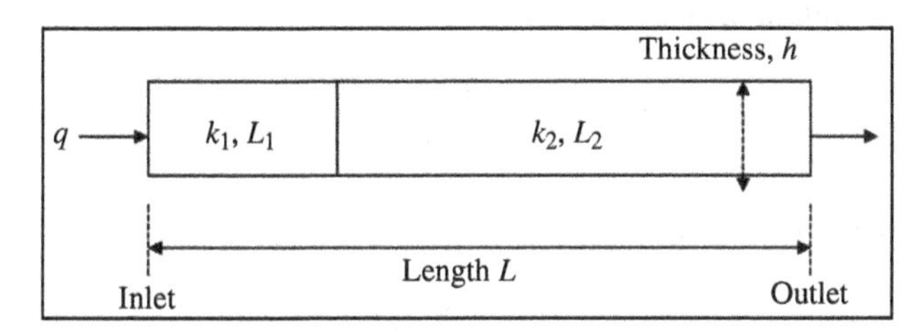

FIGURE 4.5 Serial flow through two adjacent beds of porous material.

drop for each segment. Using Darcy's law, it can be shown that the average permeability is

$$k_{\text{ave}} = \frac{L_1 + L_2}{L_1/k_1 + L_2/k_2} \tag{4.20}$$

Generalizing to multiple segments in series gives

$$k_{\text{ave}} = \frac{\sum_i L_i}{\sum_i L_i/k_i} \tag{4.21}$$

In the case of radial flow, beds in series are concentric rings around a wellbore with radius r_w. The average permeability for a series of three beds is

$$k_{\text{ave}} = \frac{\ln(r_e/r_w)}{\frac{\ln(r_e/r_2)}{k_3} + \frac{\ln(r_2/r_1)}{k_2} + \frac{\ln(r_1/r_w)}{k_1}} \tag{4.22}$$

4.3.3 Dykstra–Parsons Coefficient

The Dykstra–Parsons coefficient is one of several ways to characterize the heterogeneity of permeability with a single number. Having a single number allows simple comparisons of permeability data from different wells, reservoirs, and even numerical models.

Modern numerical models often incorporate randomly chosen permeabilities in an effort to recreate reservoir behavior. With the Dykstra–Parsons coefficient, one can compare the heterogeneity of the numerical model to that of the reservoir.

To calculate the Dykstra–Parsons coefficient V_{DP}, we must have a collection of permeability data for multiple layers of the same thickness in a reservoir. For example, we can determine permeability for each 2-ft-thick interval in a reservoir that is 40 ft thick so that we have a set of 20 permeabilities. Then, V_{DP} can be calculated as follows:

$$V_{DP} = 1 - \exp\left[-\sqrt{\ln\left(\frac{k_A}{k_H}\right)}\right] \tag{4.23}$$

where k_A is the arithmetic average

$$k_A = \frac{1}{n}\sum_{i=1}^{n} k_i \tag{4.24}$$

and k_H is the harmonic average

$$\frac{1}{k_H} = \frac{1}{n}\sum_{i=1}^{n}\frac{1}{k_i} \tag{4.25}$$

For a reservoir with homogeneous permeability, V_{DP} is 0. With increasing heterogeneity, V_{DP} increases toward 1. In most cases, V_{DP} is between 0.3 and 0.9.

Example 4.5 Dykstra–Parsons

A. A reservoir has three layers with the following permeabilities from the upper layer to the lower layer: 100 md, 5 md, and 25 md. What is the arithmetic average of permeability in md?

B. What is the harmonic average of permeability in md?

C. What is the Dykstra–Parsons coefficient?

Answer

A. $k_A = \frac{1}{n}\sum_{i=1}^{n} k_i = \frac{1}{3}(100 + 5 + 25) = 43.3\,\text{md}.$

B. $\frac{1}{k_H} = \frac{1}{n}\sum_{i=1}^{n}\frac{1}{k_i} = \frac{1}{3}\left(\frac{1}{100} + \frac{1}{5} + \frac{1}{25}\right)$ or $k_H = 12.0\,\text{md}.$

C. $V_{DP} = 1 - \exp\left[-\sqrt{\left(\ln\left(\frac{k_A}{k_H}\right)\right)}\right] = 0.678.$

4.4 DIRECTIONAL PERMEABILITY

In general, the value of permeability depends on direction. The directional dependence of permeability is often represented in the $\{x, y, z\}$ directions as $\{k_x, k_y, k_z\}$. Permeability in a porous medium is considered isotropic if permeability does not depend on direction so that $k_x = k_y = k_z$, otherwise permeability is considered anisotropic.

In many reservoirs, permeability in the horizontal plane ($k_x = k_y$) is about 10 times the permeability in the vertical direction k_z. This difference is usually more pronounced in sandstones than in limestones, and it results from layering that is visible in sandstones. Permeability parallel to the direction of deposition of the layers is greater than permeability perpendicular to the direction of deposition of the layers. In many sandstone reservoirs, the layers are horizontal or nearly so. But in other sandstone reservoirs, the layers are inclined. Layer inclination depends on the local depositional environment and geological events such as folding and tectonic activity.

A medium is considered homogeneous if permeability is isotropic at a point in the medium and the directional values of permeability do not change from one position in the medium to another. If permeability varies from one point in the medium to another, the medium is considered heterogeneous. Virtually all reservoirs exhibit some degree of anisotropy and heterogeneity, but the flow behavior in many reservoirs can be approximated as homogeneous and isotropic.

The effect of the directional dependence of permeability on fluid flow is illustrated in Figure 4.6 (see also Fanchi, 2010). Part A is a sketch of the drainage area of four production wells with isotropic permeability, and Part B is a sketch of the drainage area of four production wells with anisotropic permeability. Fluid flow into the wells is radial in Part A, while it is elliptical in Part B with flow greater in the direction of larger directional permeability.

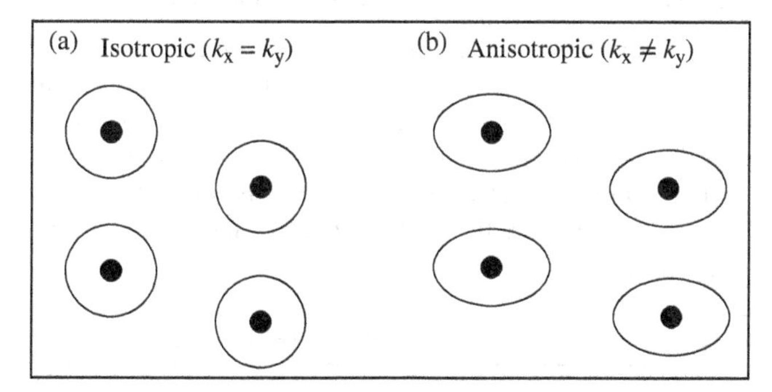

FIGURE 4.6 Illustration of the effect of permeability dependence on direction (after Fanchi, 2010).

4.5 ACTIVITIES

4.5.1 Further Reading

For more information about rock properties, see Dandekar (2013), Mavko et al. (2009), and Batzle (2006).

4.5.2 True/False

4.1 Permeability is a measure of the flow capacity of a porous medium.

4.2 The distribution of porosity is heterogeneous if porosity is the same everywhere in the reservoir.

4.3 An oil–water transition zone is usually thicker than a gas–oil transition zone.

4.4 Viscosity is used in Darcy's law.

4.5 Interstitial velocity is less than superficial velocity.

4.6 Bulk volume is the sum of grain volume and pore volume.

4.7 The Dykstra–Parsons coefficient is dimensionless.

4.8 The oil–water contact is usually found above the gas–oil contact in a reservoir with free gas, oil, and water.

4.9 The value of permeability depends on the direction of fluid flow.

4.10 Real gas pseudopressure has the unit of pressure, such as psia.

4.5.3 Exercises

4.1 Estimate volumetric flow rate using Darcy's law. Assume the permeability in a cylindrical core is 100 md, the length of the core is 6 in, the diameter of the core is 1 in., the pressure drop across the core is 10 psi, and the viscosity of liquid passing through the core is 1.1 cp. Express volumetric flow rate in bbl/day.

4.2 The pressure at an injection well is 2000 psi and the pressure at a production well is 1500 psi. The injection well and production well are separated by a distance of 1000 ft. The mobile fluid in the reservoir between the injection well and the production well has a viscosity of 0.9 cp. The net thickness of the reservoir is 15 ft and the effective width of the reservoir is 500 ft. Use the linear version of Darcy's law to fill in the following table. Assume the formation volume factor of the mobile fluid is 1 RB/STB.

Permeability (md)	Flow Rate from Injector to Producer (bbl/day)
1	
10	
100	
1000	

4.3 A core sample is 3 in. long. The flow path through the core can be observed by injecting dye into the core and then slicing the core into thin sections which can be studied under the microscope. Estimate the tortuosity of the core if the length of the flow path is approximately 4 in. and assume the flow path is confined to flow along a two-dimensional surface. Note: Tortuosity is the actual length of the flow path divided by the length of a linear flow path.

4.4 **A.** The superficial velocity of a core is approximately 1 ft/day. What is the interstitial velocity if the core has 20% porosity?

B. Effective permeability to oil is 100 md and absolute permeability is 200 md. What is the relative permeability to oil?

4.5 Hall's correlation relates porosity to formation compressibility using the relation

$$c_f = \left(\frac{1.782}{\phi^{0.438}} \right) \times 10^{-6}$$

where ϕ = porosity (fraction) and c_f = formation compressibility (1/psi). Use Hall's correlation to estimate formation compressibility at 18% porosity. Remember to include units.

4.6 A reservoir has three layers. Each layer has a thickness of 1 ft and the following permeabilities:

Layer	Permeability (md)
1	126
2	4
3	35

A. What is the arithmetic average of permeability?

B. What is the harmonic average of permeability?

C. Use these values to estimate the Dykstra–Parsons coefficient for a log-normal permeability distribution.

4.7 A cylindrical sandstone core sample has a length of 3 in. and a diameter of 0.5 in. The core is dried and weighed. The dried core weighs 25.0 g. The core is then saturated with freshwater. The water-saturated core weighs 27.5 g. What is the porosity of the sandstone core? Neglect the weight of air in the dried core and assume the density of water is 1 g/cc at laboratory temperature and pressure.

4.8 **A.** Suppose a well flows 600 STBO/D at a pressure drawdown of 25 psia. Calculate the productivity index of the well. Recall that production rate is equal to productivity index times pressure change.

B. What pressure drawdown is required to produce 100 STBO/D if the productivity index for a well is 0.23 STBO/D/psi?

5

MULTIPHASE FLOW

Single-phase flow refers to flow of a single fluid phase in rock. Other phases may be present but are immobile. For example, single-phase oil flows in undersaturated oil reservoirs even though water is present at connate water saturation. Multiphase flow refers to the simultaneous flow of two or more phases in the rock. The presence of two or more flowing phases in a pore space affects the flow of each phase and also the interaction between fluid and rock. To understand and manage multiphase flow processes, reservoir engineers have identified properties that represent the interaction between fluid phases flowing in the same pore space and the interaction between rock and fluid phases. Rock–fluid properties and their application to multiphase flow through rock are introduced in this chapter.

5.1 INTERFACIAL TENSION, WETTABILITY, AND CAPILLARY PRESSURE

Reservoir rock may contain oil, gas, and water phases dispersed throughout the pore space of the rock. The total surface area of these phases, including the solid surface of the rock, is large. As a result, interactions between the rock surface and fluid phases are important in understanding flow in the reservoir. Two surface properties affect the distribution of fluid phases in a reservoir: interfacial tension (IFT) and wettability.

Introduction to Petroleum Engineering, First Edition. John R. Fanchi and Richard L. Christiansen.
© 2017 John Wiley & Sons, Inc. Published 2017 by John Wiley & Sons, Inc.
Companion website: www.wiley.com/go/Fanchi/IntroPetroleumEngineering

IFT has units of force per unit length, which is equivalent to energy per unit area. IFT can refer to the force acting at the boundary of the interface between two phases or to the energy needed to form the area within the boundary. IFT arises because of the differences in molecular attractions that are experienced by molecules at the interface between phases. Consider, for example, a water–oil interface. In water, molecules are capable of hydrogen bonding with each other because of their polarity and shape. In oil, molecules have little or no polarity, and bonding between molecules is weak. A water molecule at the interface will feel strong attractive forces toward the other water molecules in the water phase but not from the molecules in the oil phase. This difference in attractions produces IFT. Some examples of IFT are listed in Table 5.1.

An important consequence of IFT is a difference in pressure between two adjacent phases. That pressure difference is called capillary pressure p_c. It is proportional to IFT and the inverse of curvature of the interface. For a spherical oil drop of radius r surrounded by water, capillary pressure is the difference in pressure given by

$$p_{cow} = p_o - p_w = \frac{2\sigma_{ow}}{r} \tag{5.1}$$

where σ_{ow} is the oil–water IFT. Pressure inside the drop of oil is higher than pressure in the surrounding water. Capillary pressure increases with decreasing size of pore space. Shales have very small pores and capillary pressure can approach 1000 psi.

Wettability is a result of the interactions between a solid surface and two adjacent fluid phases as shown in Figure 5.1. The contact angle is the most fundamental measure of wettability. Figure 5.1a shows water and oil in contact with a solid surface. The surface is termed water wet because the contact angle is less than 90°. In this competition for contact with the solid, the water is spreading over the solid. Water is the wetting phase and oil is the nonwetting phase. If a droplet of water is placed on a clean glass surface with surrounding air, its shape is similar to

TABLE 5.1 Examples of Interfacial Tension

Fluid Pair	IFT Range (mN/m or dyne/cm)
Air–brine	72–100
Oil–brine	15–40
Gas–oil	35–65

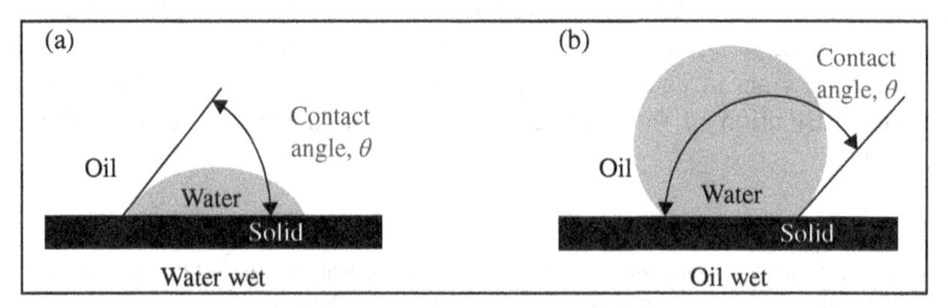

FIGURE 5.1 Wettability of a surface in contact with two phases is measured by the contact angle.

TABLE 5.2 Wetting Condition and Contact Angle

Wetting Condition	Contact Angle (°)
Strongly water wet	0–30
Moderately water wet	30–75
Neutrally wet	75–105
Moderately oil wet	105–150
Strongly oil wet	150–180

Figure 5.1a—the water wets the glass. Figure 5.1b shows a droplet of water on a surface that is oil wet. The oil almost lifts the water off the surface.

Values of contact angle for different wetting conditions are illustrated in Table 5.2 for oil and water in contact with a solid surface. Wettability can be changed by several factors, including contact with drilling fluids, fluids on the rig floor, and contact of the core with oxygen or water from the atmosphere.

Capillary pressure can be defined in terms of wetting and nonwetting phases as

$$p_c = p_{nw} - p_w \tag{5.2}$$

where p_{nw} is the pressure of the nonwetting phase and p_w is the pressure of the wetting phase. In the case of a gas–water system, gas is the nonwetting phase and water is the wetting phase. Capillary pressure in a reservoir is

$$p_c = p_{nw} - p_w = \frac{2\sigma\cos\theta}{r} \tag{5.3}$$

where σ is the IFT between wetting and nonwetting phases, r is the pore radius of the rock, and θ is the contact angle between rock and fluid in a consistent set of units. Contact angle depends on the wettability of the rock. Equation 5.3 shows that capillary pressure increases as pore radius decreases.

Example 5.1 Oil–water Capillary Pressure

Capillary pressure of an oil–water system p_{cow} is 35 psia and water-phase pressure p_w is 2500 psia. Assuming water is the wetting phase, calculate the oil-phase pressure.

Answer

$$p_o = p_{cow} + p_w = 2535 \text{ psia}$$

IFT and wettability are important because they determine the distribution of phases in porous media, at the small scale of a rock sample, and at reservoir scale. The distribution of phases at reservoir scale affects reservoir management. The small-scale phase distribution determines relative permeabilities, which affect fluid flow in the reservoir and into the well.

5.2 FLUID DISTRIBUTION AND CAPILLARY PRESSURE

Oil, gas, and water phases coexist in many hydrocarbon reservoirs. If these three phases were in a bottle at rest on a table, they would segregate with the gas on top, the water on bottom, and the oil between (provided that gas is least dense, water is most dense, and oil density falls between the other two). The gas–oil boundary and the oil–water boundary would be planar and well defined. In a reservoir, these same phases will segregate, but not completely, and the boundaries between them will be very fuzzy. These zones are known as transition zones. In an oil–water system, the transition zone separates the water zone where only water flows from the oil zone where only oil flows. Wells producing from transition zones typically produce two phases simultaneously.

Imagine that you and a pressure gauge could shrink to the scale of pores in a reservoir that is saturated with oil and water. The less dense oil resides mostly in the upper portion of the reservoir, and the more dense water is mostly in the lower portion, but the segregation of these phases is not complete. If you were to measure the pressure in immediately adjacent phases of oil and water, you will find a difference that is termed the oil–water capillary pressure, p_{cow}:

$$p_{cow} = p_o - p_w \tag{5.4}$$

As explained in the previous section, this difference results from the curvature of the interface between the phases. If you were to climb to a higher elevation in the reservoir, the pressure in the water phase would decrease at approximately 0.45 psi/ft. The pressure in the oil would decrease at a lower rate, perhaps 0.35 psi/ft, because the oil density is lower than the water density. Consequently, the capillary pressure increases at a rate of 0.10 psi/ft as you climb upward and decreases at the same rate as you climb downward in the reservoir. In your up-and-down exploration, you will likely find an elevation for which the capillary pressure is zero. Above that elevation, p_{cow} is positive; below that elevation, it is negative. In Figure 5.2a, the linear relationship between measured capillary pressure and elevation is shown. Figure 5.2b shows a trend for water saturation that you might have seen in your exploration of the reservoir. Clearly, the boundary between the oil and water is blurred.

Combining the observed capillary pressure and water saturation for each elevation in Figure 5.2 will yield Figure 5.3. Such relationships are routinely measured for rock samples taken from oil and gas reservoirs. The details of such measurements are found elsewhere. After the relationship between capillary pressure and water saturation has been measured, the distribution of fluids in a formation can be estimated.

The relationship between capillary pressure and fluid saturation depends on the direction of change of saturation. Consider a rock sample that is initially saturated with water. If oil is injected into the sample at increasing saturations, the "primary drainage" trend in Figure 5.4 might be obtained. If water is next injected into the sample, the "secondary imbibition" trend might be obtained. And finally, if oil is again injected, one might obtain a trend like the "secondary drainage" trend in Figure 5.4. The variation of the capillary pressure relationship with direction of saturation change is termed hysteresis. Historically, drainage referred to decreasing saturation of the wetting phase, and imbibition referred to increasing saturation of the wetting phase.

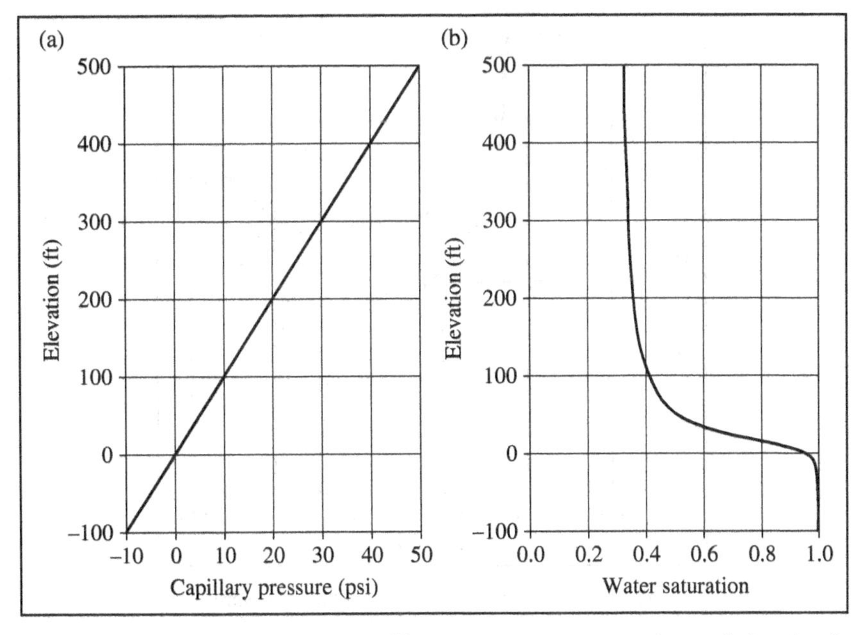

FIGURE 5.2 Relationship between capillary pressure, water saturation, and elevation for a hypothetical reservoir.

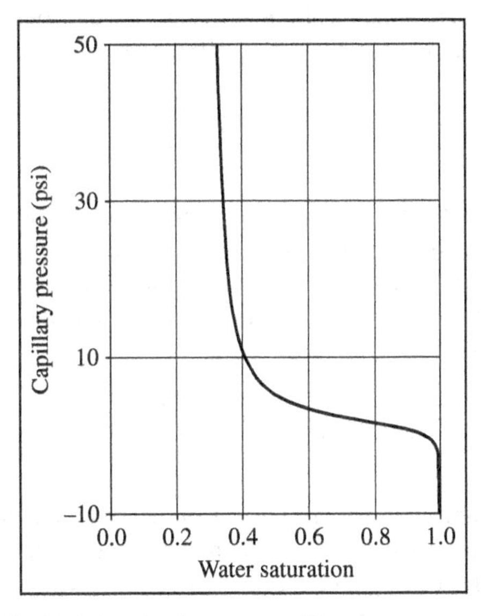

FIGURE 5.3 Relationship between capillary pressure and saturation.

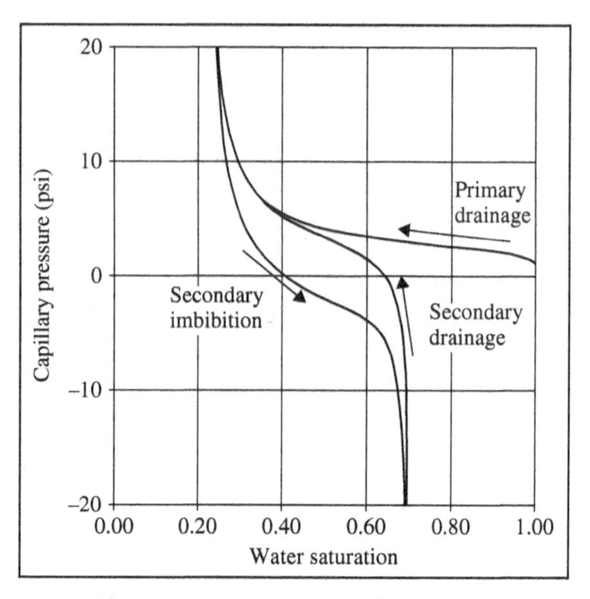

FIGURE 5.4 Hysteresis of capillary pressure.

5.3 RELATIVE PERMEABILITY

Darcy's equation for calculating volumetric flow rate q through a linear, horizontal, water-saturated permeable material is proportional to pressure gradient:

$$q_w = -0.001127 \frac{kA}{\mu_w} \frac{\Delta p}{\Delta x} \tag{5.5}$$

The variables are expressed in the following units: volumetric flow rate q (bbl/day), permeability k (md), cross-sectional area A (ft^2), pressure p (psi), fluid viscosity μ (cp), and length Δx (ft). We represent flow of water through a material that is not entirely saturated with water by introducing a new term called water relative permeability, as shown as follows:

$$q_w = -0.001127 \frac{kk_{rw}(S_w)A}{\mu_w} \frac{\Delta p}{\Delta x} \tag{5.6}$$

Water relative permeability $k_{rw}(S_w)$ is a dimensionless variable that depends on water saturation. If the permeable material also contains some oil, then we can write a similar expression for the oil flow rate using an oil relative permeability $k_{ro}(S_o)$ that depends on oil-phase saturation:

$$q_o = -0.001127 \frac{kk_{ro}(S_o)A}{\mu_o} \frac{\Delta p}{\Delta x} \tag{5.7}$$

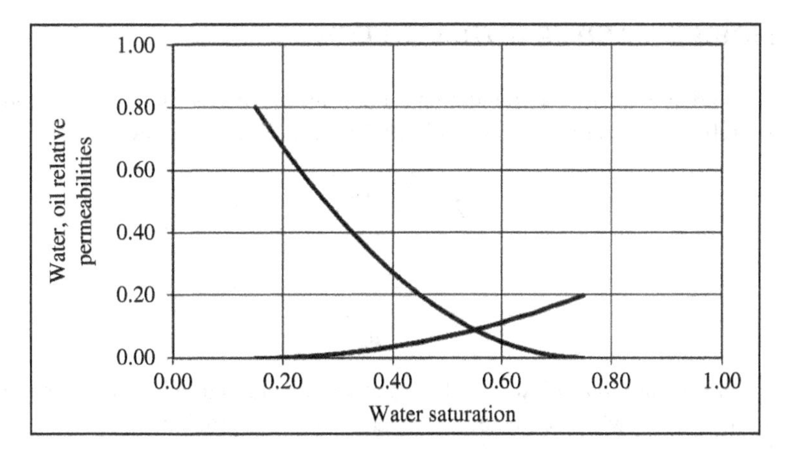

FIGURE 5.5 Example of water and oil relative permeabilities for a porous medium containing water and oil.

An example of oil and water relative permeabilities is shown in Figure 5.5. The curves are not labeled. Which curve is water relative permeability and which is oil relative permeability? Relative permeability of a phase increases as its saturation increases. Therefore, the curve that rises with increasing water saturation is water relative permeability, and the other curve is oil relative permeability.

According to the example in Figure 5.5, oil and water relative permeabilities are defined for a range of water saturations from 15% up to 75%. Irreducible water saturation S_{wirr} is the endpoint of the water relative permeability curve at $S_{\mathrm{w}} = 15\%$. Residual oil saturation S_{or} is the endpoint of the oil relative permeability curve at $S_{\mathrm{o}} = 25\%$ which corresponds to $S_{\mathrm{w}} = 75\%$ since $S_{\mathrm{o}} + S_{\mathrm{w}} = 1$. Relative permeabilities are less than one between the endpoints. At the endpoints, water relative permeability goes to zero at $S_{\mathrm{w}} = 15\%$, and oil relative permeability goes to zero at $S_{\mathrm{w}} = 75\%$. Relative permeabilities as defined by Equations 5.6 and 5.7 are less than one because the capacity for flow of either phase is hindered by the presence of the other phase.

Example 5.2 Relative Permeability

Effective permeability of a phase is the relative permeability of the phase times the absolute permeability of the porous medium. Suppose the effective permeability to air in a core is 200 md for single-phase flow of air and the relative permeability to water at 100% water saturation is 0.1. What is the effective permeability of water in the core at 100% water saturation? Use the effective permeability to air as the absolute permeability of the core.

Answer

$$k_{\mathrm{rw}} = \frac{k_{\mathrm{w}}}{200\ \mathrm{md}} = 0.1 \text{ gives } k_{\mathrm{w}} = 20\ \mathrm{md}.$$

5.4 MOBILITY AND FRACTIONAL FLOW

Equations 5.6 and 5.7 are the foundation for definitions of mobility, relative mobility, mobility ratio, and fractional flow. Mobility for oil and water are defined as

$$\lambda_o = \frac{kk_{ro}(S_o)}{\mu_o} \tag{5.8}$$

$$\lambda_w = \frac{kk_{rw}(S_w)}{\mu_w} \tag{5.9}$$

Viscous oil typically has a low mobility relative to water and will flow with a lower velocity than water when subjected to the same pressure gradient.

Relative mobilities are defined as follows:

$$\lambda_{ro} = \frac{k_{ro}(S_o)}{\mu_o} \tag{5.10}$$

$$\lambda_{rw} = \frac{k_{rw}(S_w)}{\mu_w} \tag{5.11}$$

Mobility ratio is used for assessing effectiveness of a displacement in porous media. In general, mobility ratio is the mobility of the displacing fluid divided by the mobility of the fluid being displaced. As an example, consider a water flood of an oil reservoir. Water is injected into the reservoir to displace oil and provide pressure support. In this case, the mobility ratio M is

$$M = \frac{kk_{rw}(S_{or})/\mu_w}{kk_{ro}(S_{wirr})/\mu_o} = \frac{k_{rw}(S_{or})/\mu_w}{k_{ro}(S_{wirr})/\mu_o} \tag{5.12}$$

If M is greater than 1, the displacement is "unfavorable" because a less mobile fluid is being pushed by a more mobile fluid. In a "favorable" displacement, the displacing fluid is less mobile than the displaced fluid.

Water fractional flow is the fraction of water flowing through the reservoir during a water–oil displacement process:

$$f_w = \frac{q_w}{q_w + q_o} \tag{5.13}$$

By contrast, water cut (WCT) introduced in Chapter 1 is the rate of water produced at the surface divided by total liquid production rate at the surface:

$$\text{WCT} = \left(\frac{q_w}{q_w + q_o}\right)_{\text{surface}} \tag{5.14}$$

WCT should not be confused with water fractional flow.

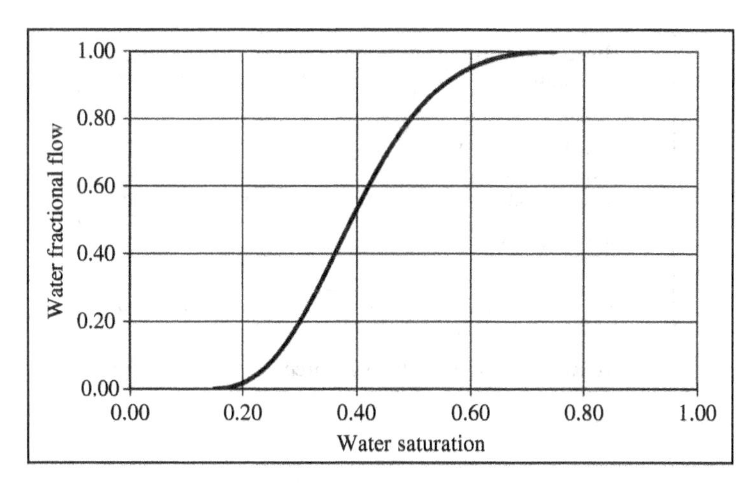

FIGURE 5.6 Fractional flow for oil–water displacement with relative permeabilities shown in Figure 5.5.

Fractional flow of water depends on mobility ratio, buoyancy effects relative to oil, and the gradients of capillary pressure between oil and water. Here, we neglect buoyancy and capillary pressure effects. Substituting Equations 5.6 and 5.7 for water and oil flow rates into Equation 5.13 gives

$$f_w = \frac{1}{1 + \dfrac{\mu_w}{k_{rw}(S_w)} \dfrac{k_{ro}(S_o)}{\mu_o}} \tag{5.15}$$

Thus, for a given pair of water and oil viscosities, fractional flow depends only on water saturation through the saturation dependence of the relative permeabilities. The fractional flow relationship for the relative permeabilities of Figure 5.5, with water viscosity of 0.9 cp and oil viscosity of 8.0 cp, is shown in Figure 5.6.

Example 5.3 Fractional Flow

A well produces 1000 STB/day oil and 100 STB/day water. What is the fractional flow of water?

Answer

$$f_w = \frac{q_w}{q_w + q_o} = \frac{100}{100 + 1000} = 0.091$$

5.5 ONE-DIMENSIONAL WATER-OIL DISPLACEMENT

Fractional flow is a useful concept for modeling water–oil displacement in one dimension. Although actual reservoirs are not one-dimensional, understanding the basics of displacements in one dimension is a good start on understanding more

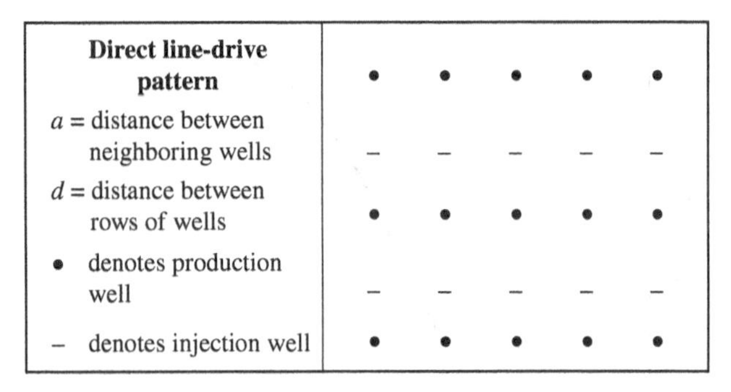

FIGURE 5.7 Well locations in direct line-drive pattern.

complex displacements. In some cases, such as direct line-drive water floods, fluid flow can be approximated as one-dimensional linear flow from an injection well to a production well. Figure 5.7 illustrates direct line-drive well pattern. Rows of injectors are alternately spaced between rows of producers.

Buckley and Leverett (1942) published a key idea for understanding one-dimensional water–oil displacement. They showed that the velocity at which any water saturation propagates through a porous medium during a water flood is proportional to the slope of the fractional flow curve. For water saturations less than 20% in Figure 5.6, the slope is relatively small. The slope increases as saturation increases to about 35%, and then the slope decreases and approaches zero as water saturation approaches 75%.

Ten years later, Welge (1952) published a method for using fractional flow curves to predict results of water floods of oil reservoirs—again in one dimension. Welge's method starts with the saturation-velocity idea of Buckley and Leverett. The outline of Welge's method is presented here. The starting saturation condition is initial water saturation S_{wi} and the corresponding oil saturation $S_o = 1 - S_{wi}$. The method consists of the following four steps:

1. *Water fractional flow.* Generate a water fractional flow curve from relative permeabilities and viscosities using Equation 5.15, which neglects gravitational and capillary pressure effects.
2. *Shock saturation.* Construct the tangent to the water fractional flow curve from the initial condition for fractional flow ($f_w = 0$) and initial water saturation (S_{wi}). The point of tangency to the fractional flow curve corresponds to the water saturation S_{wf} and water fractional flow f_{wf} immediately behind the saturation shock front.
3. *Oil production at water breakthrough.* Water breakthrough occurs when the shock front reaches the producing well. To find oil production at breakthrough, extrapolate the tangent found in step 2 to intersect with the line $f_w = 1$.

The water saturation at the intersection is the average water saturation behind the front. The pore volume of oil produced at breakthrough equals the average water saturation minus the initial water saturation. The volume of water injected equals the volume of oil recovered at breakthrough; it also equals the inverse of the slope of the tangent.

4. *Postbreakthrough production behavior.* Construct a tangent to the water fractional flow curve for water saturation greater than S_{wf}. The inverse of the slope of this tangent is the pore volume of water injected Q_{wi} when the saturation at the tangent point reaches the producing well. Extrapolate the tangent to intersect with the line $f_w = 1$. The water saturation at the intersection is the average water saturation after injecting Q_{wi} pore volume of water. The pore volume of oil produced Q_{op} equals the average water saturation minus the initial water saturation S_{wi}. Repeat step 4 as desired to produce additional oil recovery results.

Steps 2 and 3 are demonstrated in Figure 5.8. A tangent line is drawn from $f_w = 0.0$ at $S_{wi} = 0.15$. The point of tangency is $f_{wf} = 0.78$ at $S_{wf} = 0.48$. The tangent line reaches $f_w = 1.0$ at $S_w = 0.58$. Average water saturation behind the flood front is 58%. The pore volume of water injected and oil produced at water breakthrough are equal and given by $Q_{wibt} = Q_{opbt} = 0.58 - 0.15 = 0.43$.

Step 4 is illustrated in Figure 5.9 for water saturation of 58%. A line is drawn tangent to the f_w curve at $S_w = 0.58$. The slope of the dashed line in Figure 5.9 is 1.04. The corresponding pore volume of water injected Q_{wi} is the inverse of the slope, so $Q_{wi} = 0.96$. The tangent line reaches $f_w = 1.0$ at $S_w = 0.64$. The average water saturation behind the flood front is 0.64, and the pore volume of oil produced Q_{op} is $0.64 - 0.15 = 0.49$. Repeated application of step 4 generates a curve of oil production with increasing water injection. Results are shown in Figure 5.10.

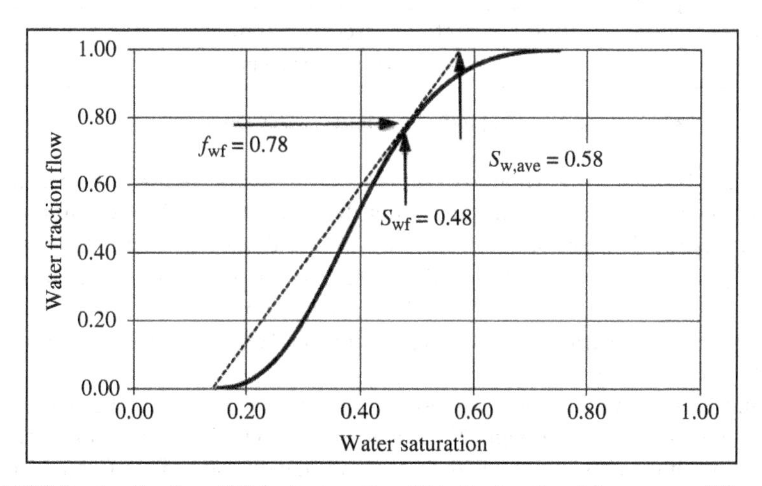

FIGURE 5.8 Application of Welge's steps 2 and 3 to the fractional flow curve of Figure 5.6.

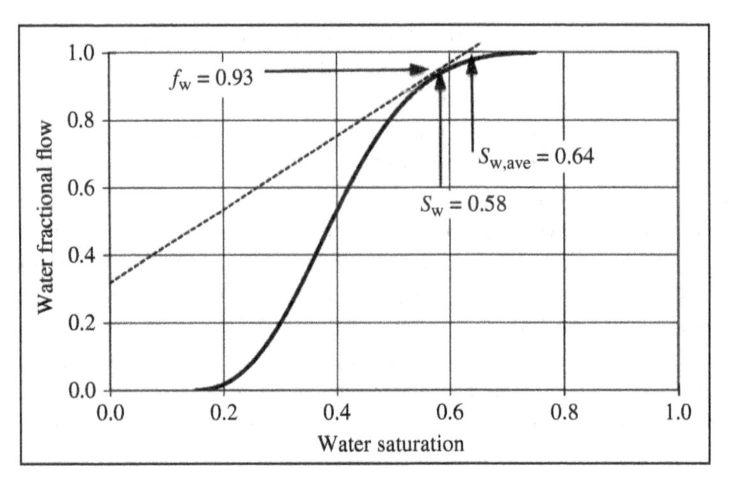

FIGURE 5.9 Application of Welge's step 4 to the fractional flow curve of Figure 5.6.

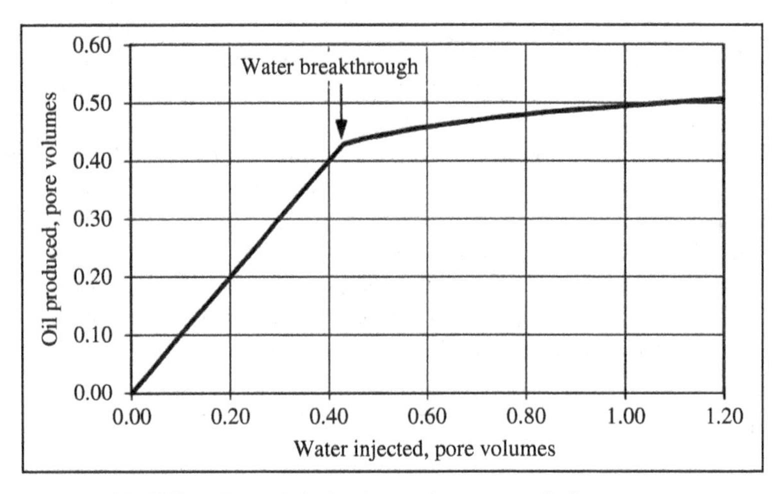

FIGURE 5.10 Oil production during water–oil displacement.

Figure 5.10 shows that volume of oil recovered equals volume of water injected up to the point of water breakthrough. After that point, continuing volumes of water injected produce less and less oil. In other words, the water fractional flow keeps increasing, while the oil fractional flow decreases. Considering economics of an actual water flood, a point is reached where the value of the oil produced equals the cost of water flood operations, including injection and disposal. Operations ordinarily cease before that point.

Figure 5.11 shows the water saturation profile midway between the start of water injection and water breakthrough. Normalized position in this figure ranges from 0 at the inlet to 1 at the outlet. Normalized position is the distance of the flood front from the injection well divided by the distance between injection well and production well. The rapid increase in water saturation at normalized position of 0.5 is often

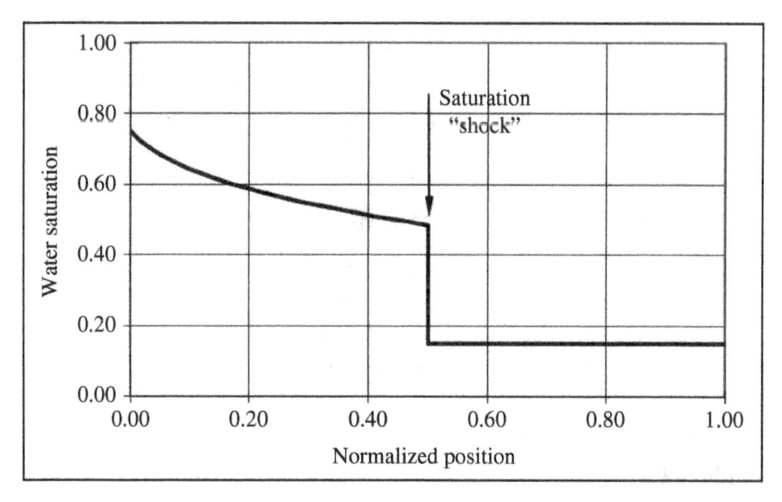

FIGURE 5.11 Water saturation profile when the front is halfway through the sample.

called a saturation shock or a shock front. Ahead of the shock front, water saturation is constant and equal to initial water saturation S_{wi}. Behind the front, water saturation gradually increases to $S_w = 1 - S_{or}$ at the inlet position. Water breakthrough corresponds to arrival of the front at the outlet position. After breakthrough, the saturation profile continues to rise, asymptotically approaching $S_w = 1 - S_{or}$.

Welge's method was primarily intended to provide a graphical means for estimating oil production for water or gas flooding. Such an approach was satisfactory at the time. Today, software is available that uses numerical techniques to apply the Buckley–Leverett–Welge method.

5.6 WELL PRODUCTIVITY

Production of fluid from a well can be quantified using the concept of well productivity. Consider the case of radial flow into a vertical well. Volumetric flow rate q_ℓ for phase ℓ is proportional to pressure differential Δp so that

$$q_\ell = PI \times \Delta p \tag{5.16}$$

where the proportionality factor is the productivity index PI. The pressure differential is the difference between reservoir pressure and flowing wellbore pressure, or

$$\Delta p = p_{res} - p_{fwb} \tag{5.17}$$

The productivity index terms are illustrated in Figure 5.12. Fluid flows from the reservoir, through perforations in the casing into the wellbore, and up the tubing to the surface. The pressure differential is greater than zero ($\Delta p > 0$) for production wells and less than zero ($\Delta p < 0$) for injection wells. In the case of fluid injection, the term injectivity index is used.

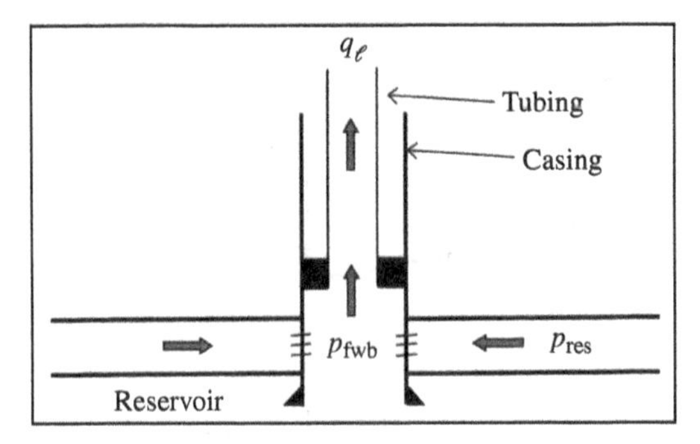

FIGURE 5.12 Productivity index terms.

Example 5.4 Well PI

A 20 psia pressure drawdown is required to produce 500 STBO/day. Use well PI to estimate the oil production rate at 10 psia pressure drawdown.

Answer

$$PI = \frac{q_o}{\Delta p} = \frac{500\,\text{STBO/day}}{20\,\text{psia}} = 25\,\text{STBO/day/pisa}$$

Therefore

$$q_o = PI \times \Delta p = (25\,\text{STBO/day/psia}) \times 10\,\text{psia} = 250\,\text{STBO/day}$$

The radial form of Darcy's law can be used to write the productivity index (PI) for radial flow into a vertical well as

$$PI = \frac{q_\ell}{\Delta p} = \frac{0.00708 k_e h_{net}}{\mu_\ell B_\ell \left(\ln(r_e/r_w) + S\right)} \tag{5.18}$$

where the variables and associated units are tabulated as follows:

PI = productivity index (STB/D/psi)
q_ℓ = volumetric flow rate of phase ℓ (STB/D)
$\Delta p = p_{res} - p_{fwb}$
p_{res} = reservoir pressure (psi)
p_{fwb} = flowing wellbore pressure (psi)
k_e = effective permeability (md) = $k_{r\ell} k_{abs}$
$k_{r\ell}$ = relative permeability of phase ℓ
k_{abs} = absolute permeability (md)
h_{net} = net thickness (ft)
μ_ℓ = viscosity of phase ℓ (cp)
B_ℓ = FVF of phase ℓ (RB/STB)

r_e = drainage radius (ft)
r_w = wellbore radius (ft)
S = dimensionless skin

Effective permeability in Equation 5.18 incorporates relative permeability into the calculation of flow rate and *PI*. Skin *S* is an indicator of formation damage in the vicinity of the well. The well is damaged if $S>0$ and stimulated if $S<0$. Skin is discussed in more detail in Chapter 12.

Productivity index is positive for both an injection well and a production well. We can verify our sign conventions for pressure differential and flow rate by noting that a production well has a positive pressure differential ($\Delta p > 0$) and a positive flow rate ($q_\ell > 0$); hence the productivity index in Equation 5.18 is positive. Similarly, an injection well has a negative pressure differential ($\Delta p < 0$) and a negative flow rate ($q_\ell < 0$), so once again the productivity index in Equation 5.18 is positive.

Example 5.5 Drainage Radius

Drainage radius can be estimated using well spacing, which is the area associated with a production well. For example, 16 equally spaced wells in a square mile occupy an area of 40 acres. The well spacing is called 40-acre spacing. We can estimate the drainage radius by approximating the area as a circular area. What is the drainage area of a well in a well pattern with 40-acre spacing?

Answer

The drainage area A_d assigned to the well is 40 acres $\approx 1742400\,\text{ft}^2$.

Drainage radius for a circular area is $r_e \approx \sqrt{\dfrac{A_d}{\pi}} = \sqrt{\dfrac{1742700}{\pi}} \approx 745\,\text{ft}$.

5.7 ACTIVITIES

5.7.1 Further Reading

For more information about rock properties, see Dandekar (2013), Mavko et al. (2009), and Batzle (2006).

5.7.2 True/False

5.1 The oil–water contact is usually found below the gas–oil contact in a reservoir with mobile gas, oil, and water.

5.2 Water cut is water production rate at the surface divided by the sum of oil production rate and water production rate at the surface.

5.3 A positive skin represents stimulation.

5.4 Capillary pressure decreases when pore radius decreases.

5.5 Welge developed a method to use fractional flow curves to predict oil recovery from a water flood.

5.6 A well producing from the transition zone between a water zone and a gas zone will produce both water and oil.

5.7 Wettability is a measure of the ability of a liquid to maintain contact with a solid surface.

5.8 The value of relative permeability to water k_{rw} is must be in the range $0 \le k_{rw} \le 1.0$.

5.9 The rate of oil production during a water flood increases after water breakthrough.

5.10 Mobility ratio is the mobility of the displacing fluid divided by the mobility of the displaced fluid.

5.7.3 Exercises

5.1 Suppose the effective permeability to air in a core is 200 md for single-phase flow of air and the relative permeability to water at 100% water saturation is 0.1. What is the effective permeability of water in the core at 100% water saturation? Use the effective permeability to air as the absolute permeability of the core.

5.2 Capillary pressure for a gas–water system p_{cgw} is 10 psia and water-phase pressure p_w is 2500 psia. If water is the wetting phase, what is the gas-phase pressure?

5.3 A well produces 100 STB/day oil and 1000 STB/day water. What is the water cut?

5.4 Suppose a well originally produces 10 000 STBO/day at a pressure drawdown of 10 psia. The well PI declined 5% a year for the first 2 years of production. What is the well PI at the end of year 2?

5.5 Consider the following oil–water relative permeability table:

S_w	k_{rw}	k_{ro}
0.30	0.000	1.000
0.35	0.005	0.590
0.40	0.010	0.320
0.45	0.017	0.180
0.50	0.023	0.080
0.55	0.034	0.030
0.60	0.045	0.010
0.65	0.064	0.005
0.70	0.083	0.001
0.80	0.120	0.000

A. What is residual oil saturation (S_{orw})?
B. What is connate water saturation (S_{wc})?
C. What is the relative permeability to oil at connate water saturation $k_{ro}(S_{wc})$?
D. What is the relative permeability to water at residual oil saturation $k_{rw}(S_{orw})$?
E. Assume oil viscosity is 1.30 cp and water viscosity is 0.80 cp. Calculate mobility ratio for water displacing oil in an oil–water system using the data in the table and mobility ratio = $\lambda_{\text{Displacing}}/\lambda_{\text{displaced}} = (k_{rw}(S_{orw})/\mu_w)/(k_{ro}(S_{wc})/\mu_o)$.

5.6 A well is producing oil in a field with 20-acre well spacing. The oil has viscosity 2.3 cp and formation volume factor 1.25 RB/STB. The net thickness of the reservoir is 30 ft and the effective permeability is 100 md. The well radius is 4.5 in. and the well has skin = 0. What is the productivity index (PI) of the well in STB/D/psi?

5.7 **A.** Interfacial tension (IFT) σ can be estimated using the Weinaug–Katz variation of the Macleod–Sugden correlation:

$$\sigma^{1/4} = \sum_{i=1}^{N_c} \sigma_i^{1/4} = \sum_{i=1}^{N_c} P_{\text{chi}} \left(x_i \frac{\rho_L}{M_L} - y_i \frac{\rho_V}{M_V} \right)$$

where σ is the interfacial tension (dyne/cm), P_{chi} is an empirical parameter known as the parachor of component i [(dynes/cm)$^{1/4}$/(g/cm^3)], M_L is the molecular weight of liquid phase, M_V is the molecular weight of vapor phase, ρ_L is the liquid phase density (g/cm^3), ρ_V is the vapor phase density (g/cm^3), x_i is the mole fraction of component i in liquid phase, and y_i is the mole fraction of component i in vapor phase. The parachor of component i can be estimated using the molecular weight M_i of component i and the empirical regression equation (Fanchi, 1990):

$$P_{\text{chi}} = 10.0 + 2.92\,M_i$$

This procedure works reasonably well for molecular weights ranging from 100 to 500. Estimate the parachor for octane. The molecular weight of octane (C_8H_{18}) is approximately $8*12+18*1=114$, and the density of octane in the liquid state is approximately 0.7 g/cc.

B. Estimate the contribution σ_i of IFT for octane in a liquid mixture with density 0.63 g/cc and molecular weight 104. Assume there is no octane in the vapor phase and the mole fraction of octane in the liquid phase is 0.9.

6

PETROLEUM GEOLOGY

The ability to manage subsurface resources depends in part on our knowledge of the formation and evolution of the Earth. We describe the geologic history and structure of the Earth and then present concepts from petroleum geology.

6.1 GEOLOGIC HISTORY OF THE EARTH

The age of the Earth is approximately 4.5 billion years old based on radioactive dating of the oldest available rock samples. The cross section of the Earth's interior is subdivided into an inner core, the mantle, and the crust (Figure 6.1). Seismic measurements have shown that the core consists of a crystalline inner core and a molten outer core. The electric and magnetic properties of the Earth and the density of the core provide evidence for identifying the metal as an alloy of iron and nickel. Iron is the dominant constituent.

The density of rock in the mantle is greater than in the crust because the mantle contains primarily basalt, a dark volcanic rock. Basalt is composed of magnesium and iron silicates. Basalt at the surface of the mantle exists in a semimolten state. This layer of semimolten basalt is called the asthenosphere. As we descend through the mantle, the rigidity of the basalt increases until the mantle acquires a rigidity in excess of the rigidity of steel.

Introduction to Petroleum Engineering, First Edition. John R. Fanchi and Richard L. Christiansen.
© 2017 John Wiley & Sons, Inc. Published 2017 by John Wiley & Sons, Inc.
Companion website: www.wiley.com/go/Fanchi/IntroPetroleumEngineering

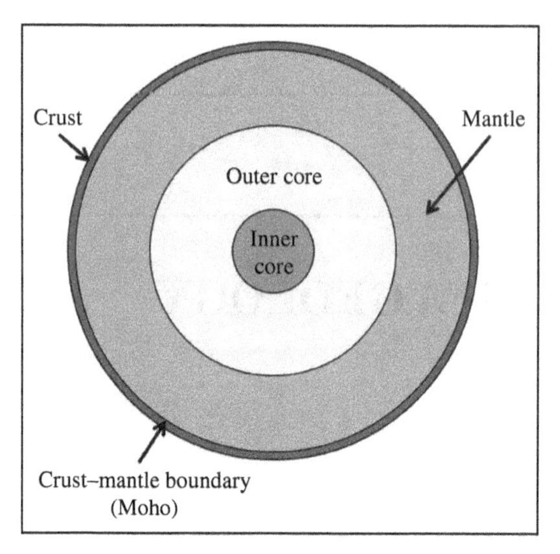

FIGURE 6.1 The interior of the Earth.

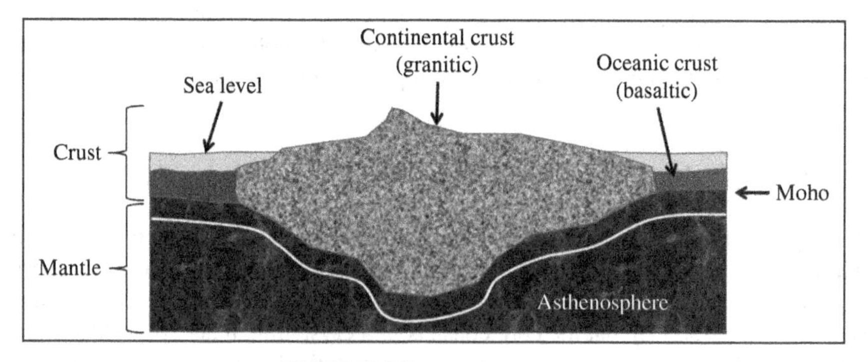

FIGURE 6.2 The lithosphere.

A relatively thin crust is above the mantle. Figure 6.2 is a sketch of the crust of the Earth. The granitic continental crust is composed principally of silica and aluminum; the oceanic crust contains silica and magnesia. Underlying the continental and oceanic crusts is a layer of basalt. The boundary between the crustal basalt and the semimolten basalt of the mantle is the Mohorovicic discontinuity, or Moho for short. The combination of crust and solid basalt above the Moho is known as the lithosphere. The upper mantle below the Moho that behaves elastically is part of the lithosphere. The asthenosphere is below the lithosphere and contains the part of the upper mantle that is ductile. The lithosphere and asthenosphere are separated by the lithosphere–asthenosphere boundary (LAB).

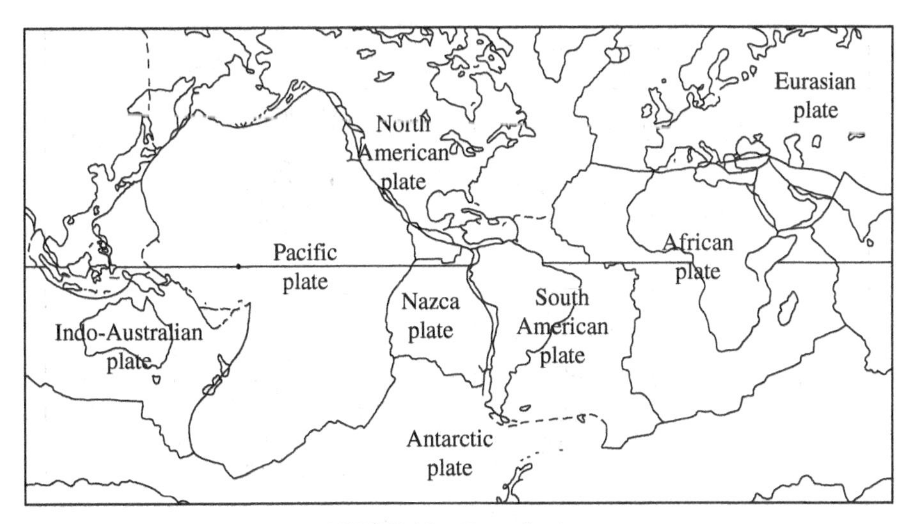

FIGURE 6.3 Tectonic plates.

The history of the lithosphere depends on the dynamic relationship that exists between the lithosphere and the asthenosphere. The lithosphere consists of a set of rigid plates floating on the semimolten asthenosphere. Crustal plates move in relation to one another at the rate of a few inches per year. Molten material in the asthenosphere can enter the lithosphere through cracks between plates. This transfer of material can be by violent volcanic eruptions, or it can be gradual through the extrusion of basaltic lavas at the boundaries between plates.

Satellite measurements of the gravitational field of the Earth show boundaries between continents and tremendous mountain ranges rising from ocean floors. The shapes of the boundaries are suggestive of vast plates, as depicted in Figure 6.3. Only the largest of the known plates are depicted in the figure. Many of these plates are associated with continental land masses.

Oceanic mountain ranges are sources of basaltic extrusion and seafloor spreading. As seafloors spread, the continental plates are forced to move. Movement of continental plates is known as continental drift and was first proposed by the German meteorologist Alfred Wegener (1880–1930 CE) in 1915. A collision of two plates can form great mountain ranges, such as the Himalayas. The boundary at the interface between two colliding plates is a convergent plate boundary. Alternatively, a collision can deflect one plate beneath another. The boundary where one plate is moving under another is a subduction zone. Material in the subduction zone can be forced down through the Moho and into the semimolten asthenosphere. The separation of two plates by the extrusion of material at the mid-Atlantic Ridge is an example of seafloor spreading. Together, seafloor spreading and subduction zones are the primary mechanisms for transferring material between the crust and upper mantle.

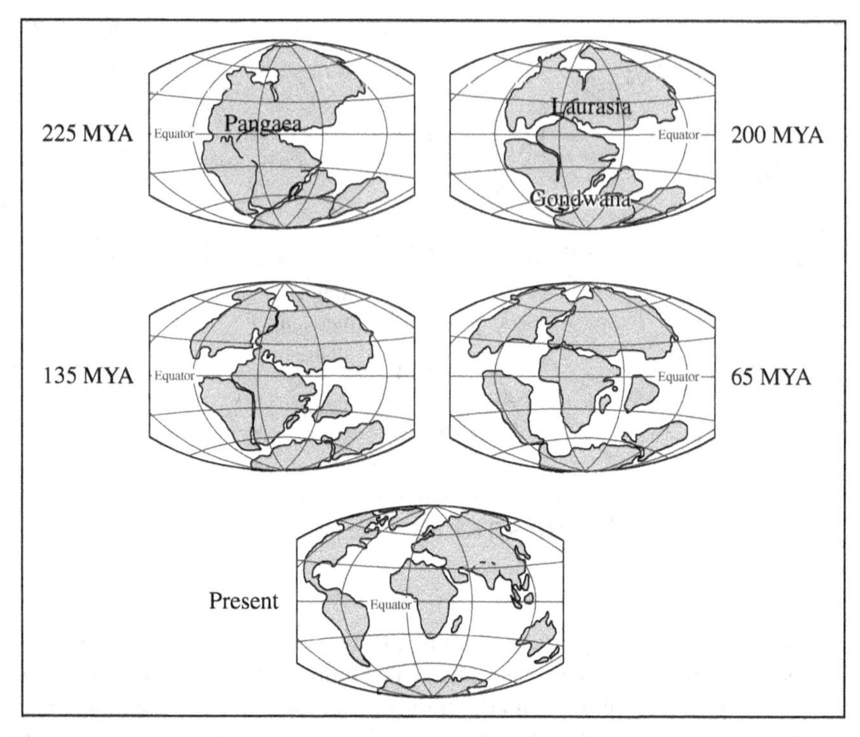

FIGURE 6.4 Tectonic plate movement. (Source: U.S. Geological Survey Historical (2013).)

The theory describing the movement of lithospheric plates is known as plate tectonics. According to plate tectonics, the land masses of the Earth have been moving for millions of years. Using radiometric dating and similarities in geologic structures, a reconstruction of lithologic history is sketched in Figure 6.4 for the past 225 million years.

The movement of tectonic plates is responsible for much of the geologic heterogeneity that is found in hydrocarbon-bearing reservoirs. The figure begins at the time that all surface land masses were gathered into a single land mass known as Pangaea. Pangaea was formed by the movement of tectonic plates, and the ongoing movement of plates led to the breakup of the single land mass into the crustal features we see today.

Table 6.1 is an abridged version of the geologic time scale beginning with the formation of the Earth. The most encompassing period of time is the eon, which is subdivided into eras and further subdivided into periods. The acronym MYBP stands for millions of years before the present. The geologic time scale is from the Geological Society of America (Gradstein et al., 2012; Cohen et al., 2015). Our understanding of the chronology of the Earth is approximate.

TABLE 6.1 Geologic Time Scale

Eon	Era	Period	Epoch	Approximate End of Interval (MYBP)
Phanerozoic	Cenozoic	Quaternary	Holocene (recent)	0.01
			Pleistocene	2.6
		Neogene	Pliocene	5.3
			Miocene	23.0
		Paleogene	Oligocene	33.9
			Eocene	56.0
			Paleocene	66.0
	Mesozoic	Cretaceous		145
		Jurassic		201
		Triassic		252
	Paleozoic	Permian		299
		Carboniferous	Pennsylvanian	323
			Mississippian	359
		Devonian		419
		Silurian		444
		Ordovician		485
		Cambrian		541
Precambrian	Proterozoic			2500
	Archean			4000
	Hadean			4600

Example 6.1 Plate Tectonics

The South American and African plates are about 4500 mi apart. If they began to separate from Pangaea about 225 million years ago, what is their rate of separation? Express your answer in meters per year.

Answer

The rate of separation is estimated as the distance of separation divided by the time it took to achieve the separation:

$$\frac{4500\,\text{mi}}{225\times10^{6}\,\text{yr}}\times\frac{5280\,\text{ft}}{\text{mi}}=0.106\frac{\text{ft}}{\text{yr}}\times\frac{0.3048\,\text{m}}{\text{ft}}=0.032\frac{\text{m}}{\text{yr}}$$

Human fingernails grow at approximately 2 cm/yr = 0.02 m/yr.

6.1.1 Formation of the Rocky Mountains

The movement of tectonic plates across the surface of the globe generated forces powerful enough to form the structures we see today. We illustrate the concepts by describing the formation of the Rocky Mountains.

Fossils of marine life—such as the spiral-shaped, invertebrate ammonites—found in the Rocky Mountains are evidence that the central plains of the United States and the foothills of the Rocky Mountains were below sea level in the past. The Western Interior Seaway, which is also known as the Cretaceous Sea or North American Inland Sea, was a sea that extended from the Arctic Ocean through the Great Plains of North America and south to the Gulf of Mexico. The inland sea split North America into two land masses: Laramidia and Appalachia.

Scientists believe the Rocky Mountains were formed by the subduction of plates beneath the western edge of the North American plate. The subduction created stress that caused buckling of the North American plate along a line of weak rock that paralleled the North American Inland Sea. The buckling uplifted the central part of the United States and formed the Rocky Mountains. The flatirons and Red Rocks near Denver, Colorado, show the degree of structural buckling.

The Rocky Mountains run north–south through much of North America and are part of the North American plate. The shapes of fossilized leaves indicate that the Rocky Mountains were once twice as high as they are today. The shape of leaf edges in fossilized leaves can be used to estimate temperature and elevation. Edges of the leaves of some plants are jagged at lower temperatures and smooth at higher temperatures. Since atmospheric temperature decreases as elevation above sea level increases, we are more likely to find leaves with jagged edges at higher elevations and cooler temperatures than leaves with smooth edges. Based on these observations and the study of fossilized leaves, the elevation of the Rocky Mountains today appears to be lower than it was in the past. The loss of material can be attributed to weathering and erosion. Much of the eroded material was collected in valleys and basins or was transported away from the region by rivers.

The Rocky Mountains have also been shaped by the movement of glaciers. A river can carve a V-shaped valley through mountains. A glacier moving through the valley can carve out the sides of a V-shaped valley so the valley acquires a U shape. The glacier can move large boulders over great distances. When the glacier melts, it leaves behind debris, such as boulders, that was carried by the glacier.

Example 6.2 Erosion

How long would it take to completely erode a mountain that is 1 mi high? Suppose the rate of erosion is 3 mm per century. Express your answer in years.

Answer

Mountain height is 1 mi = 5280 ft = 1609 m. The rate of erosion is 3 mm per 100 years or 0.03 mm per year. Thus, the time to erode a mile high mountain is about 1609 m/0.03 mm/yr or 54 million years.

6.2 ROCKS AND FORMATIONS

The three primary rock types are igneous, sedimentary, or metamorphic. Rocks can transition from one rock type to another when the physical conditions change. The transitions are part of the rock cycle shown in Figure 6.5. The rock cycle begins with the formation of magma, or molten rock, in the mantle. Convection cells in the mantle carry the magma toward the surface of the Earth where it encounters lower temperatures and pressures. At sufficiently low temperature and pressure, the magma cools and solidifies into igneous rock.

Igneous rock on the surface of the Earth is subjected to atmospheric conditions. Erosion can break the igneous rock into smaller particles which can be transported by wind, water, and ice. The particles can become finer as they collide with other objects during the transport process. Eventually, the particles will be deposited when the energy of the transporting agent dissipates.

The deposition of particles can lead to sizable accumulations of sediment. The loose sediment is subjected to increasing temperature and pressure as it is buried beneath additional layers of sediment. The combination of increased temperature, pressure, and chemical processes can transform loose sediment into sedimentary rock. If the sedimentary rocks are buried deep enough, they can encounter pressures and temperatures that can change the character of the rock in a process called metamorphism. Given enough time, pressure, and heat, rocks will melt and start the rock cycle again.

Sedimentary rocks are important because they are often the porous medium associated with commercially important reservoirs. The key attributes used to classify sedimentary rock are mineral composition, grain size and shape, color, and structure.

Rocks are composed of minerals. Each mineral is a naturally occurring, inorganic solid with a specific chemical and crystalline structure. Mineralogy is the study of the set of minerals within the rock. The source of the minerals, the rate of mineral breakdown, and the environment of deposition are important factors to consider in characterizing the geologic environment.

Grains which form sedimentary rocks are created by weathering processes at the surface of the Earth. Weathering creates particles that can be practically any size,

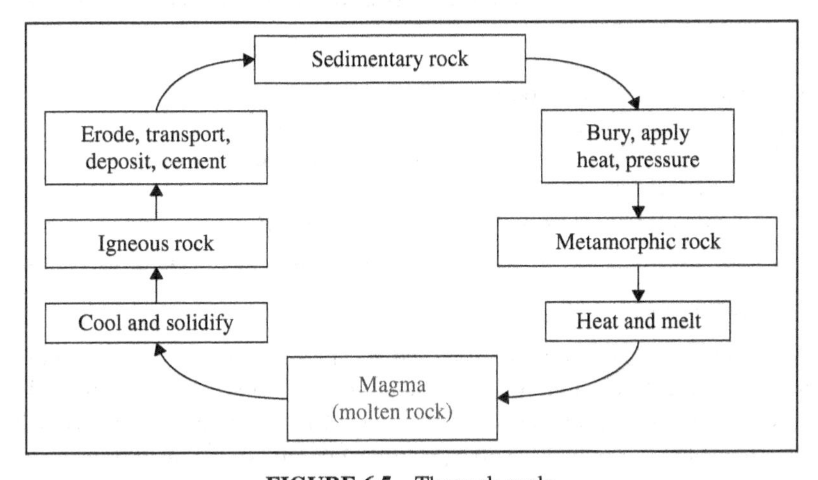

FIGURE 6.5 The rock cycle.

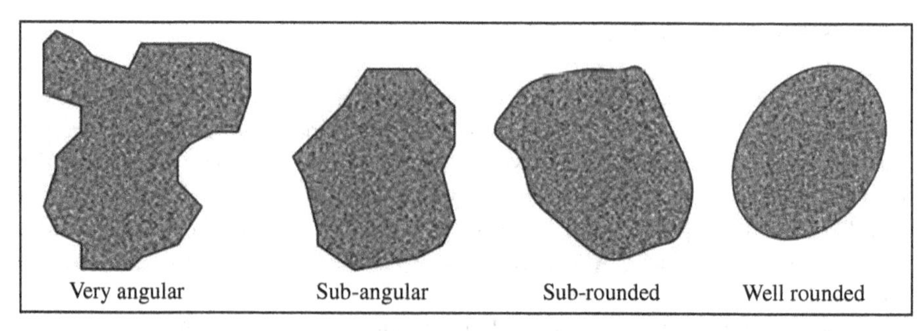

FIGURE 6.6 Grain shapes.

shape, or composition. A glacier may create and transport a boulder the size of a house, while a desert wind might create a uniform bed of very fine sand. The particles, also known as sediments, are transported to the site of deposition. Particles transported by wind or water roll and bump along the transport pathway. The edges of a grain give information about the depositional environment.

Figure 6.6 illustrates different grain shapes. Grains with sharp edges probably did not get transported very far, which suggests the grain was transported in a low-energy environment such as a slow-moving stream. Rounded grains suggest transport in a high-energy environment such as a fast-moving stream. Rocks made up of rounded grains tend to have larger permeability than rocks composed of grains that are flat or have sharp edges.

Sorting refers to the uniformity of grain size. The size of grains in well-sorted rock is relatively uniform, while there is considerable variation in grain size in poorly sorted rock. Fluids will typically flow better through well-sorted rock than poorly sorted rock. The ability to flow is characterized by a rock property called permeability.

Example 6.3 Canyon Formation

It has taken the Colorado river about 8 million years to carve a mile deep canyon as the surface of the Earth uplifted. Use this observation to estimate the rate of uplift of the Earth's surface at the Grand Canyon. Express your answer in inches per year.

Answer

Note: 1 mi = 5280 ft and 1 foot = 12 in.

The rate of uplift = $(1\ \text{mi})/(8 \times 10^6\ \text{yr}) = 1.25 \times 10^{-7}\ \text{mi/yr}$.

Convert to in./yr: $(1.25 \times 10^{-7}\ \text{mi/yr})\ (5280\ \text{ft/mi})\ (12\ \text{in./ft}) = 7.92 \times 10^{-3}\ \text{in./yr}$.

This is less than 1 in. per century.

Convert to mm/yr: $(7.92 \times 10^{-3}\ \text{in./yr})\ (2.54\ \text{cm/in.})\ (10\ \text{mm/cm}) = 0.2\ \text{mm/yr}$.

6.2.1 Formations

The environment where a rock is deposited is called the environment of deposition. As the environment, such as a shoreline, moves from one location to another, it leaves a laterally continuous progression of rock which is distinctive in character. If the progression is large enough to be mapped, it can be called a formation.

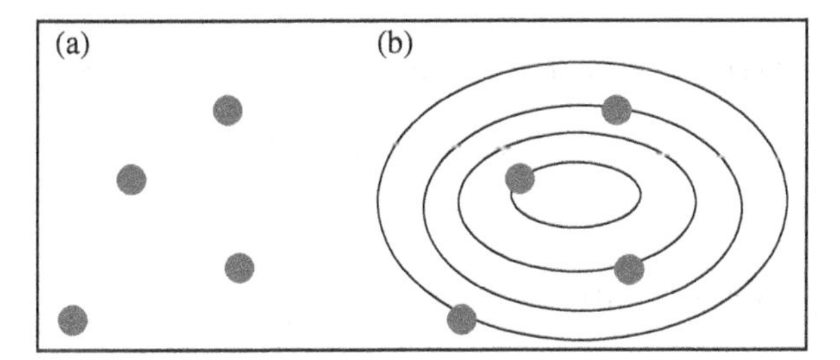

FIGURE 6.7 Preparing a map. (a) Gather data and place in spatial location and (b) contour data.

The mapping process is the point where geological and geophysical interpretations have their greatest impact on the representation of the formation. Data is gathered and plotted in their relative spatial locations as illustrated in Figure 6.7. An important function of geologic maps is to present values for a spatially distributed property at any point in a volume from a set of control point values. Control point values correspond to property values measured at wells or determined by seismic methods that apply to volume of interest. Control points can also be imposed by a modeler using soft data such as seismic indications of structure boundaries. Once the set of control points is determined and the corresponding data are specified, contours can be drawn.

The contouring step is a key point where geological interpretation is included in our understanding of the formation. The following guidelines apply to contouring:

1. Contour lines do not branch.
2. Contour lines do not cross.
3. Contour lines either close or run off the map.
4. Steep slopes have close contour lines.
5. Gentle or flat slopes have contour lines that are far apart.

The first three of these guidelines are illustrated in Figure 6.8. Discontinuities in contour lines are possible but need to be justified by the inferred existence of geologic discontinuities such as faults and unconformities. While tedious, traditional hand contouring can let a geologist imprint a vision on the data that many computer algorithms will miss.

Formations are the basic descriptive unit for a sequence of sediments. The formation is a mappable rock unit that was deposited under a dominant set of depositional conditions at one time. Formations may consist of more than one rock type. The different rock types are referred to as members if they are mappable within the formation.

The thickness of a formation is related to the length of time an environment was in a particular location and how much subsidence was occurring during that period. A formation can contain one or more layers of sedimentary rock. Each layer is referred to as a stratum and has characteristics that distinguish it from other layers. A stratigraphic column presents the column of layers, or strata.

Sedimentary layers are originally deposited in a time sequence. The uppermost stratum is the youngest, and the lowermost stratum is the oldest. Missing layers, or gaps, can occur in the stratigraphic column. A gap can occur when strata are exposed

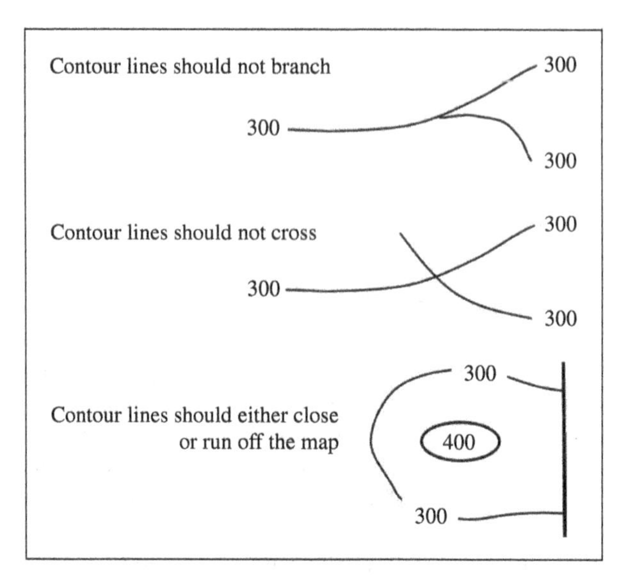

FIGURE 6.8 Examples of contour tips.

to the atmosphere. The exposed strata can be removed by a mechanism such as erosion. The strata can then be buried by later deposits of sediment.

Geologic units are often displayed using the stratigraphic column, structure map, and isopach map. The structure map shows the surface of a stratum by presenting contours of the depth to either the top or base of the stratum. An isopach map shows the thickness of a rock unit.

The characteristics of a rock sequence depend on the depositional environment. For example, sandstone may be deposited in either a fluvial (river) environment or deltaic environment. The sandstone deposited in each environment has a distinct character. Rocks in a fluvial environment can be deposited over great distances in a meandering pattern, while rocks in a deltaic environment tend to be deposited in a more compact location. The grain size of sand deposited in a fluvial environment becomes finer at shallower depths in a process called fining upward, and a deposit of coarser sand is at the base of the formation.

Example 6.4 Net and Gross Thickness

A flow unit consists of 6 ft of impermeable shale, 4 ft of impermeable mudstone, 30 ft of permeable sandstone, and 10 ft of permeable conglomerate. What is the net-to-gross ratio of the flow unit for a conventional reservoir?

Answer

Gross thickness $= H = 6\,\text{ft} + 4\,\text{ft} + 30\,\text{ft} + 10\,\text{ft} = 50\,\text{ft}$.
Net thickness $= h = 30\,\text{ft} + 10\,\text{ft} = 40\,\text{ft}$.
Net-to-gross ratio $= h/H = 40\,\text{ft}/50\,\text{ft} = 0.80$.

6.3 SEDIMENTARY BASINS AND TRAPS

A sedimentary basin is "an area of the earth's crust that is underlain by a thick sequence of sedimentary rocks" (Selley and Sonnenberg, 2015, page 377). The sediments have accumulated to a greater extent in basins than in adjacent areas. Sedimentary basins are formed in large crustal regions that are lower than surrounding regions. Deposition of sediment in these relatively low-lying regions can result in thick accumulations of sedimentary rock.

The rock in sedimentary basins can exhibit significant variability throughout the basin. Rock heterogeneity is due to different depositional environments and changes in pressure, density, and composition of deposited material. The application of tectonic forces such as folding, faulting, and fracturing can result in major changes in the orientation and continuity of rock strata. The three-dimensional distribution of rock strata that have been folded and faulted is an example of a geologic structure.

Some layers of sediment may contain organic matter. These layers are called source rock if there is enough organic matter to generate petroleum. A source rock can be characterized by a factor called the transformation ratio. The transformation ratio is the ratio of the amount of petroleum formed in the basin to the amount of material that was available for generating petroleum.

Pressure and temperature at the base of the accumulation of sediment increase as the accumulation thickens. The increasing temperature and pressure can facilitate the decay of organic matter and the formation of petroleum. The volume of petroleum generated in a sedimentary basin depends on the area of the sedimentary basin, the average total thickness of source rock, and the efficiency of transforming organic matter to petroleum.

Hydrocarbon fluids generated in source rocks are usually less dense than water. They will migrate upward and follow permeable pathways until they encounter an impermeable boundary. A permeable pathway is typically a permeable or fractured formation. The permeable pathway can be referred to as carrier rock to distinguish it from source rock. Traps are structures where hydrocarbon fluids accumulate since they can no longer migrate.

6.3.1 Traps

There are two primary types of traps: structural and stratigraphic. A structural trap is present when the geometry of the reservoir prevents fluid movement. Structural traps occur where the reservoir beds are folded and perhaps faulted into shapes which can contain commercially valuable fluids like oil and gas. Anticlines are a common type of structural trap. Folding and faulting can be caused by tectonic or regional activity. Tectonic activity is a consequence of moving plates, while regional activity is exemplified by the growth of a salt dome. Formation of a structural trap by regional activity is also known as a diapiric trap.

Stratigraphic traps occur where the fluid flow path is blocked by changes in the character of the formation. The formation changes in such a manner that hydrocarbons can no longer move upward. Types of stratigraphic traps include the thinning

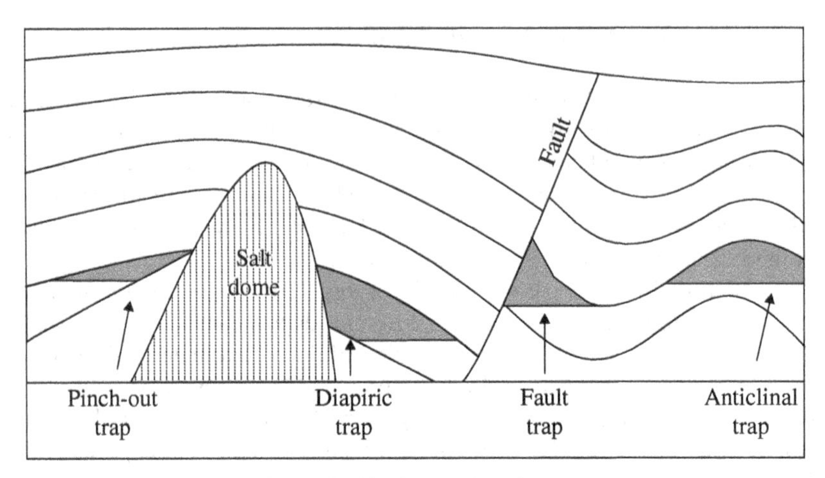

FIGURE 6.9 Examples of traps.

and eventual disappearance of a sand or porosity reduction because of diagenetic changes. Diagenesis refers to processes in which the lithology of a formation is altered at relatively low pressures and temperatures when compared with the metamorphic formation of rock. Diagenesis includes processes such as compaction and cementation.

In addition to structural and stratigraphic traps, there are many examples of traps formed by a combination of structural and stratigraphic features. These traps are called combination traps. An example of a combination trap is the Prudhoe Bay Field on the North Slope of Alaska (Selley and Sonnenberg, 2015, p. 366). It is an anticlinal trap that has been truncated and bounded by impermeable shale. Different types of traps are illustrated in Figure 6.9. The horizontal lines at the base of each trap indicate interfaces between water-saturated rock and hydrocarbon-saturated rock.

6.4 WHAT DO YOU NEED TO FORM A HYDROCARBON RESERVOIR?

Production of hydrocarbons from a subsurface reservoir is a commercial application of multiphase flow through porous media. Hydrocarbons may exist in the solid, liquid (oil), or gaseous (gas) state. Several key factors must be present for a hydrocarbon reservoir to develop.

First, a source rock for the hydrocarbon must be present. Hydrocarbons are thought to form from the decay of organic material. The sedimentary environment that contains the decaying organic material is the source rock.

Second, a permeable flow path known as carrier rock must exist from the source rock to the reservoir rock.

Third, a trap is needed to hold the hydrocarbon. The trap is a structure that is called the reservoir. Rock is considered reservoir rock if fluids can be confined in a volume of rock and fluids can be produced from the rock at economic flow rates.

FIGURE 6.10 Examples of reservoir rocks.

Two important characteristics that control the economic viability of the reservoir are porosity and permeability. Porosity helps quantify the volume of fluid that can be stored in the rock, and permeability helps quantify the rate that fluids can be produced from the rock.

Overriding all of these factors is the fourth key factor: timing. A trap must exist when hydrocarbon migrates from source rock to reservoir rock. If a trap forms at a location after the hydrocarbon has passed, there will be no hydrocarbon to capture.

Reservoir rocks are usually sandstones, carbonates, and shales. Sandstone is compacted sediment and includes consolidated rock with cemented grains, unconsolidated rock with uncemented grains, and conglomerates. A conglomerate consists of larger grains of sediment, such as pebbles or boulders, embedded in a matrix of smaller grains of sediment. Carbonate rock is produced by chemical and biochemical sources. Examples of carbonate rock are limestone (calcium carbonate) and dolomite (calcium magnesium carbonate). Shale is laminated sediment that is formed from consolidated mud or clay. Examples of reservoir rocks are shown in Figure 6.10.

6.5 VOLUMETRIC ANALYSIS, RECOVERY FACTOR, AND EUR

Different calculation procedures and data sources can be used to estimate reservoir fluid volumes. Geologists determine volume using static information in a procedure called volumetric analysis. Static information is information that does not change significantly between the time the reservoir is discovered and the time production begins in the reservoir. Static information includes reservoir volume and the original saturation and pressure distributions. By contrast, engineers use dynamic information to estimate reservoir fluid volumes. Dynamic information is information that changes with respect to time such as pressure changes and fluid production. Material balance is an engineering procedure for estimating original fluid volumes from dynamic data. Reservoir fluid volume estimates obtained from different procedures and sources of data provide a means of assessing the quality of information used by different disciplines. Calculated original fluid volumes can be combined with a recovery factor (RF) to calculate estimated ultimate recovery (EUR) for a given economic limit.

In this section, we present the equations for volumetric estimates of original oil and gas in place and then define RF and EUR.

6.5.1 Volumetric Oil in Place

The volume of original hydrocarbon in place (OHIP) in an oil reservoir is original oil in place (OOIP). It is calculated using the expression

$$N = \frac{7758\phi A h_o S_{oi}}{B_{oi}} \tag{6.1}$$

where 7758 is a unit conversion factor, N is OOIP (STB), ϕ is reservoir porosity (fraction), A is reservoir area (acres), h_o is net thickness of oil zone (feet), S_{oi} is initial reservoir oil saturation (fraction), and B_{oi} is initial oil formation volume factor (RB/STB). Initial oil formation volume factor is the volume of oil at reservoir conditions divided by the volume of oil at stock tank conditions. Associated gas, or gas in solution, is the product of solution gas–oil ratio R_{so} and N.

Example 6.5 Oil in Place

An oil reservoir has average porosity = 0.15 in an area of 6400 acres with a net thickness of 100 ft, initial oil saturation of 75%, and initial oil formation volume factor of 1.3 RB/STB. Use the volumetric OIP equation to estimate OOIP.

Answer
OOIP is

$$N = \text{OOIP} = \frac{7758\phi A h_o S_{oi}}{B_{oi}}$$

$$= \frac{7758 \times (0.15) \times (6400\,\text{acres}) \times (100\,\text{ft}) \times 0.75}{1.3\,\text{RB/STB}} \approx 430\,\text{million STB}$$

6.5.2 Volumetric Gas in Place

OHIP for a gas reservoir is original free gas in place:

$$G = \frac{7758\phi A h_g S_{gi}}{B_{gi}} \tag{6.2}$$

where 7758 is a unit conversion factor, G is original gas in place (SCF), ϕ is reservoir porosity (fraction), A is reservoir area (acres), h_g is net thickness of gas zone (ft), S_{gi} is initial reservoir gas saturation (fraction), and B_{gi} is initial gas formation volume factor (RB/SCF). Initial gas formation volume factor is the volume of gas at reservoir conditions divided by the volume of gas at standard conditions. Equation 6.2 is often expressed in terms of initial water saturation S_{wi} by writing $S_{gi} = 1 - S_{wi}$. Initial water saturation can be determined from well log or core analysis.

Example 6.6 Gas in Place

A well is draining a gas–water reservoir. The drainage area of the well is 160 acres and has a net thickness of 20 ft. Initial properties are 15% porosity, 70% gas saturation, and gas FVF of 0.0016 RB/SCF. What was the original gas in place in the drainage area?

Answer
OGIP is

$$G = \frac{7758\phi A h_g S_{gi}}{B_{gi}} = \frac{7758(0.15)(160\,\text{acres})(20\,\text{ft})0.70}{0.0016\,\text{RB/SCF}} = 1.63\times10^{9}\,\text{SCF}$$

6.5.3 Recovery Factor and Estimated Ultimate Recovery

RF is the fraction of OHIP that can be produced from a reservoir. EUR is calculated from RF and OHIP as

$$\text{EUR} = \text{OHIP}\times\text{RF} = \frac{\text{GRV}\times\bar{\phi}\times\text{NTG}\times\bar{S}_{hi}}{B_{hi}}\times\text{RF} \tag{6.3}$$

where EUR is estimated ultimate recovery (standard conditions), OHIP is original hydrocarbon in place (standard conditions), RF is recovery factor (fraction) to economic limit, GRV is gross rock volume (reservoir conditions), $\bar{\phi}$ is average porosity in net pay (reservoir conditions), NTG is net-to-gross ratio (reservoir conditions), $\bar{S}_{hi}$ is initial average hydrocarbon saturation in net pay (reservoir conditions), and B_{hi} is initial hydrocarbon formation volume factor in a consistent set of units. Formation volume factor is the ratio of reservoir volume to surface volume. The product $\text{GRV}\times\bar{\phi}\times\text{NTG}$ is reservoir pore volume. The volume of fluid produced from a reservoir is the product of RF and original fluid in place. For example, a 30% oil RF means that 30% of the OOIP can be produced. It also implies that 70% of OOIP will remain in the reservoir. EUR is the volume of fluid produced at a specified economic limit.

6.6 ACTIVITIES

6.6.1 Further Reading

For more information about geology, see Selley and Sonnenberg (2015), Hyne (2012), Bjørlykke (2010), Reynolds et al. (2008), and Gluyas and Swarbrick (2004).

6.6.2 True/False

6.1 The K–T boundary is the boundary between the Earth's crust and mantle.

6.2 A hydrocarbon reservoir must be able to trap fluids.

6.3 Formation volume factor is the ratio of surface volume to reservoir volume.

6.4 Capillary pressure must be a positive value.

6.5 Porosity is the fraction of void space in rock.

6.6 The Paleocene epoch is part of the Mesozoic era.

6.7 Mountain ranges can be formed by the collision of tectonic plates.

6.8 Reservoirs are found in sedimentary basins.

6.9 Volumetric analysis depends on dynamic information.

6.10 A microdarcy is greater than a nanodarcy.

6.6.3 Exercises

6.1 The temperature in some parts of the Earth's crust increases by about 1 °F for every 100 ft of depth. Estimate the temperature of the Earth at a depth of 2 mi. Assume the temperature at the surface is 60 °F. Express your answer in °C.

6.2 **A.** A formation contains 10 ft of impermeable shale, 30 ft of impermeable mudstone, and 60 ft of permeable sandstone. What is the gross thickness of the formation?

B. What is the net-to-gross ratio of the formation?

6.3 **A.** Suppose a reservoir is 3 mi long by 6 mi wide and has an average gross thickness of 40 ft, a net-to-gross ratio of 0.7, and a porosity of 0.18. Well logs show an average water saturation of 0.30. An oil sample has a formation volume factor of 1.4 RB/STB. We can calculate original oil in place (in STB) using the following procedure.

B. What is the bulk volume of the reservoir (in RB)?

C. Calculate the pore volume of pay in the reservoir (in RB) if porosity is 0.18 and net-to-gross ratio is 0.7.

D. If the reservoir has an oil saturation of 0.7, what is the volume of oil in the reservoir (in RB)?

E. Calculate original oil in place (in STB).

6.4 Radiometric dating is accomplished using the age equation

$$t = \frac{t_{\text{half}}}{\ln 2} \ln\left(1 + \frac{D}{P}\right)$$

where t is the age of the rock or mineral specimen, t_{half} is the half-life in years of the parent isotope, P is the current number of atoms of the parent isotope, and D is the current number of atoms of a daughter product. Use the age equation to fill in the following table for U-238 half-life of 4.5 billion years.

D/P	Age t (billion years)
0.10	
0.25	
0.50	
0.75	
1.00	
1.50	
2.00	

6.5 The European and North American plates separate at the rate of 1 in. per year in the vicinity of Iceland. How far apart would the two plates move in 10 000 years? Express your answer in feet.

6.6 Is the height of a transition zone greater for a reservoir with small pore throats than large pore throats?

6.7 Assume a formation is approximately 3 mi wide, 10 mi long, and 30 ft thick. The porosity of the sand averages almost 40%. Considering this surface sand body as a model of an underground reservoir, determine the volume of pore space in bbl. Assume net-to-gross ratio = 1.

6.8 An oil reservoir has average porosity = 0.23 in an area of 3200 acres with a net thickness of 80 ft, an initial oil saturation of 70%, and an initial oil formation volume factor of 1.4 RB/STB. Use the volumetric OIP equation to estimate OOIP.

6.9 **A.** A reservoir is 3 mi wide and 5 mi long. What is the area of the reservoir in acres?

B. The reservoir has a net thickness of 50 ft and 10% porosity. The initial gas saturation is 70% with a gas formation volume factor of 0.005 RB/SCF. What is the gas in place in SCF?

7

RESERVOIR GEOPHYSICS

A picture of the large-scale structure of the reservoir can be obtained using vibrations called seismic waves that propagate through the earth. The study of vibrations in the earth is one aspect of geophysics, which is the study of the physical properties and processes associated with the earth and surrounding space. This chapter reviews the process for acquiring and analyzing vibrational data in porous media, discusses the resolution of vibrational data, and introduces modern applications.

7.1 SEISMIC WAVES

Seismic waves are vibrations that propagate from a source and through the earth until some of the vibrational energy is reflected back to the surface where it is detected by receivers (Figure 7.1). Sources include explosions, mechanical vibrators created by vibrating trucks, and expulsion of compressed air from seismic air guns in marine environments.

Receivers include geophones on land and hydrophones offshore. A geophone converts ground motion to a voltage, and a hydrophone is a microphone that detects underwater sound. Receivers record travel time from the source to the receiver during the data acquisition phase of a seismic survey.

Geophones and hydrophones are seismometers. A seismometer detects vibrational energy and sends a record of the vibrations to a seismograph, which displays the signal as a seismogram. The seismogram records the motion of the seismometer

Introduction to Petroleum Engineering, First Edition. John R. Fanchi and Richard L. Christiansen.
© 2017 John Wiley & Sons, Inc. Published 2017 by John Wiley & Sons, Inc.
Companion website: www.wiley.com/go/Fanchi/IntroPetroleumEngineering

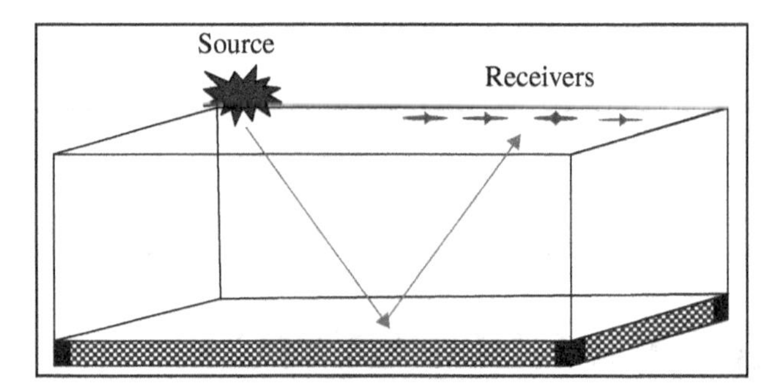

FIGURE 7.1 Energy propagation as vibrations in the subsurface.

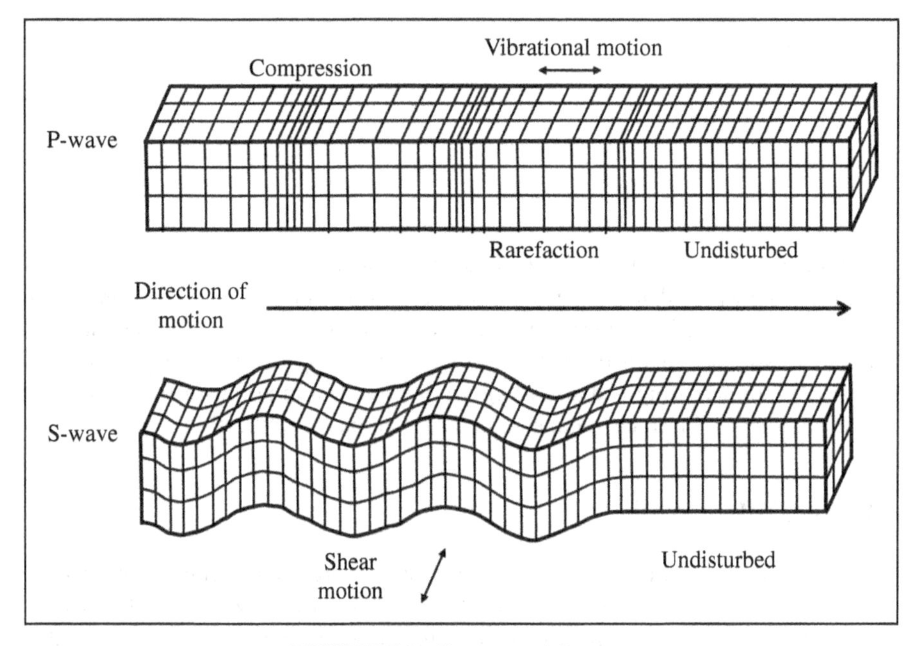

FIGURE 7.2 P-wave and S-wave.

as a function of time. The graph from a single seismometer is called a trace. A seismic section displays a collection of traces from different seismometers.

The vibrational energy propagates as seismic waves through the earth. Two common types of seismic vibrations are compressional (P-) waves and shear (S-) waves. P-waves are longitudinal waves that propagate as compressions and rarefactions in the direction of wave motion illustrated in Figure 7.2. S-waves are vibrations that move perpendicular to the direction of wave motion. S-waves are also known as shear waves because the particles of the disturbed medium are displaced in a shearing motion that is perpendicular to the direction of wave motion. P-waves and S-waves are body waves because they travel through the body of a medium. Surface waves are waves that travel along the surface of a medium.

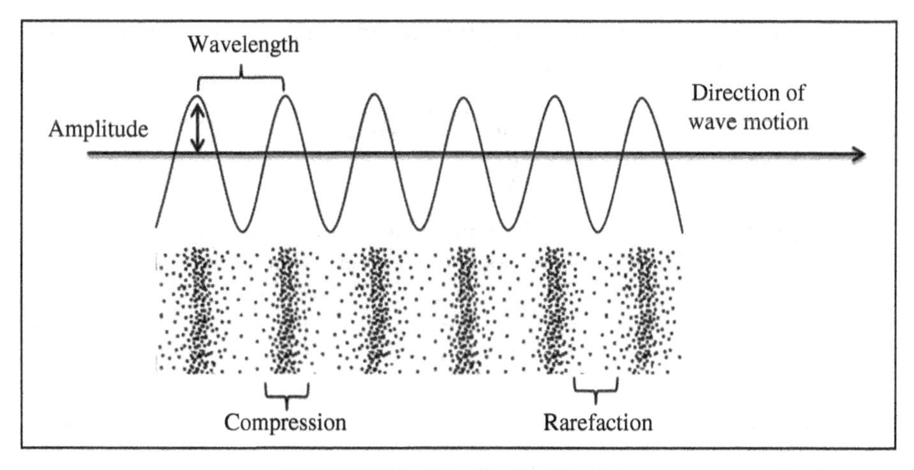

FIGURE 7.3 Longitudinal P-wave.

Waves are characterized by amplitude, wavelength, and frequency (Figure 7.3). The wavelength is the length of the wave from one point on the wave to an equivalent point on the wave. Wave frequency is the number of waves passing a particular point during a specified interval of time. Typical seismic frequencies range from 5 to 100 Hz (Bjørlykke, 2010, page 380). Vibrational energy from the source propagates as a packet of waves known as a wavelet. The wavelet has a dominant wavelength λ_d and a dominant frequency f_d. The amplitude of the wavelet varies from zero to total amplitude and back to zero over the length of the wavelet.

Example 7.1 Seismic Wave Travel Time

A seismic wave propagates through the crust of the earth at a speed of 5 km/s. The seismic wave propagates vertically downward to a depth of 2000 m and is reflected back to the surface. How long does it take for the seismic wave to make the round trip?

Answer

The travel time downward is 2000 m/(5000 m/s) or 0.4 s. The time to make the round trip is called two-way travel (TWT) time and is two times 0.4 s so that TWT = 0.8 s.

Seismic velocities depend on the physical properties of rock. The relevant rock properties are shear modulus, bulk modulus of rock grains, and mass density of the medium. Shear modulus is the ratio of shear stress to shear strain. Bulk modulus refers to the bulk deformation of a rock subjected to confining pressure.

The P-wave was originally called the primary wave because the P-wave arrived before the S-wave. The S-wave was called the secondary wave because it arrived after the primary wave. The velocity of P-wave propagation V_P in an elastic, homogeneous, and isotropic medium is

$$V_P = \sqrt{\frac{\left(K_B + (4/3)G\right)}{\rho}} \tag{7.1}$$

where K_B is bulk modulus, G is shear modulus, and ρ is the mass density of the medium. The corresponding velocity of S-wave propagation V_S is

$$V_S = \sqrt{\frac{G}{\rho}} < v_P \tag{7.2}$$

Example 7.2 P-Wave and S-Wave Velocities

Dolomite has bulk modulus $K = 95$ GPa, shear modulus $G = 46$ GPa, and density $\rho = 2800$ kg/m^3. Calculate compressional wave velocity in m/s and shear wave velocity in m/s.

Answer

A consistent set of units is obtained by first converting GPa to Pa (1 GPa = 10^9 Pa and 1 Pa = 1 N/m^2 = 1 kg/m·s^2): $K = 85$ GPa $= 85 \times 10^9$ Pa and $G = 55$ GPa $= 55 \times 10^9$ Pa.

Compressional wave velocity is

$$V_P = \sqrt{\frac{K + (4/3)G}{\rho}} = \sqrt{\frac{85 \times 10^9 + (4/3)55 \times 10^9}{2800}} = 7519 \text{ m/s}$$

Shear wave velocity is

$$V_S = \sqrt{\frac{G}{\rho}} = \sqrt{\frac{55 \times 10^9}{2800}} = 4008 \text{ m/s}$$

7.1.1 Earthquake Magnitude

An earthquake occurs when the crust slips along a subsurface fault. Seismic wave vibrations caused by the slippage are detected by a seismograph that records the amplitudes of ground movement beneath the instrument. Charles F. Richter developed a scale (Richter, 1935) to measure the magnitude of earthquakes in southern California using a Wood–Anderson torsion seismometer (Wood and Anderson, 1925). A copper mass suspended by a thin wire would rotate when the seismometer moved. The seismogram was made when a beam of light was reflected onto photosensitive paper from a mirror on the copper mass.

The Richter scale expresses the local magnitude M_L of a vibration on a Wood–Anderson seismograph as

$$M_L = \log \frac{A}{A_0} = \log A - \log A_0 \tag{7.3}$$

where A is the maximum amplitude measured on the seismograph and A_0 is a correction factor for a particular region like southern California. Both A, A_0 were expressed in mm. Richter assigned $M_L = 3$ to an amplitude $A = 1$ mm on a Wood–Anderson seismograph that was recorded 100 km from the source of an earthquake.

Example 7.3 Richter Scale

What is the correction factor A_0 when $M_L = 3$ and $A = 1$ mm on a Wood–Anderson seismograph that was recorded 100 km from the source of an earthquake?

Answer

We rearrange Equation 7.3 to find the correction factor A_0:

$$A_0 = A \cdot 10^{-M_L} = 1\text{ mm} \cdot 10^{-3} = 0.001\text{ mm}$$

The Richter scale tended to underestimate the size of large earthquakes. A more fundamental measure of the magnitude of an earthquake is the seismic moment M_0 defined in terms of the movement of one fault block relative to another. If we define G as the shear modulus (Pa) of the rock, d as the distance one fault block slips relative to another fault block (m), and S as the estimated surface area that is ruptured at the interface between the two fault blocks (m^2), then seismic moment M_0 is

$$M_0 = G \cdot d \cdot S \tag{7.4}$$

The unit of seismic moment M_0 is N·m in SI units. The seismic moment M_0 is a measure of the strength of the motion of the earthquake that connects seismographic measurements to the physical displacement of fault blocks. The energy E released by the earthquake is based on the empirical relationship

$$E \approx \frac{M_0}{20000} \tag{7.5}$$

The SI unit of E is Joules. The physical units of seismic moment M_0 (N·m) and energy E (J) are used to denote that M_0 and E are different physical quantities.

The magnitude of the earthquake is based on empirical relationships between seismic moment M_0 and moment magnitude M_w. For example, the moment–magnitude relationship (Hanks and Kanamori, 1978)

$$M_w = \frac{2}{3}\left[\log(M_0) - 9.1\right] \tag{7.6}$$

relates seismic moment M_0 and moment magnitude M_w, where M_w is dimensionless and M_0 is in N·m.

TABLE 7.1 Classification of Earthquakes

Class	Magnitude	Comment
>8.0	Great	Can destroy communities near the epicenter
7.0–7.9	Major	Serious damage
6.0–6.9	Strong	Significant damage in populated areas
5.0–5.9	Moderate	Slight damage to structures
4.0–4.9	Light	Obvious vibrations
3.0–3.9	Minor	Weak vibrations
2.0–2.9	Very minor	Hardly felt

Table 7.1 presents a classification of earthquakes by magnitude. Moment magnitudes can be negative. For example, Warpinski et al. (2012) pointed out that microseismic events associated with fracturing operations in shale have negative moment magnitudes on the order of −2 to −3. The fracturing operations are known as hydraulic fracturing and are used to fracture very low permeability shale. The fractures are kept open with proppants to provide higher permeability fluid flow paths to the wellbore.

7.2 ACOUSTIC IMPEDANCE AND REFLECTION COEFFICIENTS

Seismic waves are detected at receivers after some of the vibrational energy of the incident seismic wave is reflected by a geologic feature. Some of the vibrational energy from the incident seismic wave is transmitted and some is reflected when the incident seismic wave encounters a reflecting surface. Seismic reflection occurs at the interface between two regions with different acoustic impedances. Acoustic impedance Z is the product of bulk density and seismic wave velocity in a medium. If the velocity is compressional (P-wave) velocity, acoustic impedance of P-wave velocity is compressional impedance:

$$Z_{\mathrm{P}} = \rho_{\mathrm{B}} V_{\mathrm{P}} \tag{7.7}$$

The bulk density of a rock–fluid system depends on rock matrix grain density ρ_{m}, fluid density ρ_{f}, and porosity ϕ:

$$\rho_{\mathrm{B}} = (1-\phi)\rho_{\mathrm{m}} + \phi\rho_{\mathrm{f}} \tag{7.8}$$

Fluid density for an oil–water–gas system is

$$\rho_{\mathrm{f}} = \rho_{\mathrm{o}} S_{\mathrm{o}} + \rho_{\mathrm{w}} S_{\mathrm{w}} + \rho_{\mathrm{g}} S_{\mathrm{g}} \tag{7.9}$$

where ρ_ℓ is fluid density of phase ℓ and S_ℓ is saturation of phase ℓ. Subscripts o, w, and g stand for oil phase, water phase, and gas phase, respectively. Shear impedance is acoustic impedance calculated using shear velocity:

$$Z_{\mathrm{S}} = \rho_{\mathrm{B}} V_{\mathrm{S}} \tag{7.10}$$

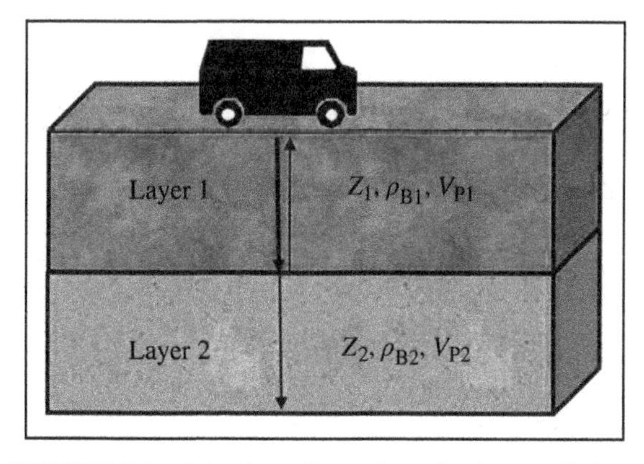

FIGURE 7.4 Seismic attributes for reflection coefficient.

Figure 7.4 depicts an incident wave as a bold arrow directed downward in layer 1. The energy in the incident wave is split between the reflected wave in layer 1 and the transmitted wave in layer 2. The reflection coefficient RC at the interface between layer 1 and layer 2 with acoustic impedances Z_1 and Z_2, respectively, is the dimensionless ratio

$$\mathrm{RC} = \frac{Z_2 - Z_1}{Z_2 + Z_1} \tag{7.11}$$

Equation 7.11 assumes zero offset between source and receiver. Offset is the distance between source and receiver. The zero offset reflection coefficient corresponds to an incident plane wave propagating in a direction that is perpendicular to a horizontal reflecting interface.

The seismic reflection coefficient RC is zero at the interface between two formations with equal acoustic impedances. Nonzero values of reflection coefficients are obtained when an incident wave is reflected at an interface between two media with different acoustic impedances. The acoustic impedance changes if there is a change in either bulk density or wave velocity as the wave travels from one medium into another.

Example 7.4 Seismic Reflection Coefficient

Consider two rock layers with the following properties:

Upper layer 1 has density = 2800 kg/m^3 and P-wave velocity = 3.0 km/s.
Lower layer 2 has density = 2600 kg/m^3 and P-wave velocity = 2.8 km/s.

A. Calculate P-wave acoustic impedance for each layer.

B. Calculate reflection coefficient for a downward wave reflected at the interface between layers 1 and 2.

Answer

A. P-wave acoustic impedance for upper layer

$$Z_{\mathrm{U}} = (\rho V_{\mathrm{P}})_{\mathrm{U}} = 2800\frac{\mathrm{kg}}{\mathrm{m}^3} \times 3.0\frac{\mathrm{km}}{\mathrm{s}} = 8400\frac{\mathrm{kg}}{\mathrm{m}^3}\frac{\mathrm{km}}{\mathrm{s}}$$

and P-wave acoustic impedance for lower layer

$$Z_{\mathrm{L}} = (\rho V_{\mathrm{P}})_{\mathrm{L}} = 2600\frac{\mathrm{kg}}{\mathrm{m}^3} \times 2.8\frac{\mathrm{km}}{\mathrm{s}} = 7280\frac{\mathrm{kg}}{\mathrm{m}^3}\frac{\mathrm{km}}{\mathrm{s}}$$

B. Reflection coefficient

$$\mathrm{RC} = \frac{Z_{\mathrm{L}} - Z_{\mathrm{U}}}{Z_{\mathrm{L}} + Z_{\mathrm{U}}} = \frac{7280 - 8400}{7280 + 8400} = -0.0714$$

7.3 SEISMIC RESOLUTION

Seismic wave reflecting surfaces are due to changes in acoustic impedance between two adjacent media. Seismic resolution is the ability to distinguish between two reflecting surfaces that are close together. The quality of seismic resolution depends on the orientation of the reflecting surfaces and interference effects, as we discuss in this section.

7.3.1 Vertical Resolution

Vertical resolution tells us how close two horizontal reflecting surfaces can be and still be separable. The two reflecting surfaces are the top and bottom of a thin layer. Maximum vertical resolution is expressed in terms of the dominant wavelength of a seismic wavelet λ_{d}, which is the ratio of wavelet velocity v_{i} to the dominant frequency f_{d}:

$$\lambda_{\mathrm{d}} = \frac{v_{\mathrm{i}}}{f_{\mathrm{d}}} = v_{\mathrm{i}} T_{\mathrm{d}} \tag{7.12}$$

The dominant period T_{d} is the inverse of dominant frequency f_{d}. As an illustration, suppose the wavelet velocity is 10 000 ft/s and the dominant period is 40 ms or 0.040 s, then the dominant frequency f_{d} is 25 Hz and the dominant wavelength is 400 ft.

Maximum vertical resolution δz_{V} is one fourth of the dominant wavelength λ_{d}:

$$\delta z_{\mathrm{V}} = \frac{\lambda_{\mathrm{d}}}{4} = \frac{v_{\mathrm{i}}}{4 f_{\mathrm{d}}} \tag{7.13}$$

The maximum vertical resolution in the illustration earlier is approximately 100 ft. If the separation between two reflecting surfaces is less than the maximum vertical resolution, it is difficult or impossible to distinguish each of the reflecting surfaces.

The maximum vertical resolution can be increased by reducing the dominant frequency of the seismic wavelet.

7.3.2 Lateral Resolution

Lateral resolution is the ability to distinguish two points that are separated in the lateral, or horizontal, plane. The maximum horizontal resolution is estimated by considering constructive interference of waves from the Fresnel zone. A Fresnel zone is the part of a reflecting surface that reflects seismic energy back to a detector. The radius r of the Fresnel zone depends on the depth z from the reflecting interface to the detector and the dominant wavelength λ_d. A criterion based on analysis of wave interference says that two reflective surfaces can be distinguished when they are separated by a distance of at least ¼ wavelength. The geometry of the Fresnel zone for a reflector–receiver system is illustrated in Figure 7.5.

The radius of the Fresnel zone for estimating the maximum horizontal resolution is

$$r = \frac{1}{2} \times \frac{\lambda_d}{2} = \frac{\lambda_d}{4} \tag{7.14}$$

The maximum ray path of a wave arriving at the detector from the Fresnel zone is given by the Pythagorean relation

$$z^2 + r^2 = \left(z + \frac{\lambda_d}{4} \right)^2 \tag{7.15}$$

Expanding Equation 7.15 and solving for r gives

$$r^2 = \frac{\lambda_d z}{2} + \frac{\lambda_d^2}{16} \tag{7.16}$$

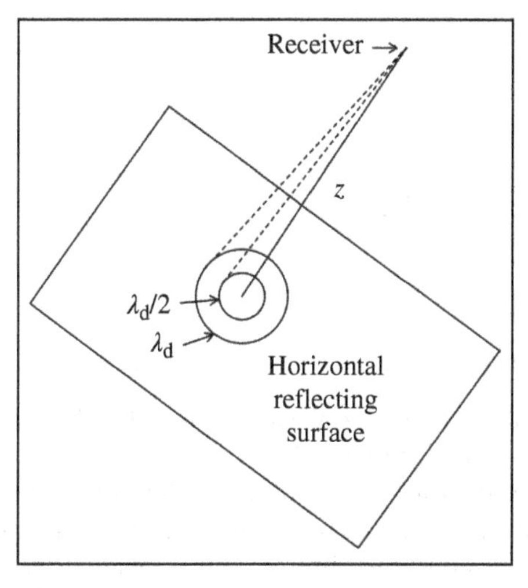

FIGURE 7.5 Geometry of Fresnel zone.

Depth in terms of velocity and two-way travel time Δt is $z = v_i \Delta t/2$, and the dominant wavelength is related to dominant frequency by $\lambda_d = v_i/f_d$. Substituting these relations in Equation 7.16 gives

$$r = \frac{v_i}{2}\left[\frac{\Delta t}{f_d} + \frac{1}{4f_d^2}\right]^{1/2} \tag{7.17}$$

The radius r_F of the first-order Fresnel zone is obtained by keeping only first-order terms in dominant frequency to obtain an estimate of maximum horizontal resolution δz_H:

$$r_F = \frac{v_i}{2}\sqrt{\frac{\Delta t}{f_d}} = \delta z_H \tag{7.18}$$

The Fresnel radius r_F is about 775 ft for wavelet velocity of 10 000 ft/s, dominant frequency of 25 Hz, and two-way travel time Δt of 0.6 s. Thus, two horizontal features can be resolved if they are separated by at least 775 ft. If we increase the dominant frequency, we can resolve horizontal features on a smaller separation because the maximum horizontal resolution decreases.

Example 7.5 Seismic Resolution of Two Reflective Surfaces

A. A wave propagating at 1800 m/s has a frequency of 60 Hz. What is the wavelength of the wave?

B. Assume two reflective surfaces can be distinguished when they are separated by a distance of at least ¼ wavelength. Use this criterion and the wavelength in Part A to estimate the minimum separation needed to distinguish two reflective surfaces.

Answer

A. $\lambda = \frac{1800\ \text{m/s}}{60\ \text{Hz}} = 30\ \text{m}$

B. $\text{Separation} = \frac{\lambda}{4} = \frac{30\ \text{m}}{4} = 7.5\ \text{m}$

7.3.3 Exploration Geophysics and Reservoir Geophysics

Historically, the primary role of geophysics in the oil and gas industry was exploration. Today, geophysical technology has value during both the exploration and development stages of the life of a reservoir. Exploration geophysics is used to provide a picture of subsurface geologic features such as stratigraphy and structure. Reservoir geophysics is conducted in fields where wells have penetrated the target horizon. The resolution associated with reservoir geophysics is more quantitative than the resolution associated with exploration geophysics.

Wittick (2000) and Pennington (2001) observed that the difference in resolution between exploration and reservoir geophysics is due to the role of calibration. Ordinarily, no wells are available during the exploration process, so it is not possible to calibrate seismic data measurements with well log measurements. By contrast, well log data can be acquired during the development process to provide measurements in the wellbore that can be used to calibrate seismic surveys conducted at the surface. Consequently, reservoir geophysics can have more information available to improve quantitative estimates of reservoir properties if appropriate well log data is used to calibrate seismic data.

The process of using calibrated seismic information to predict reservoir properties is called seismic inversion. Seismic inversion is an attempt to correlate seismic attributes like acoustic impedance to rock properties. Seismic attributes are cross-plotted against groupings of rock properties. Examples of cross-plots for a formation with permeability K, oil saturation S_o, and net thickness h_{net} include acoustic impedance versus porosity ϕ, seismic amplitude versus flow capacity (Kh_{net}) or rock quality (ϕKh_{net}), and seismic amplitude versus oil productive capacity ($S_o\phi Kh_{net}$).

One of the first examples of seismic inversion that included a field test was provided by De Buyl et al. (1988). They predicted reservoir properties at two wells using seismic inversion and then compared actual results to predicted results. A similar comparison was made between measurements at wells and predictions prepared using only well logs. Predictions made with seismic inversion were at least as accurate as predictions made with well log data only and were more accurate in some cases.

7.4 SEISMIC DATA ACQUISITION, PROCESSING, AND INTERPRETATION

Subsurface geologic features are mapped in reflection seismology by measuring the time it takes an acoustic signal to travel from the source to a seismic reflector and then to a receiver. Three steps are required to analyze seismic measurements: data acquisition, data processing, and interpretation.

7.4.1 Data Acquisition

Seismic surveys are conducted to acquire seismic data. A 2-D seismic survey uses a vibrational source and a single line of receivers to prepare a cross-sectional image of the subsurface. A 3-D seismic survey uses a line of sources with a 2-D array of receivers to prepare a 3-D image of the subsurface.

The recorded seismic trace is a function of travel time and combines source signal with the sequence of seismic reflectors known as reflectivity sequence. Reflectors are determined by changes in acoustic impedance and the corresponding reflection coefficient. The seismic trace also includes noise, which is always present in realistic data acquisition. In practice, traces from a single source are recorded at multiple receivers. A family of seismic traces is called a gather.

Time-lapse seismology, also known as 4-D seismic, is the comparison of two 3-D seismic surveys taken in the same geographic location at different times. The fourth dimension in 4-D seismic is time. The difference between the two 3-D seismic surveys should show changes in the same rock volume due to changes resulting from operations. The structure should be the same while the pressure and saturation distributions change. Thus, 4-D seismic can be used to identify the movement of fluids between wells, improve the quality of reservoir characterization, and highlight bypassed reserves in reservoirs where a signal can be detected.

7.4.2 Data Processing

Data processing is used to prepare seismic data for interpretation. One of the most important data processing tasks is to transform travel time to depth. The time-to-depth conversion makes it possible to view seismic traces as functions of depth and compare them to geological and engineering measurements which are typically expressed as functions of depth. The conversion of travel time measurements to depths depends on the velocity of propagation of the acoustic signal through the earth. Seismic velocity varies with depth and depends on the properties of the media along the travel path. The collection of velocities used in the time-to-depth conversion is called the velocity model, and the process is known as depth migration. If evidence becomes available that suggests the velocity model needs to be changed, it may be necessary to process seismic data again with the revised velocity model.

Vertical seismic profiles (VSP) or checkshots in wellbores can improve the quality of velocity models. A checkshot is conducted by discharging a vibration source at the surface and recording the seismic response in a borehole receiver. The checkshot is a VSP with zero offset if the vibration source is vertically above the receiver. The checkshot is a VSP with offset if the vibration source is offset relative to the receiver. A reverse VSP is obtained by placing the source in the wellbore and the receiver on the surface. The velocity model can be validated or refined by comparing well log and core sample data to VSP and checkshot data.

Data processing is also used to maximize the signal to noise ratio of the seismic data. There is always noise in seismic data. Noise arises from such sources as ground roll, wind, vibrations from other operations, and interference with other seismic waves that intersect the path of the signal. A variety of mathematical techniques have been developed to maximize the signal and minimize the noise.

7.4.3 Data Interpretation

Seismic data that have been subjected to data processing are ready for interpretation. Processed seismic data are reviewed in conjunction with data from other disciplines to provide a better understanding of the fluid content, composition, extent, and geometry of subsurface rock. Some applications of the interpretation process include development of geologic models, preparing an image of the geologic structure, identifying faults and folds, and designing wellbore trajectories.

7.5 PETROELASTIC MODEL

Seismic attributes can be related to one another using a petroelastic model. One example of a petroelastic model, the integrated flow model (IFM) (Fanchi, 2009, 2010), is used in reservoir simulation software to conduct studies of fluid flow in porous media. The IFM is based on the assumptions that temperature does not significantly affect rock properties, and rock properties are elastic. The elasticity assumption is reasonable in regions where rock failure does not occur over the range of pressure and temperature encountered during the life of a reservoir. The IFM is presented here.

7.5.1 IFM Velocities

The seismic attributes compressional velocity (V_P), shear velocity (V_S), and associated acoustic impedances (Z_P, Z_S) are calculated in the IFM using the relations

$$V_P = \sqrt{\frac{S^*}{\rho^*}}, \quad S^* = K^* + \frac{4G^*}{3} \tag{7.19}$$

$$V_S = \sqrt{\frac{G^*}{\rho^*}} \tag{7.20}$$

$$Z_P = \rho^* V_P, \quad Z_S = \rho^* V_S \tag{7.21}$$

where S^* is stiffness, K^* is bulk modulus, G^* is shear modulus, and ρ^* is bulk density. Bulk density accounts for both pore volume occupied by fluids and volume of rock grains in the bulk volume. The equation for bulk density is

$$\rho^* = (1-\phi)\rho_m + \phi\rho_f \tag{7.22}$$

where ϕ is porosity, ρ_m is rock matrix grain density, and ρ_f is fluid density. Fluid density for different fluids occupying pore space is

$$\rho_f = \rho_o S_o + \rho_w S_w + \rho_g S_g \tag{7.23}$$

where ρ_ℓ is fluid density of phase ℓ and S_ℓ is saturation of phase ℓ. Subscripts o, w, and g stand for oil, water, and gas, respectively.

The ratio V_P/V_S of compressional velocity to shear velocity is

$$\frac{V_P}{V_S} = \sqrt{\frac{K^* + (4G^*/3)}{G^*}} = \sqrt{\frac{4}{3} + \frac{K^*}{G^*}} \tag{7.24}$$

The ratio V_P/V_S is greater than $\sqrt{4/3}$ since the moduli K^* and G^* are greater than zero.

Seismic velocities can be estimated from correlations that depend on rock type. For example, Castagna et al. (1985) presented a correlation for seismic velocities in sandstone. The velocities depend on porosity and clay content C:

$$V_{\mathrm{P}} = 5.81 - 9.42\phi - 2.21C \tag{7.25}$$

and

$$V_{\mathrm{S}} = 3.89 - 7.07\phi - 2.04C \tag{7.26}$$

where the seismic velocities are in km/s, C is expressed as the fraction of clay volume, and porosity ϕ is a fraction.

7.5.2 IFM Moduli

Gassmann's equation (Gassmann, 1951) was introduced to provide an estimate of saturated bulk modulus, which is the bulk modulus of a rock saturated with fluids. Gassmann's equation is widely used in rock physics because it is relatively simple and does not require much data. Saturated bulk modulus K^* in the IFM is approximated as a form of Gassmann's equation. It is given by

$$K^* = K_{\mathrm{IFM}} + \frac{\left[1 - \left(K_{\mathrm{IFM}}/K_{\mathrm{m}}\right)\right]^2}{\left(\phi/K_{\mathrm{f}}\right) + \left(1 - \phi/K_{\mathrm{m}}\right) - \left(K_{\mathrm{IFM}}/K_{\mathrm{m}}^2\right)} \tag{7.27}$$

where K_{IFM} is dry frame bulk modulus, K_{m} is rock matrix grain bulk modulus, and K_{f} is fluid bulk modulus. The term "dry" means no fluid is present in the rock pore space. The rock would not be considered "dry" if the pore space was filled with gas, including air; instead, the rock would be considered gas saturated. The variables in Equation 7.27 can change as reservoir conditions change.

Fluid bulk modulus is the inverse of fluid compressibility c_{f}:

$$K_{\mathrm{f}} = \frac{1}{c_{\mathrm{f}}} = \frac{1}{\left(c_{\mathrm{o}}S_{\mathrm{o}} + c_{\mathrm{w}}S_{\mathrm{w}} + c_{\mathrm{g}}S_{\mathrm{g}}\right)} \tag{7.28}$$

where fluid compressibility is the saturation weighted average of phase compressibility. The bulk modulus of rock saturated with pore fluid can be calculated in terms of seismic velocities by rearranging Equations 7.19 and 7.20:

$$K^* = \rho_{\mathrm{B}}\left[V_{\mathrm{P}}^2 - \frac{4}{3}V_{\mathrm{S}}^2\right] \tag{7.29}$$

Similarly, effective shear modulus is

$$G^* = \rho_{\mathrm{B}}V_{\mathrm{S}}^2 \tag{7.30}$$

Saturated bulk modulus K^* equals grain modulus K_{m} when porosity goes to zero.

7.6 GEOMECHANICAL MODEL

The mechanical behavior of rocks in the subsurface can be modeled using geomechanical models. Parameters used in geomechanical modeling include Poisson's ratio (ν), Young's modulus (E), uniaxial compaction (Δh), and horizontal stress (σ_H). This set of geomechanical parameters can be estimated using the IFM. Poisson's ratio and Young's modulus depend on the frequency of the vibration. Dynamic Poisson's ratio and dynamic Young's modulus are obtained from measurements of compressional and shear velocities. Static Poisson's ratio and static Young's modulus are measured in the laboratory using measurements on cores. Static measurements more accurately represent mechanical properties of reservoir rock. Consequently, it is worthwhile to be able to transform from dynamic to static properties.

Dynamic Poisson's ratio (ν_d) is given in terms of compressional (P-wave) velocity and shear (S-wave) velocity as

$$\nu_d = \frac{0.5V_P^2 - V_S^2}{V_P^2 - V_S^2} \tag{7.31}$$

Once an estimate of dynamic Poisson's ratio is known, it can be combined with shear modulus to estimate dynamic Young's modulus (E_d):

$$E_d = 2(1+\nu_d)G^* \tag{7.32}$$

Static Poisson's ratio (ν_s) is calculated from dynamic Poisson's ratio using the algorithm

$$\nu_s = a\nu_d^b + c \tag{7.33}$$

where a, b, c are dimensionless coefficients. Coefficients a, b are functions of effective pressure p_e, which can be estimated from the IFM as

$$p_e = p_{con} - \alpha p \tag{7.34}$$

where p is pore pressure. Confining pressure p_{con} at depth z is estimated from the overburden pressure gradient γ_{OB} as

$$p_{con} = \gamma_{OB} z \tag{7.35}$$

The Biot coefficient α is the ratio of the change in pore volume to the change in bulk volume of a porous material in a dry or drained state (Mavko et al., 2009). The Biot coefficient α is estimated in the IFM from the Geertsma–Skempton correction

$$\alpha = 1 - \left(\frac{K_{IFM}}{K_m}\right) \tag{7.36}$$

Static Poisson's ratio is equal to dynamic Poisson's ratio when $a = 1,\ b = 1,\ c = 0$.

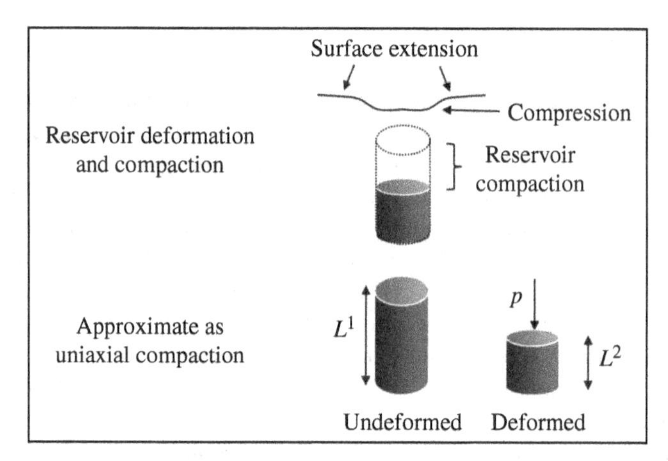

FIGURE 7.6 Schematic of reservoir compaction features. (Source: Fanchi (2010). Reproduced with permission of Elsevier-Gulf Professional Publishing.)

Static Young's modulus (E_s) is calculated from dynamic Young's modulus using the algorithm

$$E_s = a' E_d^{b'} + c' \tag{7.37}$$

where a', b' are dimensionless coefficients and c' has the same unit as shear modulus. If the functional dependence of a', b' on effective pressure p_e is not known, static Young's modulus is set equal to dynamic Young's modulus when $a' = 1, b' = 1, c' = 0$.

Surface extension, compression, and reservoir compaction shown in the upper half of Figure 7.6 occur when a reservoir is deformed. Deformation can occur when reservoir fluids are withdrawn. The effects of deformation can be approximated using the uniaxial compaction model sketched in the lower half of Figure 7.6. Uniaxial compaction Δh is the compaction of an object along one axis and is estimated in the IFM from static Poisson's ratio as

$$\Delta h = \frac{1}{3}\left[\frac{1+\nu_s}{1-\nu_s}\right]\phi c_\phi h_{\text{net}}\left(p - p_{\text{init}}\right) \tag{7.38}$$

where ϕ is porosity, c_ϕ is porosity compressibility, h_{net} is net thickness, p_{init} is initial pore pressure, and p is pore pressure.

Horizontal stress in a formation is estimated in the IFM as

$$\sigma_H = \frac{\nu_s}{1-\nu_s}\left(p_{\text{con}} - \alpha p\right) + \alpha p \tag{7.39}$$

for pore pressure p, confining pressure p_{con}, static Poisson's ratio ν_s, and Biot coefficient α. Vertical stress in the formation is approximated as confining pressure p_{con}.

7.7 ACTIVITIES

7.7.1 Further Reading

For more information, see Selley and Sonnenberg (2015), Tiab and Donaldson (2011), Johnston (2010), Bjørlykke (2010), Pennington (2007), Schön (1996), and Sheriff (1992).

7.7.2 True/False

7.1 Seismic measurements use vibrations to provide images of the subsurface.

7.2 The S-wave is faster than a P-wave.

7.3 A single seismic survey is sufficient to perform time-lapse seismic analysis.

7.4 Shear modulus is the ratio of shear strain to shear stress.

7.5 Compressional impedance is the product of bulk density and compressional velocity.

7.6 The unit of strain is length.

7.7 Seismic resolution is the ability to distinguish between two features.

7.8 A velocity model is used to transform seismic travel time to depth.

7.9 The seismic reflection coefficient RC at the interface between two formations with equal acoustic impedances is 0.

7.10 Gassmann's equation can be used to estimate saturated bulk modulus.

7.7.3 Exercises

7.1 A sandstone sample has bulk modulus $K = 20.0\,\text{GPa}$, shear modulus $G = 19.9\,\text{GPa}$, and density $\rho = 2500\,\text{kg/m}^3$. Calculate compressional wave velocity V_P in m/s and shear wave velocity V_S in m/s where

$$V_P = \sqrt{\frac{K + (4G/3)}{\rho}}; \quad V_S = \sqrt{\frac{G}{\rho}}$$

To get velocity in m/s, express K and G in Pa and density in kg/m^3. Hint: first convert GPa to Pa ($1\,\text{GPa} = 10^9\,\text{Pa}$ and $1\,\text{Pa} = 1\,\text{N/m}^2 = 1\,\text{kg/m·s}^2$).

7.2 A shale sample has bulk modulus $K = 20.8\,\text{GPa}$, shear modulus $G = 13.4\,\text{GPa}$, and density $\rho = 2600\,\text{kg/m}^3$. Calculate compressional wave velocity V_P in m/s and shear wave velocity V_S in m/s where

$$V_P = \sqrt{\frac{K + (4G/3)}{\rho}}; \quad V_S = \sqrt{\frac{G}{\rho}}$$

To get velocity in m/s, express K and G in Pa and density in kg/m^3. Hint: first convert GPa to Pa (1 GPa = 10^9 Pa and 1 Pa = 1 N/m^2 = 1 kg/m·s^2).

7.3 **A.** Calculate the ratio of P-wave velocity to S-wave velocity (V_P/V_S) when V_P = 18 736 ft/s and V_S = 10 036 ft/s.
B. How long does it take a P-wave to travel 10 000 ft?
C. How long does it take an S-wave to travel 10 000 ft?
D. What is the difference in arrival times between the P-wave and S-wave travel times in Parts B and C?

7.4 **A.** Suppose the correlations $V_P = 5.81 - 9.41\phi - 2.21C$ and $V_S = 3.89 - 7.07\phi - 2.04C$ can be used to estimate P-wave velocity and S-wave velocity for a particular sandstone. The velocities are in km/s, porosity is a fraction, and C is clay volume fraction. Estimate P-wave velocity for sandstone with 20% porosity and clay volume fraction C = 0.05. Express your answer in km/s.
B. Estimate S-wave velocity for sandstone with 20% porosity and clay volume fraction C = 0.05. Express your answer in km/s.
C. Use Parts A and B to calculate the ratio of velocities V_P/V_S.

7.5 P-wave velocity is V_P = 12 500 ft/s in a medium. What is the two-way travel (TWT) time for the P-wave to travel vertically downward from the surface to a depth of 10 000 ft and then back to the surface?

7.6 Young's modulus (E) and Poisson's ratio (ν) are related to bulk modulus (K) and shear modulus (G) by the equations $K = E/(3(1-2\nu))$ and $G = E/(2+2\nu)$. Calculate bulk modulus (K) and shear modulus (G) when $E = 4.4 \times 10^6$ psia and $\nu = 0.25$.

7.7 Bulk modulus K and shear modulus G are related to Young's modulus E and Poisson's ratio ν by the equations $E = 9KG/(3K+G)$ and $\nu = (3K-2G)/(2(3K+G))$. Suppose bulk modulus $K = 4.19 \times 10^6$ psia and shear modulus $G = 2.28 \times 10^6$ psia. Calculate Young's modulus E and Poisson's ratio ν.

7.8 Compressibility is the reciprocal of modulus. Calculate fluid compressibility for brine that has a modulus of 2.97×10^5 psia. Express your answer in psia^{-1}.

7.9 **A.** An earthquake is caused by the slippage of a fault block. The shear modulus of the rock is 15 GPa, and the fault block slips 1 m relative to the adjacent fault block. The ruptured area is 2650 m^2. What is the seismic moment in N·m?
B. Ho.w much energy was released?
C. What is the moment magnitude of the earthquake?

8

DRILLING

Reservoir fluids are accessed by drilling a well and then preparing the well for the production or injection of fluids. In this chapter, we discuss drilling rights and related issues, describe rotary drilling rigs and the basics of the drilling process, and survey different types of wells. Well completions are discussed in Chapter 10.

8.1 DRILLING RIGHTS

Extraction of oil and gas from a subsurface formation requires access to the resource. Drilling is one step in that direction. But before drilling, an operator must have the right to place equipment on the surface of the drilling site. In the United States, mineral rights give access to subsurface minerals and fluids and allow reasonable operations at the surface for extraction.

An operator in the United States must lease the mineral rights from the owner of the rights before drilling begins. In exchange for the lease, the mineral rights owner can receive an upfront bonus in addition to regular royalty payments for the minerals as they are produced. An example of a royalty is 12.5% of gross revenue from production minus taxes. The operator receives the right to explore and drill, produce oil and gas, and sell produced oil and gas to the market.

US oil and gas leases include two or more terms. For example, a lease may have a primary and secondary term. The primary term of the lease may be 1–5 years long

Introduction to Petroleum Engineering, First Edition. John R. Fanchi and Richard L. Christiansen.
© 2017 John Wiley & Sons, Inc. Published 2017 by John Wiley & Sons, Inc.
Companion website: www.wiley.com/go/Fanchi/IntroPetroleumEngineering

and is considered the exploration period. The secondary term of the lease covers the production period. If the lessee (the production company) does not do any drilling during the primary term of the lease, the lease may expire and the lessee would have to renegotiate to be allowed to produce the resource. If a government is the lessor, it has the opportunity to sell the lease to a new company if the lease expires. Companies typically prioritize projects based on terms of their leases. Leases may specify how much fluid must be produced to constitute "production" to satisfy the legal agreement.

The cost for drilling and completing a well can be millions of US dollars. Financing project costs often involve a joint operating agreement (JOA) with multiple participants. The JOA defines the rights and duties of each participant. It specifies the operator that will be in charge of day-to-day operations and defines how production is to be shared. The operator does not have to be a majority shareholder of the agreement.

Consider a situation where several operators produce from the same formation on several adjacent leases. The operators could benefit financially in joining those several leases into one unit with one operator. A big part of a unit agreement is prorating production to the various members based on the fraction of reserves that is attributed to each of the separate leases.

Outside the United States, the owners of mineral rights are normally the governments of the countries. The balance of control between governments and the multinational companies that aspire to extract oil and gas resources has been evolving since the early twentieth century. Today, the countries exert much more control over their mineral rights. The nature of agreements that moderate the balance of control between owner and operator continues to evolve.

8.2 ROTARY DRILLING RIGS

Drilling rigs have changed extensively since the first commercial oil well in Titusville, Pennsylvania, was drilled with a cable tool rig. Cable tool rigs lift and lower a bit to pound a hole in rock formations. As needed, pounding would be stopped so water and debris could be bailed from the hole with a "bailer" on a cable, and then the pounding resumed. Cable tool rigs could routinely drill from 25 ft per day up to 60 ft per day. Cable tool drilling, which is also known as percussion drilling, was used for all US fields in the 1800s, but this method is slow, it does not prevent unstable rock from collapsing into the wellbore, and it does not effectively control subsurface pressure. Consequently, the uncontrolled production of fluids, known as a blowout, was common.

Rotary drilling was introduced in the late nineteenth century and became the primary drilling method by the early twentieth century. A modern rotary drilling rig is shown in Figure 8.1. A rotary rig has several systems: a power system, a hoisting system to raise and lower the drill pipe, a rotation system to rotate the drill pipe, and a circulation system to circulate drilling fluid or "mud." In addition, a rotary rig has a system for controlling the well during emergencies.

Drilling rig personnel include the "company man" (operator representative), the drilling contractor crew, and service personnel. The drilling contractor crew includes a tool pusher, a driller, a derrickhand, and roughnecks who work as floorhands.

FIGURE 8.1 Modern drilling rig.

The chain of command begins with the tool pusher and proceeds to the driller, derrickhand, and floorhands. Service personnel include contractors who work with drilling mud, drill bits, cement, casing, downhole tools, and so on.

The size of a rotary rig depends on the weight and power requirements associated with the depth of the well. A larger rig is used for a deeper well. Onshore rigs can either be moved in pieces and assembled on location or mounted on a truck. Offshore rigs come in different forms depending on water depth. Barges can be used for shallow water or swamps. Jack-ups can be used in water that is relatively shallow to a few hundred feet of water. Semisubmersibles are used in a couple thousand feet of water. Drillships are used in several thousand feet of water.

8.2.1 Power Systems

A variety of power systems are used on rigs, including mechanical systems, diesel-electric silicon-controlled rectifiers (SCR), and AC systems. Using the oldest power system, mechanical rigs have the internal combustion engine connected with clutches and transmissions to the draw works, pumps, and so forth. Diesel-electric SCR systems, which replaced most mechanical systems, use internal combustion engines to generate electricity that is converted to DC with SCR. The DC electricity powers motors to run the draw works, rotary table or top drive, pumps, and so on. Diesel-electric AC systems with variable frequency drives (VFD) are preplacing the SCR systems. With VFD, motor control is superior to that with DC power. Instead of diesel for fuel, some rigs may use natural gas, and some rigs in populated areas have connected directly to the electric power grid.

Overall power efficiency of an engine can be calculated from the input power and output shaft power. The shaft power P_{sp} is

$$P_{sp} = \omega T \tag{8.1}$$

where ω is the angular velocity of the shaft and T is the output torque. The overall power efficiency η_{sp} is

$$\eta_{sp} = \frac{\text{power output}}{\text{power input}} \tag{8.2}$$

Power output is shaft power P_{sp}. Power input P_{in} for internal combustion engines depends on the rate of fuel consumption $\dot{m}_f$ and the heating value of fuel H so that

$$P_{in} = \dot{m}_f H \tag{8.3}$$

Therefore, the overall power efficiency is

$$\eta_{sp} = \frac{P_{sp}}{P_{in}} = \frac{\omega T}{\dot{m}_f H} \tag{8.4}$$

Example 8.1 Shaft Power

A diesel engine rotates a shaft at 1000 rev/min to provide an output torque of 1500 ft·lbf. What is the shaft power in hp?

Answer

Angular velocity $\omega = 2\pi \times \text{RPM} = 2\pi \times 1000 \frac{\text{rev}}{\text{min}} = 6283 \frac{\text{rad}}{\text{min}}$

Power output

$$P_{sp} = \omega T = \left(6283 \frac{\text{rad}}{\text{min}}\right) \times 1500\,\text{ft}\cdot\text{lbf} = 9.42 \times 10^6 \frac{\text{ft}\cdot\text{lbf}}{\text{min}}$$

$$P_{sp} = 9.42 \times 10^6 \frac{\text{ft}\cdot\text{lbf}}{\text{min}} \times \frac{1\,\text{hp}}{33000\,\text{ft}\cdot\text{lbf}} = 286\ \text{hp}$$

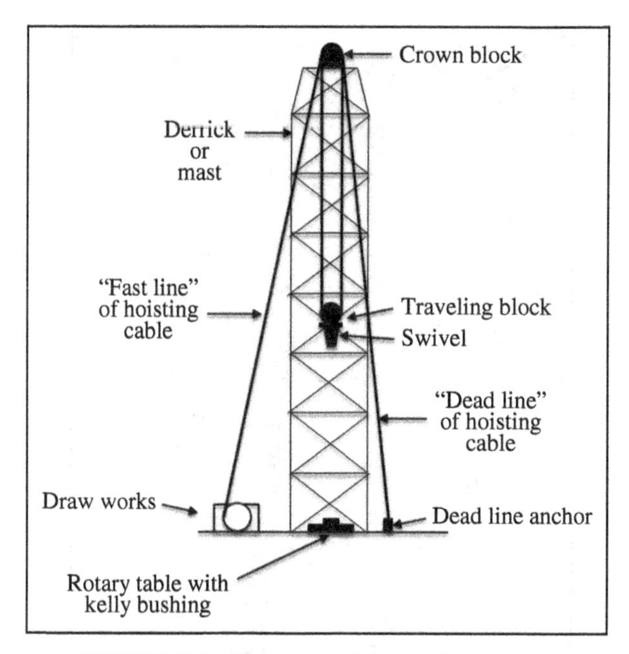

FIGURE 8.2 Illustration of the hoisting system.

8.2.2 Hoisting System

Figure 8.2 shows key components of the hoisting system. The hoisting system is used mostly to move the strings of drill pipe or casing up and down in the wellbore. The derrick, or mast, provides tall, mechanical support. The cable from the spool in the draw works loops over the crown block at the top of the derrick and under the traveling block hanging below the crown block. The swivel underneath the traveling block can be connected to the drill pipe. The weight of the string of drill pipe plus friction forces on the pipe can be as much as one million pounds, all supported by the derrick.

To repair or replace parts of the drill string, the crew must hoist, or "trip," it out of the hole. During a trip, stands of pipe are stored between the derrick floor and the monkey board, as shown in Figure 8.3. The monkey board is where the derrickhand is stationed to guide the pipe. A stand of pipe is two or three pipe joints that are screwed together. Two joints in a stand are referred to as "doubles." Three joints in a stand are "triples." Double derricks are tall enough for "doubles," and triple derricks are tall enough for "triples."

8.2.3 Rotation System

The schematic in Figure 8.4 shows key components of the rotary system. Cable from the draw works runs through the crown block and traveling block and ends at the anchor. The traveling block connects to the swivel, which connects to the kelly. The kelly is a square or hexagonal pipe that mates with the kelly bushing (KB) on the

FIGURE 8.3 Derrick with pipe in rack.

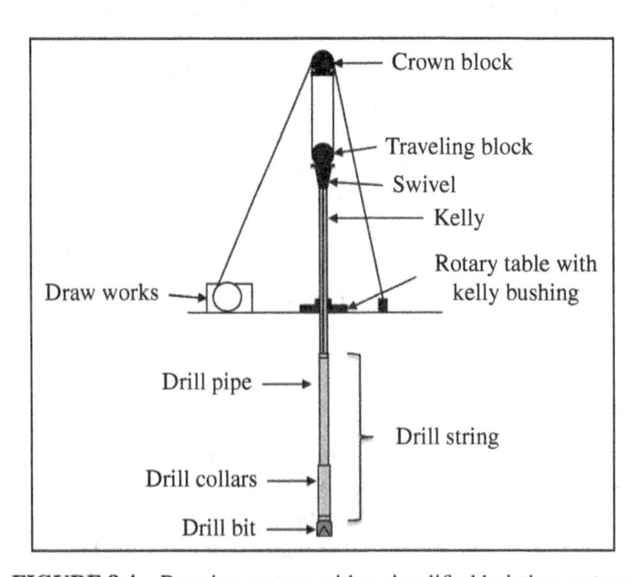

FIGURE 8.4 Rotation system with a simplified hoisting system.

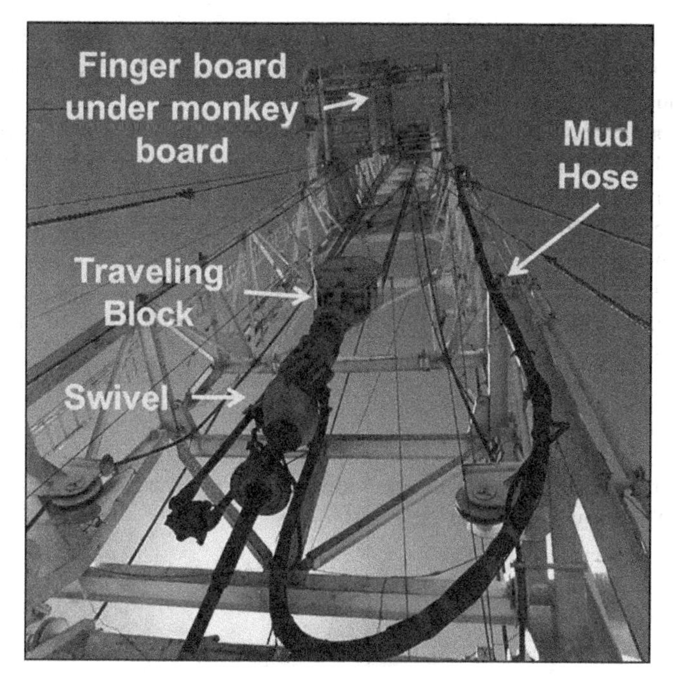

FIGURE 8.5 View up a derrick.

rotary table. The mud hose connects to the side of the swivel as shown in Figure 8.5. Mud from the pump flows through the hose and swivel before entering the top of the kelly. The swivel does not rotate, but it allows the kelly and the attached pipe to rotate while being suspended in the borehole. The bottom end of the kelly is connected below the rig floor to the drill string, which includes the drill pipe, drill collars, and the bit. The rotary table spins the KB which rotates the kelly and, by extension, the drill string. The spinning drill bit breaks up the rock into rock cuttings.

Newer rig systems use a top drive to rotate the pipe string. The top drive can be hydraulically or electrically powered. The top drive in rotary drilling eliminates the need for a kelly and rotary table. The top drive reduces the amount of manual labor during trips and the associated hazards of working on the derrick floor. Top drive rigs are designed to work with a smaller footprint than other drilling rigs, which reduces its environmental impact, especially in urban environments where the space available for well sites is limited.

8.2.4 Drill String and Bits

The drill string consists of the following components: the drill pipe, drill collars, the drill bit, and optional attachments. Drill collars are heavy-walled drill pipes that place weight on the drill bit during actual drilling and that keep the drill pipe in tension to prevent bending and buckling of the drill pipe. Drill collars are part of the bottom-hole assembly (BHA), which includes everything between the drill pipe and tip of the drill bit. The drill bit is used to grind, break, or shear the rock at the bottom of the well. Optional attachments on the drill string may include stabilizers, jars, junk baskets, mud motors, shock absorbers, and so forth.

Drill pipe and collars are rated by their size (outer diameter), weight per unit of length, grade (the steel material and manufacturing process), and connections used to screw the string together. The ratings are specified by international standards. Joints of drill pipe and drill collar are compared in Figure 8.6. Connections from one pipe to another are called box and pin connections.

The drill string is subjected to torsional stress caused by the torque associated with the rotary drilling. Shear or torsional stress τ of a pipe subjected to a torque is given by (Hibbeler, 2011)

$$\tau = \frac{Tr_{\text{outer}}}{J} \tag{8.5}$$

where T is the torque, r_{outer} is the outer radius of the pipe, and J is the polar moment of inertia. The polar moment of inertia is

$$J = \frac{\pi}{2}\left(r_{\text{outer}}^4 - r_{\text{inner}}^4\right) \tag{8.6}$$

for a pipe with inner and outer radii r_{inner} and r_{outer}. The angle of twist is

$$\phi = \frac{TL}{GJ} \tag{8.7}$$

where ϕ is the angle of twist, L is the length of the pipe (or pipe string), and G is the shear modulus.

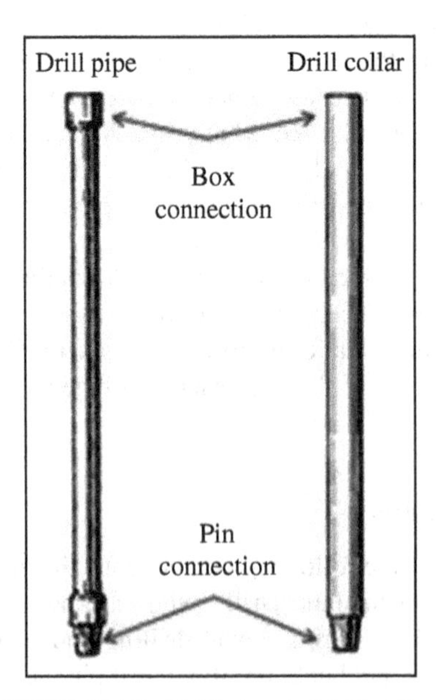

FIGURE 8.6 Drill pipe and drill collar.

Example 8.2 Torsion in a Drill String

Torsion in a drill string is caused by a twisting moment (aka rotary torque *T*). Calculate the angle of twist (in degrees) for a 9.144 m length of pipe subjected to a rotary torque = 5400 N · m. The shear modulus of elasticity is 75 GPa, and the polar moment of inertia of the drill string is $1.0 \times 10^{-5}\,\mathrm{m}^4$.

Answer

$$\text{Angle of twist } \phi = \frac{TL}{GJ} = \frac{5400\,\mathrm{N \cdot m}(9.144\,\mathrm{m})}{75 \times 10^9\,\mathrm{Pa}\left(1.0 \times 10^{-5}\,\mathrm{m}^4\right)} = 0.0658\ \mathrm{rad} \times \frac{360°}{2\pi} = 3.77°$$

Rotary bits are either roller-cone (tricone) bits or drag bits. Materials used for cutting surfaces of bits depend on what type of formation the drill bit will encounter. Roller-cone bits can have steel teeth or tungsten carbide buttons as shown in Figure 8.7. Steel teeth are sufficient for softer formations, while tungsten carbide buttons are used for harder formations. Diamond is used in two different forms for drill bits: whole diamond and fused diamond grit, also known as polycrystalline diamond compact (PDC). PDC bits have PDC discs bonded to tungsten carbide posts mounted on the surface of a bit. PDC bits are good for drilling hard formations. Diamond-impregnated bits have whole diamonds bonded to the surface of a bit. These bits can be used for the hardest formations.

In addition to cutting surfaces, bits must have nozzles for scouring cuttings from the rock surface. The nozzles allow drilling fluids to pass through the drill string and drill bit into the wellbore. Roller-cone bits also have bearings. Bits operate in harsh and abrasive conditions that produce significant wear to cutting surfaces, nozzles, and bearings, eventually ending in bit failure as indicated by declining rate of

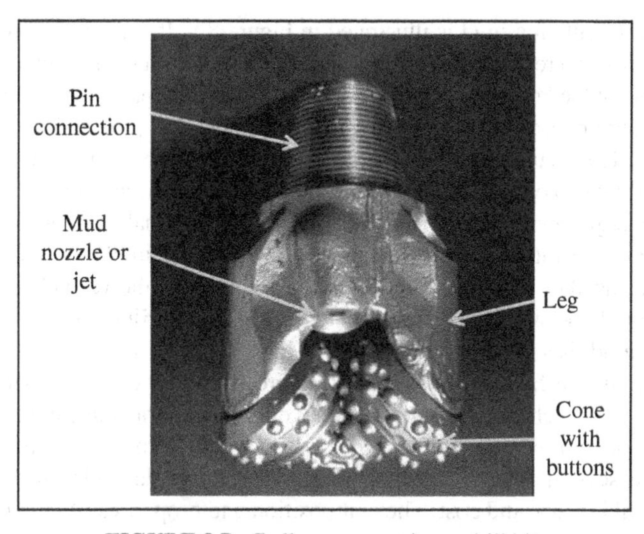

FIGURE 8.7 Roller-cone or tricone drill bit.

penetration (ROP). Replacing a bit is costly, so wise operators attempt to extend bit life by managing mud composition and circulation rate, as well as bit revolutions per minute (RPM) and weight on bit (WOB).

WOB is measured using sensors on the rig floor and in some measurement while drilling (MWD) tools in BHA. With the drill string off bottom and rotating, the rig-floor WOB sensor is zeroed. Then as the rotating drill string is lowered to touch bottom, the sensor reads the actual WOB. Drillers perform tests to find the WOB that maximizes the ROP. Bit manufacturers specify a maximum WOB for avoiding damage. High WOB and low rotation speed can cause the bit to stick and slip causing damaging vibrations. Low WOB and high rotation speed can cause the bit to whirl without cutting yet causing more damaging vibrations. The stable zone for "smooth drilling" has moderate WOB and rotation speed. The optimum WOB and rotation speed depend on the type of bit, the BHA, the mud weight, and the rock being drilled.

Replacing the bit (or any part of the BHA) is a time-consuming and costly process. It requires pulling the entire drill string out of the wellbore (tripping out), changing the drill bit, and lowering the drill string back into the wellbore (tripping in). Depending on the skill of the crew, trip time is about one hour per thousand feet of pipe. For an 8000 ft drill string, the trip time would be about 8 hr each way. And to complicate matters, pressure kicks are more likely to occur during tripping.

Efficient drilling is reflected in the cost per foot drilled. To minimize that cost, rig downtime must be kept to a minimum. Downtime includes time for repair of power generators and mud pumps, trips for bit replacement or other changes to the drill string, or delays for stuck pipe or lack of inventory. With experience drilling in an area, operators learn how to maximize ROP and bit life. Managing a rig to minimize cost is a complex task that requires a wide variety of skills, including organizational skills.

8.2.5 Circulation System

A system for circulating mud is illustrated in Figure 8.8. It is a continuous system so that the operator can reuse the mud. The mud is mixed in an aboveground tank and then pumped through the kelly and down the drill pipe. The mud passes through nozzles on the drill bit and up through the annular gap between the drill pipe and the rock wall of the wellbore. Rock cuttings are carried up with the mud to a shaker on the surface where a well site geologist can evaluate samples. The rock cuttings provide information about the geologic environment. The mud falls through the shaker to the first in a series of mud tanks, separated by baffles. Pumps (not shown) circulate mud through the desander and desilter to remove undesired solids. Finally, the reconditioned mud is returned through the swivel and kelly and to the top of the drill string.

Drilling mud has several functions. It lifts cuttings and contents of drilled formations to the surface, controls formation pressure, lubricates the drill string and bit, cools the bit, mechanically supports the wellbore, and transmits hydraulic power. The mud can also prevent movement of fluids from one formation to another.

There are several types of drilling mud. The choice depends on performance, environmental impact, and cost. The composition and properties of mud used to drill

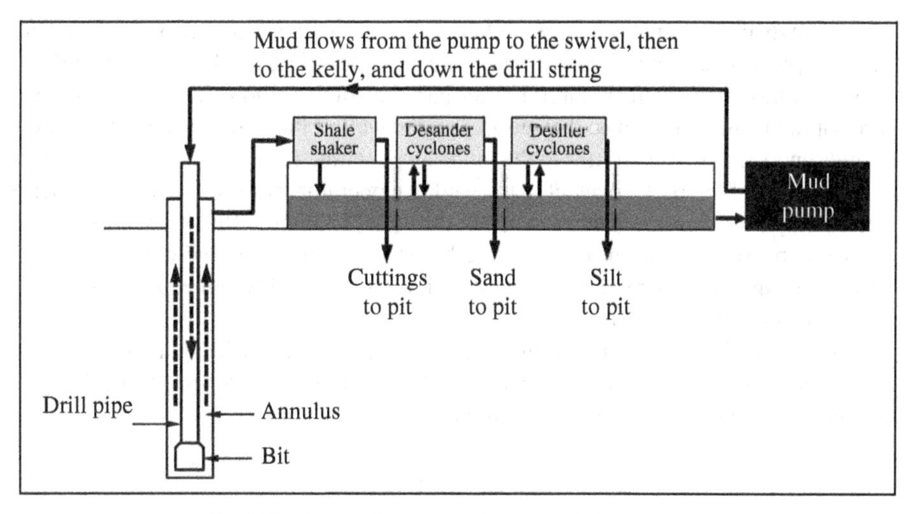

FIGURE 8.8 Illustration of the circulation system.

TABLE 8.1 Drilling Mud Density

Component	Density
Water	8.3 lbs/gal (1 g/cm^3 or 62.4 lb/ft^3)
Unweighted mud	9–10 lbs/gal
Barite-weighted mud	11–20 lbs/gal

the first well in a new region are usually an experiment. Over time and with experience, the mud can be modified to make it more effective.

Water is the primary component in water-based drilling mud, the most often used mud. Bentonite clay is added to increase the cutting-carrying capacity of the water. Water-soluble chemicals are added to adjust the properties of the suspended clay. Other solids, such as barite, are added to increase the density of the mud. Water-based mud can cause subsurface clays to swell and adversely affect reservoir properties. Table 8.1 illustrates the density of two types of mud and compares the mud density to water density.

Diesel oil is used in oil-based drilling mud. Designed to be stable at high temperature, oil-based systems are used to protect shale and can help keep the drill pipe from sticking. They can also be used for native-state coring, which is an attempt to retrieve a rock sample in its original state. Oil-based systems are messy and can have a negative environmental impact if spilled.

Synthetic oil can be used for a synthetic-based drilling mud. Synthetic mud act like oil-based mud but are more environmentally friendly. They are expensive and tend to be used in environmentally sensitive areas.

In some cases foam, air, and mist can be used to drill a well. This increases the ROP but it does not control movement of formation fluids into the wellbore. Furthermore, this type of drilling fluid provides very little mechanical support for the wellbore.

In most drilling cases, the operator chooses to pump mud of density sufficient that wellbore pressure exceeds formation pressure. This "overbalanced" condition pushes wellbore fluids into the formation. Larger particles in the mud cannot penetrate the pores of the formation, and so they collect on the wall of the bore forming mud cake. Liquids and very small particles in the mud can penetrate the formation and can alter the properties of the rock adjacent to the well. An operator may choose to drill underbalanced by pumping low-weight mud. In this case, formation fluids will move into the wellbore. If mud weight is too low, a blowout or sloughing and collapse of the formation can occur. If mud weight is too high, unintended fractures and mud loss into the formation can occur.

If a well is left inactive or shut-in, the solid particles in the drilling mud can settle to the bottom of the well and may harden. This process is called "sagging." If drill pipe is still in place, the hardening mud may lead to sticking.

Example 8.3 Mud Volume in Pipe

What volume of mud is needed to fill a section of pipe with inner diameter ID = 3.5 in. and length $L = 6000\,\text{ft}$? Express your answer in ft^3.

Answer

$$V = \pi r^2 L, \quad \text{ID} = 0.2917\,\text{ft}, r = \frac{\text{ID}}{2} = 0.1458\,\text{ft}$$

$$V = \pi r^2 L = 3.14159\,(0.1458\,\text{ft})^2\ 6000\,\text{ft} = 400.9\,\text{ft}^3$$

8.2.6 Well Control System

Mud of proper weight is the first defense against blowouts. Other parts of the well control system are the blowout preventer (BOP), the kill line, and the choke line and choke manifold. The BOP is used to shut in the well in emergencies. As shown in Figure 8.9, the BOP is a stacked sequence of two to four hydraulically actuated valves or preventers. The top preventer is an annular preventer, which functions much like a rubber sleeve for measuring blood pressure. The annular preventer can squeeze around the drill pipe or the kelly to close the annular space. The next preventer is a pair of pipe rams that slide from opposite sides of the BOP to close around a pipe. The half-circle sealing elements on the pipe rams must have the same diameter as the pipe in order to properly seal. Blind rams are designed to close an open hole and cannot shut the well if the pipe is in the hole. The final preventer is a shear ram. Shear rams are designed to cut any pipe in the hole and seal the well. The drilling crew installs the BOP on top of the casing head, which is attached to the top of the surface casing after it is cemented in place.

In addition to the BOP, a kill line and a choke line are connected at the wellhead. The kill line can be used to perform well integrity tests and to inject high-density mud into the wellbore to block fluid flow up the wellbore. The choke line has a choke

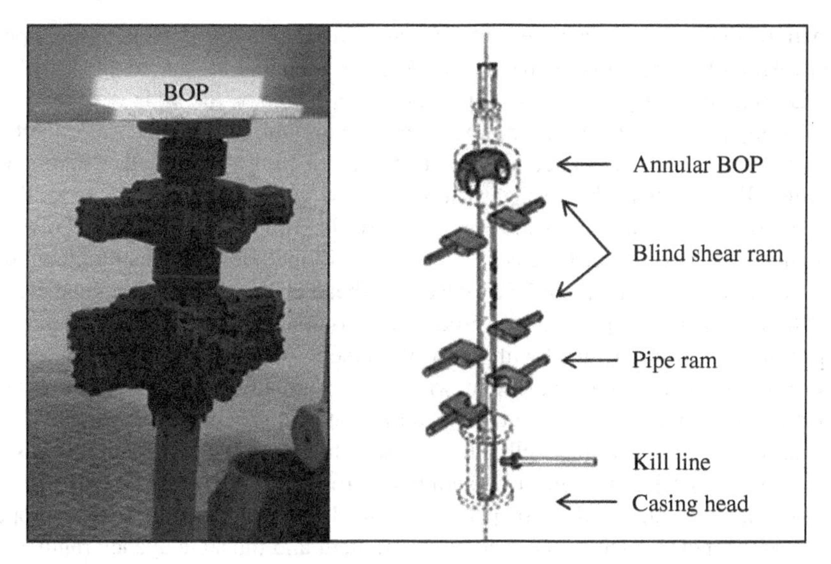

FIGURE 8.9 Blowout preventer.

manifold with various chokes in line. The choke line can be used to control wells that encounter higher pressures than the drilling fluid can contain. In this case, the well is out of balance because the formation pressure exceeds the hydrostatic pressure exerted by the drilling fluid and the chokes release pressurized fluids in a controlled manner. To resume normal drilling, the drilling crew circulates heavier mud down the drill pipe and up the annulus. The hydrostatic pressure exerted by the heavier mud should exceed the formation pressure and normal drilling can resume.

8.3 THE DRILLING PROCESS

The process of drilling begins months, and often years, before a drilling rig arrives on location. Here, the following five stages of the process will be considered: planning, site preparation, drilling, open-hole logging, and setting production casing. Planning is the longest of these five stages, and open-hole logging and setting of casing are the shortest, often just 1–3 days for each.

8.3.1 Planning

Planning begins with identification of target formations and their depths. The data used to identify the target could include data from offset (nearby) wells, seismic data, and other geologic insight. The data from offset wells includes all the drilling, logging, completion, and production records.

Well design starts after the target is selected. For some targets, a vertical well is the correct choice; but for others, a directional well may be needed. That choice depends on surface topography, surface buildings, lakes, and the subsurface formations that will be penetrated. The design will include drilling and casing operations. An authorization for expenditure (AFE) is prepared after a detailed drilling plan is complete. In general, the depth and complexity of the well will have the greatest impact on total cost. Complexity refers to variations in formation properties for each formation that is encountered during the drilling of the well. Completion costs usually dominate the cost of drilling shallow wells. Completion and drilling costs are often comparable for medium-depth wells. The cost of drilling deep wells is usually dominated by drilling costs. The AFE will include tangible and intangible costs, dry hole costs, completed well costs, overhead charges, and contingencies.

With the AFE completed, the production company must obtain stakeholder support from anyone involved with the well. Stakeholders include asset team members, different levels of management within the company, other partners, royalty owners, surface rights owners, regulators, public interest groups, and anyone that has a say on what happens to the well. Different stakeholders have different concerns or interests in the project that must be addressed. To obtain permission to drill in the United States, an application for permission to drill must be submitted to proper governing agencies, such as the federal government and the state agency regulating oil and gas development. Getting that permission can be a slow and iterative process, including site visits. Drilling operations must wait until permission is obtained.

Most production companies do not drill wells using their own personnel and equipment, so they must find a drilling contractor. The drilling contract will specify the start or spud date of the well and drilling costs. The drilling contractor will drill the well according to contract specifications. There are three types of drilling contracts. A turnkey contract will have a fixed price for drilling and equipping a well. All of the risk in this case falls on the drilling contractor to meet the terms of the contract. Footage contracts are based on cost per foot to a total depth (TD). In this case, the production company and the drilling contractor share the risk of drilling the well. Day-rate contracts are based on cost per day to drill and complete the well, and all risk falls on the production company.

8.3.2 Site Preparation

The next step after obtaining all permissions is to prepare a site for the drilling operation. This step includes building a road to the location, clearing and building the location, drilling for and setting conductor pipe, drilling mouse and rat holes, and setting anchors (usually four) for supporting rigs.

The size of onshore locations varies from about 2 to 5 acres, largely depending on the amount of room needed for equipment during completion operations. Topsoil on location is pushed to one side and saved for later site restoration activities. Berms are often built to control spills. One or two pits for used drilling and completion fluids are excavated and lined as required by governing agencies.

Conductor pipe is the first and largest diameter casing to be cemented into place for a well. It serves as a foundation for the start of the drilling operation. A small drilling rig, often truck-mounted, drills the hole and sets the casing. Conductor pipe diameter ranges from 18 in. to more than 3 ft; its length is 40–80 ft. Two other smaller holes, the rathole and the mousehole, are often drilled within 10 ft of the conductor hole. These holes are for temporary storage of pipe during the drilling process.

8.3.3 Drilling

After the location is fully prepared, the drilling rig, associated equipment, housing, and materials are moved onto the location and "rigged up," or MIRU for move in and rig up. For remote areas, this move (or mobilization) may be 5–10% of the total well cost.

The first task for the drilling rig is to drill to the depth required for the surface casing, usually 500–2000 ft as specified by the agency that permitted the well. The surface casing has two functions: first, it protects water in aquifers near the surface from contamination; second, it provides mechanical support for the well. The drilling fluid for the surface hole is typically freshwater. When the required depth is reached, the surface casing is lowered into place and cemented by pumping cement down the casing and up the annulus between the casing and the surrounding formations to the surface. The last step in cementing is to push a cement plug with drilling fluid down the casing until it reaches the bottom of the casing.

After the cement has cured, a casing head is attached to the top of the surface casing, and the BOP is attached to the top of the casing head. The BOP is used to shut the well in emergencies.

To continue drilling, the drilling crew feeds the BHA and drill pipe through the top of the BOP and into the surface casing, tripping down to the top of the cement plug. The bit on the BHA must be small enough to enter the surface casing. After starting circulation of drilling mud, the crew can drill through the plug and cement and past the bottom of the surface casing.

For many wells, the next drilling objective is the depth of the target formation. While drilling to this depth, the crew will adjust the composition of the mud as needed to clean the hole and maintain pressure control. Throughout the drilling process, a company employee submits daily reports to management of drilling activities and costs. In some cases, the drilling plan may need adjustment if an unanticipated event occurs.

At some point after setting surface casing, a separate contractor arrives on location to create a continuous tabular record, or log, of the drilling process and results for a well. This contractor is the mud logging company and its employee on location is a mud logger, usually a geologist. The mud logger installs hardware and software to monitor operations as requested by the production company. The hardware usually includes computer displays of the log in the doghouse on the drilling rig, in the office of the company employee, and in the mud logger's workspace.

The mud log consists of four to six columns of information. In the first column of the table, the ROP of the bit is recorded, typically in minutes per foot. Other operating parameters (such as rotation rate, WOB, pumping rate, mud weight and viscosity, and mud composition) are included as notes in this column. In a second column, lithology (sandstone, shale, limestone, etc.) of cuttings are recorded by the on-site geologist at 10 ft intervals. The geologist also enters detailed descriptions of cuttings collected at the shale shaker in another column. This column includes notes on oil or gas shows. Composition of hydrocarbon gas that evolves from the mud is continually recorded in the last column. The organization and extent of the mud log varies from company to company. As the first source of data from the subsurface formations, the mud log is vital for drilling management, hydrocarbon exploration, and completion designs.

When drilling reaches the target depth, the drilling crew circulates mud until the hole is clean and then trips the drill pipe out of the hole. The well is ready now for the next stage of the drilling process.

Example 8.4 Drill String

How many stands of drill pipe are needed to prepare a drill string that is 10 000 ft long? A stand is three pipes, and each pipe has a length of 30 ft. Assume one stand of drill collar and neglect elongation of pipe.

Answer

Length of drill string = 10 000 ft = length of drill collar + length of drill pipe
Length of drill collar = 90 ft

So

Length of drill string = 10 000 ft = 90 ft + length of drill pipe
Length of drill pipe = 9910 ft
One stand of drill pipe = 90 ft
Therefore we need 9910 ft/(90 ft/stand) = 110.1 stands of drill pipe.
110 stands of drill pipe would be short, so 111 stands would be needed.

8.3.4 Open-Hole Logging

After drilling to target depth, the operator needs to make a decision: set production casing and complete the well or plug and abandon the well. Before making this decision, the operator gathers more data on the formations exposed by drilling. The lowest cost data are obtained by logging the well with tools deployed on a wireline composed of steel cable wrapped around multiple strands of electrically conductive wire. Logging at this stage of the drilling process is open-hole logging because there is no casing in the hole.

When properly interpreted, open-hole logs show lithology, porosity, and water saturation in the rock around the hole from bottom to top of the logged interval. The log also includes the diameter of hole measured by the caliper from bottom to top. Hole diameter is important for two reasons. First, interpretability of the other logs rapidly declines if the hole diameter is not constant. And second, the diameter is used to estimate the amount of cement needed to case or plug the well. More details of log interpretation are discussed in the next chapter.

The wireline logging contractor arrives on location with a wireline logging truck and one or two pickup trucks loaded with 10–20 ft long logging tools and lubricator segments. While there is a wide assortment of tools available, the most commonly used are the gamma-ray tool, formation density tool, neutron density tool, and a caliper. When these tools are assembled, the total length of the logging string can be 80–120 ft. The lubricator is a pipe of sufficient length and diameter to hold the assembled logging string. The logging crew assembles the lubricator with the cable running through the stuffing box at its top and then through to its open bottom, attaches the cable to the top of the logging string, pulls the logging string into the lubricator, and uses a drill rig hoist to raise the lubricator to flange to the top of the BOP.

With the lubricator attached to the top of the BOP, the logging crew carefully lowers the tool to the bottom of the well, using the weight indicator for signs of tool sticking. The length of hanging cable is also measured as it unwinds from the drum of the wireline truck. One of the first major measurements made with the wireline string is total depth of the well. This depth is compared to the depth recorded by the driller based on length of the drill string. With total depth verified and any differences resolved, the crew slowly raises the tool and records all of the data received from the tools. Data is not collected on the downward trip but on the upward trip because the upward trip gives more reliable depth measurements. The time needed for logging varies from about 6 to 24 hr, depending on the depth range that is logged and the allowed rising speed of the logging tools. During this time, the drilling crew must be alert for signs of kicks that can occur while logging, especially on the upward stroke.

In addition to open-hole logging, operators may use other tests to assess the exposed formations. Using the drill string, the drill stem test (DST) can measure the productive capacity of formations. Using wireline, the repeat formation tester (RFT) can measure pressure in formations and collect small volumes of formation fluids. Both of these tests can end with a tool stuck in the hole. Indeed, the RFT is called the "repeat fishing tool" by some field hands.

When open-hole logging and other tests are done, the operator must decide how to proceed. Even though a substantial amount of data has been obtained, the risk of a wrong choice has not been eliminated.

8.3.5 Setting Production Casing

If the operator chooses to abandon the well, the abandonment procedure specified in the original permit for the well must be followed. This process will involve setting plugs and pumping cement to seal the well. If, on the other hand, the operator chooses to case and complete the well, then a crew will set production casing with the drilling rig following procedures similar to those used for setting surface casing. Whether plugged or cased, it is time to rig down and move out (RDMO) to the next drilling location.

Figure 8.10 shows a wellbore diagram for a vertical well. Casing prevents fluid movement between formations and it prevents collapse of the wellbore. Surface casing fits inside the conductor pipe and protects freshwater formations. Some wells have one or more strings of intermediate casing inside the surface casing, each with a smaller diameter. Intermediate casing protects formations above the target zone. Intermediate casing may go most of the way to total depth. Production casing has

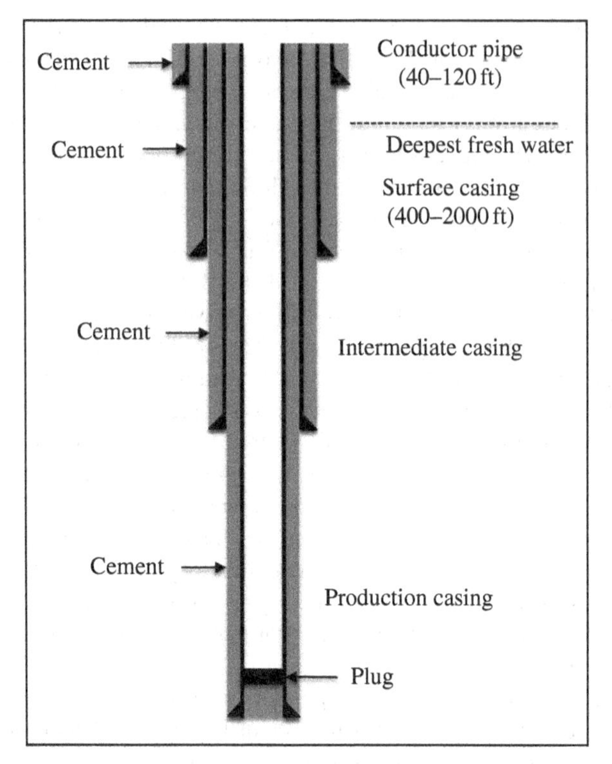

FIGURE 8.10 A wellbore diagram for a vertical well.

the smallest diameter. Production or injection tubing must be able to fit inside the production casing.

Casing is set with cement to attach the casing to the formation and to prevent fluid movement behind the casing. Cement leaks or fractures in the cement are possible. The cement must be tested to ensure it is functioning properly before the well is completed. During the completion stage, perforations (holes) must be punched through casing at the target formation to allow fluids to flow into the production tubing.

Casing is tested by the manufacturer to determine its ability to withstand the forces that will act on it. Tension forces pull at the top of the casing, collapse forces push against the outside of the casing, and burst forces from fluids in the casing push against the inside. The casing and casing joints must not split or separate during high-pressure fluid injection because the breach would lead to contamination of formations.

Example 8.5 Volume in the Annulus

Calculate the volume in the annulus between a section of pipe with outer diameter OD = 4.5 in., length $L = 1500$ ft, and the casing with inner diameter = 7 in. Express your answer in ft^3.

Answer
Radius of casing = r_c and radius of pipe = r_p:

$$\begin{aligned} V &= \pi r_c^2 L - \pi r_p^2 L \\ &= \pi \left(r_c^2 - r_p^2\right) L \\ &= 3.14159\left(0.292\,\text{ft}^2 - 0.188\,\text{ft}^2\right) \times 1500\,\text{ft} \approx 235.2\,\text{ft}^3 \end{aligned}$$

8.4 TYPES OF WELLS

Wells are drilled for exploration or development. Exploration wells include wildcat wells that are used to test a geologic trap that has never been produced, test a new reservoir in a known field, or extend known limits of a producing reservoir. A discovery well is a wildcat that discovers a new field. Wells for estimating field size include appraisal wells, delineation wells, and step-out wells. Development wells are drilled in the known extent of a producing field. An infill well is drilled between producing wells in an established field.

Wells may be categorized also by their function. A production well can produce oil, gas, and water or thermal energy as in geothermal wells. An injection well is used to inject water or gas and includes wells for water disposal and sequestration. It can also be used to inject steam for flooding of heavy oil reservoirs.

8.4.1 Well Spacing and Infill Drilling

The distance between wells and the area drained by a well correlate to well spacing (Figure 8.11). In the United States, a typical spacing is expressed in terms of acres. For example, the term "40 acre spacing" refers to vertical wells that can drain an area of 40 acres. A square mile contains 640 acres, so there are 16 areas with 40 acres each in a square mile. Therefore, the number of wells that can be drilled in a square mile with 40 acre spacing is 16.

Vertical oil wells tend to be drilled closer to one another than vertical gas wells, because gas viscosity (usually near 0.02 cp) is much less than oil viscosity. For example, vertical gas wells may be drilled on 160 acre spacing, which means that only four wells must be drilled to drain a square mile. If additional wells are needed to adequately drain a reservoir, more wells can be drilled in the space between existing wells. This is called infill drilling.

Example 8.6 Well Spacing

How many vertical production wells are needed to drain a field that covers 20 mi^2? Assume 40 acre spacing.

Answer
16 wells are needed to drain 1 mi^2 (1 mi^2 = 640 acres) on 40 acre spacing. Therefore, the number of vertical production wells needed is 20 mi^2 × 16 wells/mi^2 = 320 wells.

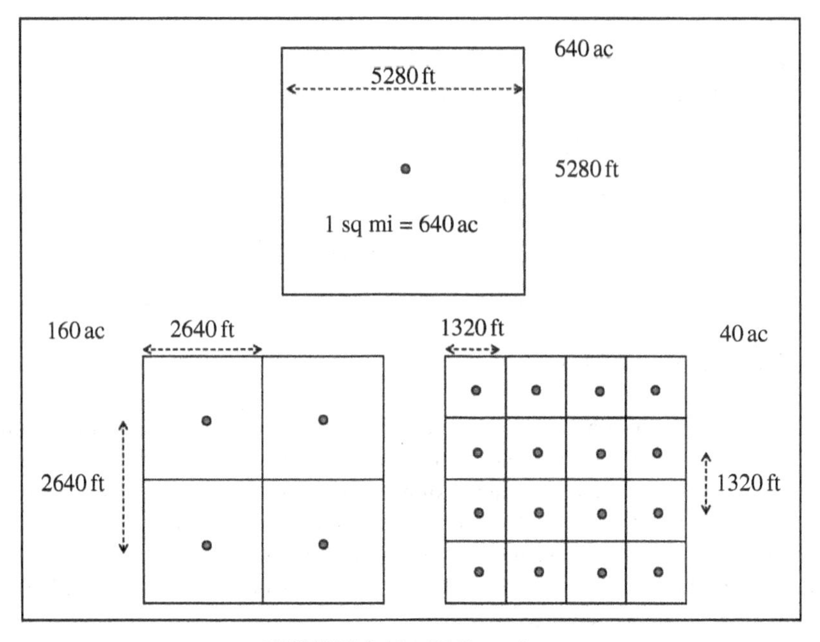

FIGURE 8.11 Well spacing.

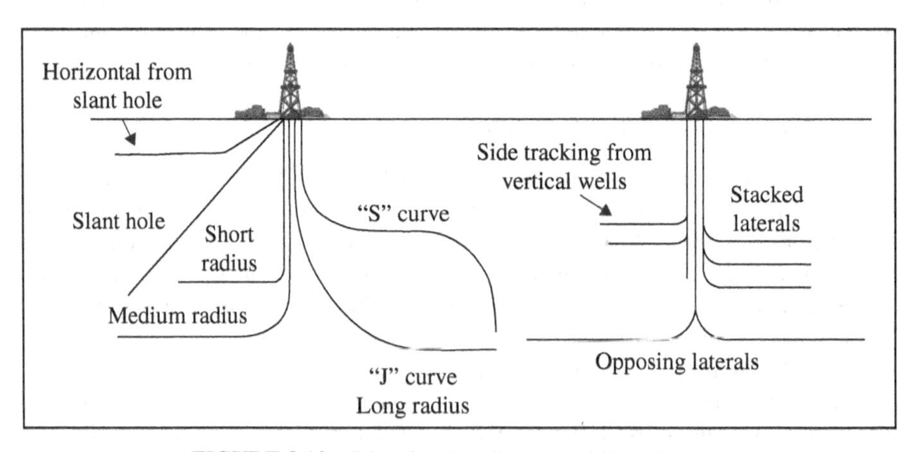

FIGURE 8.12 Directional wells and multilateral wells.

8.4.2 Directional Wells

Modern technology makes it possible to drill wells with a variety of trajectories as shown in Figure 8.12. Two common shapes for directional wells are the "S" and the "J" curves. Both shapes have a kickoff point (KOP), which is the depth at which the wellbore trajectory begins to deviate from vertical. An "S"-curve wellbore is designed to offset the drilling position before returning to a vertical orientation into the target formation. The "J" curve gives a deviated wellbore trajectory into the target formation.

Directional wells can be drilled to follow the orientation of a geologic formation. This improves access to the formation when production tubing is perforated within the formation. Horizontal wells are examples of this. The portion of a horizontal well nearest the vertical section of the well is the heel, and the end of the horizontal well is the toe. The horizontal section is often called the lateral or reach.

A sidetrack diverts the direction of the well. This can be done for many reasons, such as avoiding a piece of broken drill pipe or to reach a different reservoir.

A multilateral well has multiple segments branching from the original well. The original well is called the trunk, and the segments are called laterals or branches. Multilateral wells are used in offshore environments where the number of well slots is limited by the amount of space available on a platform. They are also used to produce fluids from reservoirs that have many compartments. A compartment in a reservoir is a volume that is isolated from other parts of the reservoir by barriers to fluid flow such as sealing faults. Before modern directional drilling, a reservoir that was broken up into several compartments required a large number of vertical wells to drain each section. Today, a smaller number of wells can be used to intersect separate reservoir compartments.

Technology for drilling directional wells has advanced rapidly in the last 30 years. One key advance is MWD technology. MWD technology transmits directional information from the BHA by creating pressure pulses that travel through the drilling mud to the surface and are decoded there. The front portion of the BHA typically includes directional sensors, mud-pulsing controls, a mud motor, a bend of 0–4°, and a drill bit. MWD gives the drilling team frequent updates about the inclination from vertical and the azimuth or direction in the horizontal plane of the BHA. With this information, the team can adjust the drilling direction, often referred to as geosteering. When geosteering, there are three angle choices: build angle orients the well toward the horizontal, drop angle orients it toward the vertical, and maintain angle continues the existing well trajectory.

With an appropriately instrumented BHA, MWD also can provide rotational speed and rotational speed variations, the type and severity of BHA vibrations, temperature, torque and weight on the bit, and the mud flow rate. Drillers use this information to adjust operations to maximize the ROP and to decrease wear on the bit. The BHA can also include formation logging tools that provide data in a process called logging while drilling (LWD). LWD tools work with the MWD system to transmit information to the surface.

Example 8.7 Drill Time

The top of a formation of interest is at a depth of 5000 ft and the ROP of the drill bit is 250 ft/day. How long will it take to drill to the top of the formation?

Answer

$$\text{Time to drill} = 5000\,\text{ft} / \left(250\,\text{ft/day}\right) = 20\,\text{days}$$

8.4.3 Extended Reach Drilling

Extended reach drilling (ERD) refers to drilling at one location with the objective of reaching a location up to 10km away that may otherwise be inaccessible or very expensive to drill. ERD wells are justified in several circumstances. For example, an ERD well may be able to access an offshore location from an onshore location, it may be drilled from a well site location that minimizes environmental impact, or it may be drilled from a well site location that minimizes its impact in a densely populated area.

Horizontal, extended reach, and multilateral wellbores that follow subsurface formations are providing access to more parts of the reservoir from fewer well locations. This provides a means of minimizing environmental impact associated with drilling and production facilities, either on land or at sea. Extended reach wells make it possible to extract petroleum from beneath environmentally or commercially sensitive areas by drilling from locations outside of the environmentally sensitive areas. In addition, offshore fields can be produced from onshore drilling locations and reduce the environmental impact of drilling by reducing the number of surface drilling locations.

8.5 ACTIVITIES

8.5.1 Further Reading

For more information about drilling, see Hyne (2012), Denehy (2011), Mitchell and Miska (2010), Raymond and Leffler (2006), and Bourgoyne et al. (1991).

8.5.2 True/False

8.1 A well integrity test can be conducted by altering mud density.

8.2 Surface casing is used to protect potable water.

8.3 Production casing has a smaller diameter than production tubing.

8.4 Drill collars are used to prevent the drill pipe from kinking and breaking by putting weight on the bottom of the drill string.

8.5 Drill cuttings are not removed from drilling mud before it is recirculated into the well.

8.6 Both "S"-curve and "J"-curve well trajectories have kickoff points.

8.7 A top drive is a motorized power swivel located below the traveling block and used to turn the drill string.

8.8 Well spacing can be reduced by infill drilling.

8.9 Delineation wells are drilled before discovery wells.

8.10 An AFE is used to obtain permission to spend money.

8.5.3 Exercises

8.1 **A.** Suppose a pipe string weighs 15 lbf/ft of length. If the pipe is used to form a string of pipe that is 10 000 ft long, what is the weight of the pipe string?

B. Suppose a pipe string is 10 000 ft long. 6000 ft of the pipe string is made of pipe that weighs 15 lbf/ft of length. 4000 ft of the pipe string weighs 18 lbf/ft of length. What is the weight of the pipe string?

8.2 **A.** What volume of mud is needed to fill a section of pipe with inner diameter ID = 4.5 in. and length = 30 ft? Express your answer in ft^3 and gal.

B. Calculate the volume in the annulus between a section of pipe with outer diameter OD = 7 in., length L = 1200 ft, and the casing with inner diameter = 10 in. Express your answer in ft^3.

8.3 The KB (kelly bushing) is 30 ft above the ground. A well is drilled to a total vertical depth of 6000 ft below the ground. How many joints of pipe are needed to reach from the KB to the bottom of the wellbore? Assume each joint of pipe is 30 ft long.

8.4 **A.** The top of a reservoir is at a depth of 4000 ft SS (SS = subsea or below sea level). The KB on the drilling rig is 5280 ft above sea level. What is the depth to the top of the reservoir relative to the KB? Express your answer in ft.

B. Estimate confining pressure p_{con} at the depth in Part A using an overburden pressure gradient of 0.43 psia/ft. Express your answer in psia.

8.5 We want to drill a well with a total length of 4000 ft. We know from previous experience in the area that the drill bit will be effective for 36 hr before it has to be replaced. The average drill bit will penetrate 20 ft of rock in the area for each hour of drilling. We expect the average trip to replace the drill bit to take about 8 hr. A trip is the act of withdrawing the drill pipe, replacing the drill bit, and then returning the new drill bit to the bottom of the hole. Given this information, estimate how long it will take to drill the 4000 ft. Hint: prepare a table as follows:

Incremental Time (hr)	Incremental Length (ft)	Cumulative Time (hr)	Cumulative Length (ft)

8.6 Consider the following types of pipe: conductor pipe, surface casing, intermediate casing, production casing, and production pipe. Which string is cemented last?

8.7 If a gas well can drain 160 acres, how many gas wells are needed to drain 3 mi^2?

8.8 Bend radius is the distance required for a string of pipe to make a 90° turn. Suppose the trajectory of a horizontal well requires a quarter of a mile to go from the kickoff point at the end of the vertical segment of the trajectory to the beginning of the horizontal segment. How many joints of pipe are needed to establish the curvature from the vertical segment to the horizontal segment if each joint of pipe is 30 ft long?

8.9 Suppose a diesel engine rotates a shaft at 1100 rev/min by consuming diesel at a rate of 28 gal/hr to provide an output torque of 1560 ft·lbf. Assume the density of diesel is 7.0 lbm/gal with a heating value of 18 000 BTU/lbm.

A. What is the output power P_{sp} in hp?

B. What is the input power P_{in} in hp?

C. What is the overall efficiency η_{sp} of the engine?

8.10 **A.** Given the following information, calculate the number of joints of pipe needed to complete the drill string:

Pipe weight	654 lbf each
Pipe length	30 ft each
Length of drill string	6000 ft = 1828.8 m
Number of joints of pipe	
Total pipe weight	lbf

B. Calculate the total pipe weight for the number of joints of pipe in Part A. Express your answer in lbf.

C. Calculate the polar moment of inertia of the drill string $J = (\pi/2)\left(r_{outer}^4 - r_{inner}^4\right)$ for outer radius = 2.25 in. and inner radius = 1.75 in. Express your answer in m^4. Hint: first convert inches to m.

D. The rotary torque is 4000 ft·lbf. What is the rotary torque in N·m?

E. Calculate the shear stress $\tau = Tr_{outer}/J$. Express your answer in MPa.

F. Assume the elastic modulus E of drill string steel is 200 GPa and its Poisson's ratio ν is 0.3. Calculate the shear modulus of elasticity $G = E/(2+2\nu)$. Express your answer in GPa.

G. Convert the length of drill string L to meters and use this result to calculate the angle of twist ϕ in degrees, where $\phi = TL/(GJ)$.

H. What is the angle of twist per joint of pipe? Use result from Part A.

9

WELL LOGGING

Well logging is an important contributor to formation evaluation. The objectives of formation evaluation include assessing resource size, supporting the placement of wells, and interpreting reservoir performance during development. Well logs provide valuable information about the formation within a few feet of the wellbore. Formation data can be obtained by examining drill cuttings, core samples, and fluid properties. Other information can be measured by instruments in a well logging tool that is lowered into a wellbore. Logging tools can be designed to measure such physical quantities as electromagnetic and sonic wave signals that pass through a section of the formation and detect elementary particles emitted by formation rock. The data are used to estimate formation properties. This chapter introduces the well logging environment, describes different types of logs, explains why combinations of logs are used, and discusses techniques and limitations of well log interpretation.

9.1 LOGGING ENVIRONMENT

A well log is produced by lowering an electric logging tool on a cable into a well and measuring various responses on the instruments as the tool is raised back to the surface. The tool instruments can measure temperature, wellbore diameter, electrical resistivity, radioactivity, sonic vibrations, and more. The tool length varies from 40 to 140 ft, depending on the logging plan.

Introduction to Petroleum Engineering, First Edition. John R. Fanchi and Richard L. Christiansen.
© 2017 John Wiley & Sons, Inc. Published 2017 by John Wiley & Sons, Inc.
Companion website: www.wiley.com/go/Fanchi/IntroPetroleumEngineering

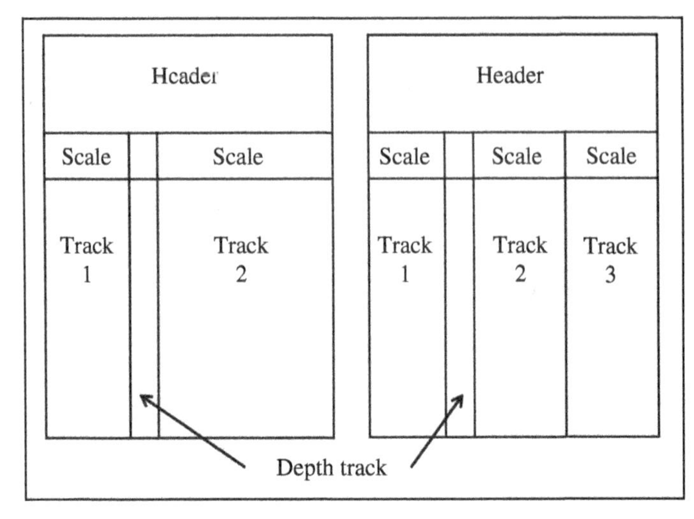

FIGURE 9.1 Well log format.

Logging measurements are recorded as a function of depth, as illustrated in Figure 9.1. The left-hand side of the figure illustrates a well log with two measurement tracks and another with three measurement tracks. Both well logs have a track for depth. The measurement tracks display measurement values relative to a scale that is suitable for the measurement. More than one measurement may be displayed in each measurement track. The header contains information about the well (such as name, location, and owner), the logging run (such as date, logging company, and tool description), and the scales for the measurement at the top of the tracks.

Performance of logging tools depends on the logging environment. The environment includes such factors as rock type and structure, fluids in the well and adjacent formation, temperature and pressure. Some features of the logging environment are discussed below.

9.1.1 Wellbore and Formation

The wellbore and adjacent formations are a complex setting for logging. The desired borehole shape is a cylinder with diameter equal to or slightly larger than the drill bit. In practice, the shape of the borehole may differ substantially from a cylinder because of borehole wall collapse or the presence of cavities in the formation. Wellbore roughness, or rugosity, is especially challenging for logging measurements. The caliper log measures the actual shape of the borehole using a caliper tool with flexible arms that push against the borehole wall. The arms move in or out as the shape of the borehole wall changes.

The caliper log has two main uses. First, it is an indicator of the quality of the other logs. When the borehole diameter varies from the bit diameter, the quality of the other logs rapidly declines. Second, the diameter in the caliper log is used to estimate amount of cement needed to cement casing.

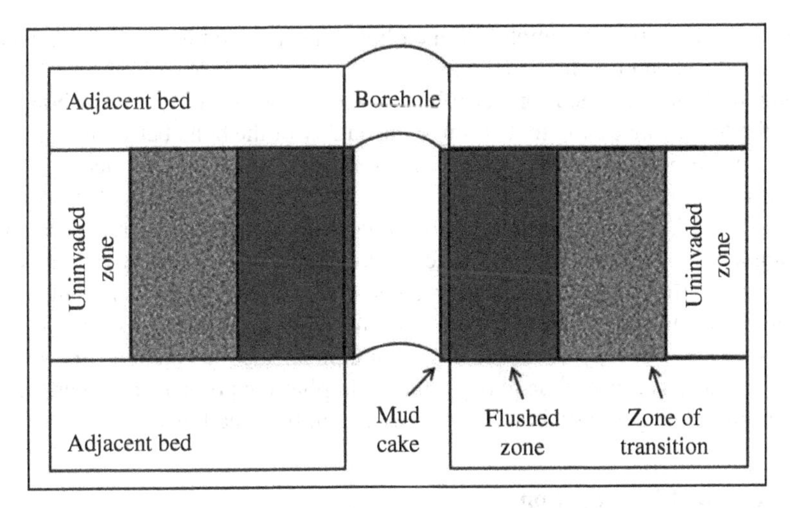

FIGURE 9.2 Schematic of invasion zones.

An idealized representation of the wellbore and formation is illustrated in Figure 9.2. It is used to make analysis of well log response more tractable. The figure shows four zones near the wellbore: the mud cake, the flushed zone, the zone of transition, and the uninvaded zone. If pressure in the borehole during drilling is greater than pressure in the formation, the pressure difference will drive drilling mud into the permeable formation. Larger particles in the drilling mud will be filtered out at the rock face of the borehole and create a mud cake adjacent to the borehole. The liquid with any small particulates that pass through the mud cake is called the mud filtrate. The mud cake can reduce the flow of fluids between the formation and the well. The formation damage caused by mud cake and filtrate can be quantified by well testing that yields a parameter called "skin."

During drilling, portions of fluids originally in the flushed zone are displaced by the invading mud filtrate and pushed into the transition zone. During logging, fluids in the flushed zone include mud filtrate with any suspended solids, some native brine, and any remaining oil and gas. Reservoir rock and fluid properties in the uninvaded zone have not been altered from their original state by fluids from the drilling operation.

9.1.2 Open or Cased?

Open-hole logging refers to logging before setting casing in the well. Sidewall coring, or the removal of a core of rock from the borehole wall, requires an uncased hole. Most logging tools are designed for uncased holes. Open-hole logging, which is the most widely used logging method, makes it possible to assess the commercial viability of a formation before spending the money to complete a well. Open-hole logs can be used to determine fluid contacts in a formation, obtain geological properties such as porosity and rock type, make pore pressure measurements, and obtain reservoir fluid samples. Examples of open-hole logs include resistivity logs, nuclear logs, sonic logs, and borehole imaging.

Some tools are designed for cased-hole logging. For example, cement-bond logs are used after setting casing to determine the quality of the bond between casing and cement. If the bond is not sufficient, then remedial operations are needed before proceeding with any completion operations. After the cement-bond log, the next cased-hole log is a gamma radiation log for correlating depths of target formations for completion operations in cased holes. Temperature logs and flow-rate logs may be used to identify sources of fluids in completed wells. A spinner flow meter can measure flow rates at different locations in the wellbore. Carbon–oxygen logs use gamma-ray spectroscopy to measure carbon content in hydrocarbons and oxygen content in water. A low carbon to oxygen ratio implies the presence of water, while a high carbon to oxygen ratio implies the presence of hydrocarbons.

9.1.3 Depth of Investigation

Some logging tools measure properties in the first few inches of the formation, while other tools measure properties deeper into the formation. This depth of investigation is usually characterized as shallow, medium, or deep and can range from a few inches to several feet. The lithology, or mineral composition, of the formation can be determined by shallow measurements. One purpose of resistivity logs is to estimate brine saturation in the formation, as an indication of the presence of oil and gas in the formation. For these estimates to be useful, deep measurements that penetrate beyond the transition zone are needed.

9.2 LITHOLOGY LOGS

Lithology logs indicate rock type. Most hydrocarbon accumulations are found in sedimentary rocks. The most important conventional reservoir rocks are sedimentary rocks classified as clastics and carbonates (Figure 9.3). A clastic rock is composed of clasts, or fragments, of preexisting rocks or minerals. Sandstones are compacted sediment, while shales or mudrock is laminated sediment. Carbonates are produced by chemical and biochemical sources. A single well can encounter several different types of rocks, as illustrated by the stratigraphic column in the figure. The stratigraphic column depicts the layering of rock strata in a column of rock. We consider three types of lithology logs in this section: the gamma-ray log, the spontaneous potential (SP) log, and the photoelectric effect (PE) log.

9.2.1 Gamma-Ray Logs

A gamma-ray tool detects gamma-ray emissions from radioactive isotopes. Gamma-ray logs imply the presence of shale when there is a high gamma-ray response. Clean (shale-free) sands or carbonates tend to have a low gamma-ray response. Depth of investigation is about 1.5 ft.

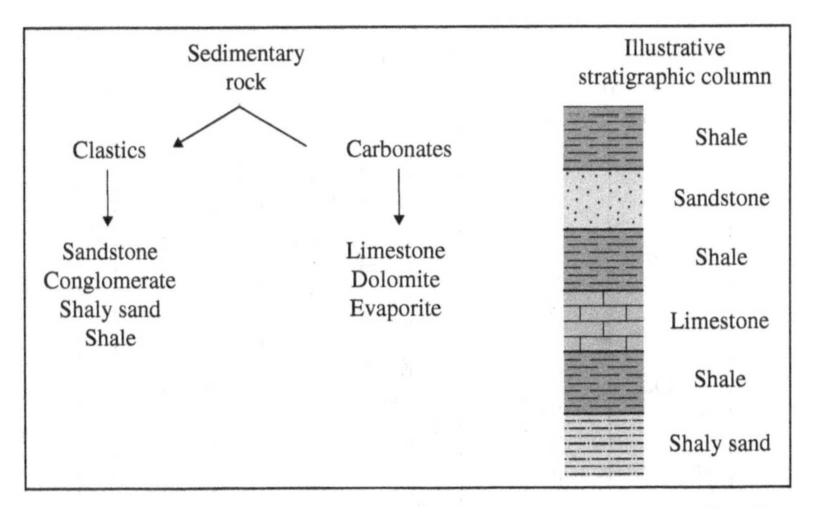

FIGURE 9.3 Common reservoir rock types and an illustrative stratigraphic column.

Natural radioactivity includes alpha, beta, and gamma rays. Alpha rays are helium nuclei, beta rays are electrons, and gamma rays are photons. Alpha and beta rays are low energy. Gamma rays emitted by nuclei have energies in the range from 10^3 eV (electron volts) to 10^7 eV and can penetrate several feet of rock. Naturally occurring radioactive materials (NORM) are minerals that contain radioactive isotopes of the elements uranium, thorium, potassium, radium, and radon. The decay of naturally occurring radioactive isotopes produces gamma rays.

Coals and evaporites, such as NaCl salt and anhydrites, usually emit low levels of gamma radiation. Anhydrites are composed of anhydrous (water-free) calcium sulfate and are usually formed as an evaporite when a body of confined seawater evaporates. Some of the highest levels of gamma radiation are observed in shales and potash (potassium chloride). A segment of a gamma-ray log is illustrated in Figure 9.4.

9.2.2 Spontaneous Potential Logs

The SP log is one of the oldest logging methods. The SP log is an electric log that records the direct current (DC) voltage difference, or electrical potential, between two electrodes. One electrode is grounded at the surface, and the other electrode on the logging tool moves along the face of formations in the wellbore.

The range of the voltage on an SP logging track is typically up to 200 mV. The SP voltage for shales in a particular well is fairly constant and forms a "shale baseline" for interpretation of the SP log for that well. The SP for shale-free formations is negative by 50–100 mV relative to the shale baseline and forms a "sand baseline." These two boundaries provide a rule for interpolating the amount of clay or shale in a formation based on its SP.

The SP results from diffusion processes in porous rock driven by differences in the ionic composition of aqueous mud filtrate and *in situ* brine. As a result, the SP log cannot function with air- or oil-based drilling mud. In porous formations with less

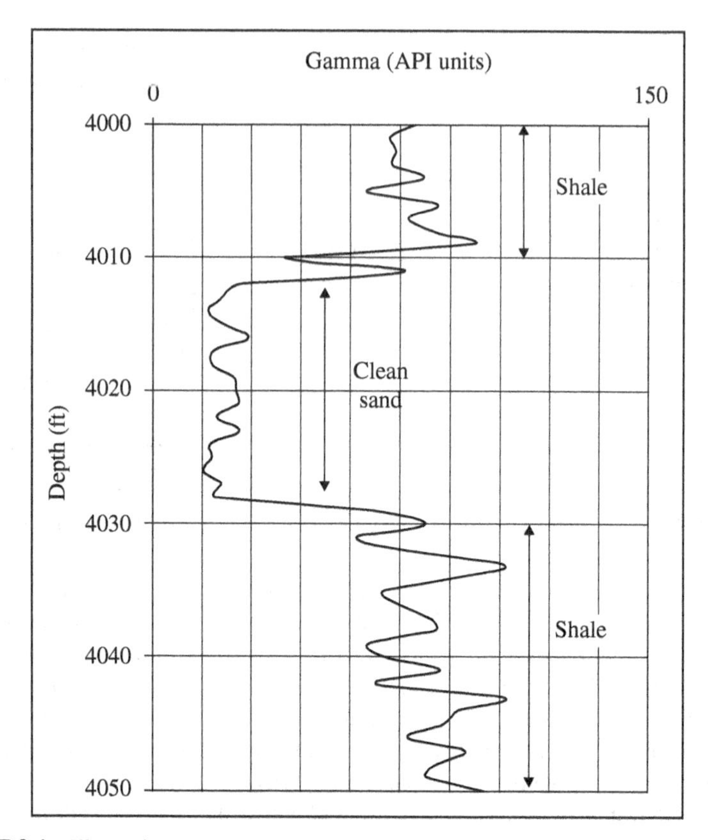

FIGURE 9.4 Illustration of gamma-ray (GR) log response. Compare with the porosity logs in Figure 9.5 and the resistivity logs in Figure 9.6.

clay or shale, the voltage is lower because the ions can diffuse more freely. Thus, deflection of the SP from the shale baseline to lower voltage indicates increasing permeability. If the concentration of salts dissolved in the mud filtrate equals the concentration in the *in situ* brine, then the SP is zero. For most wells, the filtration concentration is less than the *in situ* brine concentration, and the shale baseline is more positive than the sand baseline. However, the shale baseline can drift as the composition of the *in situ* brine changes. The change of temperature with depth can also cause baseline drift.

An increase or decrease in the SP curve indicates the boundary between one rock type and another. Boundaries between rock types that appear along the SP curve can be used to estimate net and gross thickness of a log interval. In some cases, the shape of the SP curve can yield information about the depositional environment when combined with other logs. The shape of the SP curve is used for correlating structure among wells.

The presence of oil and natural gas in the pore space with formation water reduces the number of ions in the pore space. Consequently, the SP curve changes value at the hydrocarbon–water contact, that is, the boundary between rock filled with formation water (the water zone) and rock containing both hydrocarbon and formation water (the hydrocarbon zone).

TABLE 9.1 Photoelectric Factors

Lithology	Approximate PEF (b/e)
Sandstone (quartz)	2
Dolomite	3
Shale	3.5
Salt	4.5
Limestone or anhydrite	5

Example 9.1 SP Log

On an SP log, if the shale baseline is +40 mV and the sand baseline is −80 mV, what is the shale content of a formation with SP of −30 mV?

Answer

The difference between baselines is 40 mV − (−80 mV) = 120 mV. The difference from the sand baseline to the formation SP is −30 mV − (−80 mV) = 50 mV, so the shale content is (50 mV)/(120 mV) = 0.42 or 42%.

9.2.3 Photoelectric Log

The PE gives another perspective on lithology. The PE refers to the absorption of a low energy gamma ray by inner orbital atomic electrons. The extent of this absorption varies with atomic number of the atom. The PE logging tool is part of the density logging tool. The PE tool measures the absorption or falloff of low energy gamma-ray energy. The falloff in gamma-ray energy is quantified as a PE factor (PEF). The PEF depends primarily on the atomic number of the formation. Hydrocarbons and water contribute very little to the PEF because their atomic numbers are low compared to rocks in the formation. On the other hand, the atomic numbers of different rock types can be correlated to PEF so that PEF serves as a good indicator of lithology or rock type as listed in Table 9.1. PEF is not very sensitive to porosity. The PE logging track is often scaled from 0 to 10 in units of barns per electron (b/e). A barn is a measure of cross-sectional area and 1 barn equals 10^{-28} m^2. Depth of investigation for the PE logging tool is typically 1 to 2 ft.

9.3 POROSITY LOGS

Porosity logs include density logs, acoustic logs, and neutron logs. These logs are described in this section.

9.3.1 Density Logs

A density logging tool uses intermediate energy gamma rays emitted from a radioactive source in the tool to induce Compton scattering in the formation. In this process, gamma-ray photons from the tool collide with electrons in the formation and transfer some of their energy to the electrons. The remainder of the energy is scattered as lower-energy gamma-ray photons. The density logging tool then detects some of these scattered photons. The gamma-ray count rate at the tool depends on formation electron density, which is proportional to formation bulk density ρ_b. Combining ρ_b from the density log with rock matrix density ρ_{ma} and fluid density ρ_f yields porosity ϕ:

$$\phi = \frac{\rho_{ma} - \rho_b}{\rho_{ma} - \rho_f} \tag{9.1}$$

Formation bulk density ρ_b can be obtained by rearranging Equation 9.1:

$$\rho_b = (1-\phi)\rho_{ma} + \phi\rho_f \tag{9.2}$$

Results of the density tool are presented in three tracks on a log: bulk density, density correction, and porosity. The density correction track shows corrections to the bulk density due to irregularities in wellbore diameter. It is an indication of the quality of the bulk density response. For density corrections greater than 0.2 g/cm^3, the bulk density values are suspect. The depth of investigation for the density tool is typically 1–2 ft.

Example 9.2 Density Log

A density log shows that the bulk density ρ_b of a formation is 2.30 g/cc. The density of the rock matrix ρ_{ma} is 2.62 g/cc and the density of the fluid ρ_f in the formation is 0.869 g/cc. Calculate porosity.

Answer

Use $\phi = \frac{\rho_{ma} - \rho_b}{\rho_{ma} - \rho_f} = \frac{2.62 - 2.30}{2.62 - 0.869} = 0.183$ or 18.3%

9.3.2 Acoustic Logs

The elapsed time to propagate a sound wave from a source on a logging tool through the formation and back to a receiver on the tool depends on formation porosity. The tool that sends and receives sound waves is called an acoustic log or sonic log.

Sound waves are vibrations in a medium like air or rock. The speed of sound in a medium depends on the density of the medium. In the case of rock, density depends on the density of rock matrix and the density of the fluids occupying the pore space in the rock. Fluid density is usually much smaller than matrix density. Consequently, the bulk density of a formation with relatively large pore volume is less than the bulk density of a formation with relatively small pore volume if the fluids in the pore volume are the same in both formations.

An estimate of the speed of sound v in a formation is given by Wyllie's equation. If we write the speed of sound in fluids occupying the pore volume of a rock as v_f and the speed of sound in the rock matrix as v_{ma}, Wyllie's equation is

$$\frac{1}{v} = \frac{\phi}{v_f} + \frac{(1-\phi)}{v_{ma}} \tag{9.3}$$

where ϕ is rock porosity. We can write the speed of sound in each medium as the distance traveled divided by the transit time. If we assume the distance traveled is the same in each medium, then Wyllie's equation can be written as

$$\Delta t = \phi \Delta t_f + (1-\phi)\Delta t_{ma} \tag{9.4}$$

where Δt is transit time in the bulk volume, Δt_f is transit time in the interstitial fluid, and Δt_{ma} is transit time in the rock matrix. Porosity in the bulk volume is obtained by rearranging Equation 9.4:

$$\phi = \frac{\Delta t - \Delta t_{ma}}{\Delta t_f - \Delta t_{ma}} \tag{9.5}$$

In addition to providing a measurement of porosity, acoustic logs can be used to calibrate seismic measurements. The comparison of acoustic and seismic transit time measurements in a formation can improve the accuracy of converting seismic transit times to depths.

The depth of investigation for sonic logs is typically 1–4 ft.

Example 9.3 Acoustic Log

An acoustic log measured the transit time of $75\,\mu s = 75 \times 10^{-6}\,s$ in a formation. The formation matrix has a transit time of $50\,\mu s = 50 \times 10^{-6}\,s$, and the fluid transit time is $185\,\mu s = 185 \times 10^{-6}\,s$. Use Equation 9.5 to calculate porosity.

Answer

Use $\phi = \frac{\Delta t - \Delta t_{ma}}{\Delta t_f - \Delta t_{ma}} = \frac{75-50}{185-50} = 0.185$ or 18.5%

9.3.3 Neutron Logs

The neutron log provides information about the formation by emitting neutrons from a radioactive source in the logging tool. The emitted neutrons lose energy as a result of collisions with atomic nuclei in the formation to become slower thermal neutrons. Slower neutrons can be captured by nuclei in the formation. Hydrogen nuclei are especially effective at capturing thermal neutrons. Reservoir fluids (oil, water, and hydrocarbon gas) contain hydrogen while formation rock usually does not. Consequently, the detection of captured neutrons by the neutron logging tool indicates porosity.

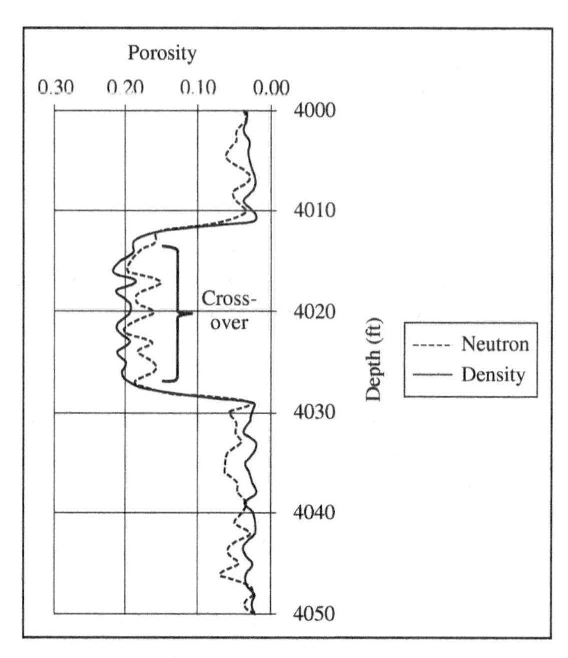

FIGURE 9.5 Illustration of crossover of porosities from density and neutron logs. Compare with the GR log in Figure 9.4 and the resistivity logs in Figure 9.6.

Pore space filled with natural gas has a relatively small hydrogen density compared to liquid water and oil. Therefore, lower porosity on the neutron log can indicate occupation of pore space by natural gas. Neutron and density porosity are often plotted on the same track. Neutron porosity is typically higher than density porosity except when natural gas occupies part of the pore space. In this case, neutron porosity is less than density porosity. This "crossover" from above to below the density porosity indicates the presence of natural gas and is illustrated in Figure 9.5.

9.4 RESISTIVITY LOGS

Formation resistivity is measured using resistivity logs. Rock grains in the formation are usually nonconductive, so formation resistivity depends primarily on the electrical properties of the fluid contained in the pore space. Hydrocarbon fluids are usually highly resistive because they do not contain ions in solution. Formation water, by contrast, contains ions in solution that can support an electrical current and have relatively small resistivity. Resistivity logs can be used to distinguish between brine and hydrocarbon fluids in the pore spaces of the formation. A resistivity log is illustrated in Figure 9.6.

Conrad and Marcel Schlumberger and Henri Doll first applied resistivity logs to the evaluation of a formation in 1927. The technology has evolved considerably since then. Here we introduce electrical properties of the ionic environment, the relationship between formation resistivity and wetting-phase saturation, and then discuss two types of resistivity logs: electrode logs and inductions logs.

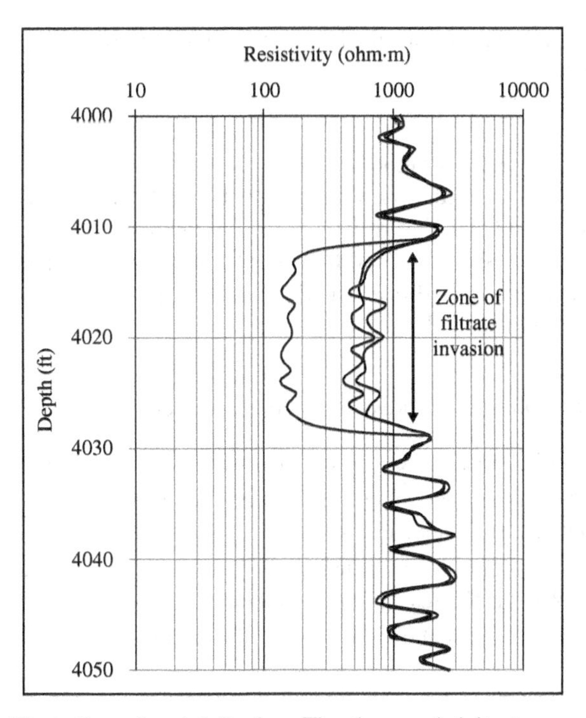

FIGURE 9.6 Illustration of resistivity log. The three resistivity traces are for shallow, medium, and deep resistivity measurements. Compare with the GR log in Figure 9.4 and the porosity logs in Figure 9.5.

In situ brine in a formation contains several inorganic salts that are ionized in water. The presence of ions in brine greatly enhances electrical conductivity. Pure water is a poor conductor of electricity. The presence of dissolved salts significantly increases the conductivity of water. The conductance of a material is inversely proportional to its resistance. Conductance S in siemens is given by

$$S = 1/R \tag{9.6}$$

where R is resistance in ohms. The units are related by 1 Siemen $= 1\,\text{mho} = 1/\text{ohm}$. Resistance R (ohms) is the proportionality constant between voltage V (Volts) and current I (Amperes) in the linear relationship

$$V = IR \tag{9.7}$$

where Equation 9.7 is Ohm's law.

Resistivity is often used to quantify the ionic environment. If we consider a uniform conducting medium with electrical resistance R_e (ohms), length L (m), and cross-sectional area A (m^2), resistivity of the conducting medium ρ (ohm·m) is

$$\rho = R_e \frac{A}{L} \tag{9.8}$$

Resistivity ρ (ohm·m) is inversely proportional to electrical conductivity σ (mho/m = S/m):

$$\sigma = 1/\rho \tag{9.9}$$

where 1 S/m = 1 siemen/m = 1 mho/m.

A high conductivity fluid has low resistivity. By contrast, a low-conductivity fluid like oil or gas would have high resistivity. A tool that can measure formation resistivity indicates the type of the fluid that is present in the pore space. Resistivity in a pore space containing hydrocarbon fluid will be greater than resistivity in the same pore space containing brine.

Example 9.4 Ohm's Law, Resistivity, and Conductivity

A. 0.006 C of charge moves through a circuit in 0.1 s. What is the current?

B. A 6 V battery supports a current in a circuit of 0.06 A. Use Ohm's law to calculate the resistance of the circuit.

C. Suppose the circuit has a length of 1 m and a cross-sectional area of 1.8 m². What is the resistivity of the circuit?

D. What is the electrical conductivity of the circuit?

Answer:

A. $\text{Current [Ampere]} = \dfrac{\text{[Coulomb]}}{\text{[Second]}} = \dfrac{0.006\,\text{C}}{0.1\,\text{s}} = 0.06\,\text{A}$

B. Ohm's law: $\text{Voltage}\,(V) = \text{current} \times \text{resistance} = I(\text{A}) \times R(\text{ohm})$ or $V = IR$.

Calculate $R = V/I = 6\text{ V}/0.06\text{ A} = 100\text{ ohm}$

C. $\rho = R_e \dfrac{A}{L} = 100\text{ ohm}\dfrac{1.8\text{ m}^2}{1\text{ m}} = 180\text{ ohm}\cdot\text{m}$

D. $\sigma = \dfrac{1}{\rho} = \dfrac{1}{180\text{ ohm}\cdot\text{m}} = 0.0056\text{ mho/m} = 5.6\text{ mmho/m}$ where mmho = milli-mho

The resistivity R_0 of a porous material saturated with an ionic solution is equal to the resistivity R_w of the ionic solution times the formation resistivity factor F of the porous material, thus

$$R_0 = FR_w \tag{9.10}$$

Formation resistivity factor F is sometimes referred to as formation factor. It can be estimated from the empirical relationship

$$F(a,m) = a\phi^{-m} \tag{9.11}$$

where ϕ is porosity, m is cementation exponent, and a is tortuosity factor. The cementation exponent m depends on the degree of consolidation of the rock and varies

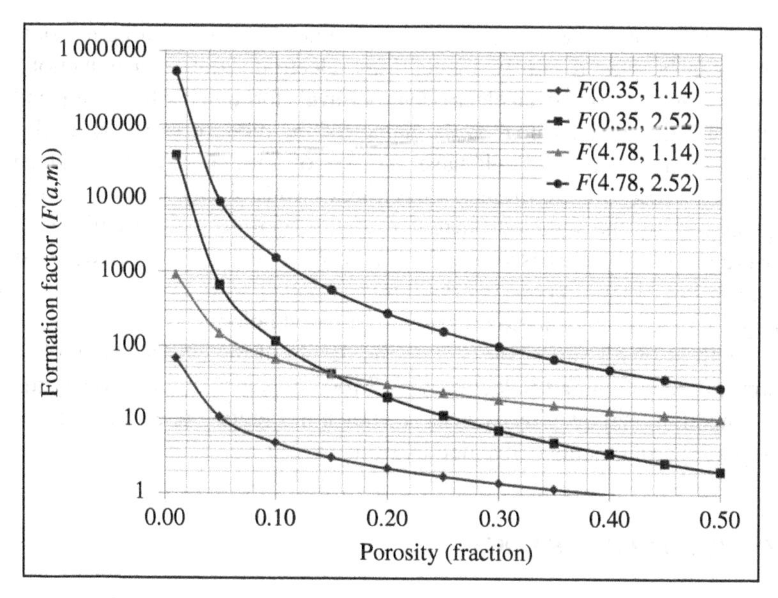

FIGURE 9.7 Sketch of formation resistivity factor for sands.

from 1.14 to 2.52. Tortuosity factor a depends on pore geometry. Formation resistivity factor $F(a,m)$ is plotted as a function of porosity for sands in Figure 9.7. In the case of sands, the empirical parameters m and a vary from 1.14 to 2.52 and from 0.35 to 4.78 (Bassiouni, 1994), respectively.

Archie's equation is an empirical relationship for the formation resistivity R_t of a porous medium that is partially saturated by an electrically conducting wetting phase with saturation S_w. Archie's equation is

$$R_t = R_0 S_w^{-n} \tag{9.12}$$

where the empirical parameter n is the saturation exponent. If the porous medium is completely saturated by the wetting phase so that $S_w = 1$, then $R_t = R_0$ from Archie's equation. Archie's equation for wetting-phase saturation

$$S_w = \left(\frac{FR_w}{R_t} \right)^{1/n} \tag{9.13}$$

is obtained by combining Equations 9.10 through 9.12. The wetting-phase saturation is often water saturation.

Example 9.5 Resistivity Log

A consolidated sandstone has tortuosity factor = 0.81, cementation factor = 2, and formation factor = 90. What is the porosity of the rock?

Answer

Solve $F(a,m) = a\phi^{-m}$ for $\phi = (a/F)^{1/m}$ to get $\phi = (0.81/90)^{1/2} = 0.095$

There have been many variations in the design of resistivity logging tools. The variations can be placed into two categories: electrode tools and induction tools. Current from an electrode in an electrode tool travels through the formation and back to another electrode. "laterolog" is the name for the modern variation of electrode tools. By contrast, induction tools have two sets of wire coils: coils for generating magnetic fields that induce current in the formation and coils for sensing the magnetic field produced by the induced current in the formation. Historically, electrode tools were used for water-based drilling fluids, while induction tools were used for oil-based drilling fluids. Those restrictions have faded with modern designs. Some laterologs can measure formation resistivity of cased holes.

The depth of investigation varies with the design of the electrodes in a laterolog or design of the coils in an induction tool. Most tools can simultaneously measure shallow, medium, and deep resistances as shown in Figure 9.6. Some tools, "array" tools, report measurements at five depths from about 1 up to 7 ft.

9.5 OTHER TYPES OF LOGS

In the previous sections, we have highlighted the most commonly used logs. A few other logs are introduced here. These logs illustrate the variety of logs that are being developed and used to meet industry challenges.

9.5.1 Borehole Imaging

Borehole imaging techniques are designed to produce centimeter-scale images of the rocks comprising the borehole wall. They include optical, acoustic, and electrical imaging techniques. Consequently, borehole imaging is not used in a cased hole. Borehole images can be used to identify fractures and fracture orientation, observe structural and stratigraphic dip, detect breakouts in the borehole wall, analyze small-scale sedimentary features, and assess net pay in thin beds.

9.5.2 Spectral Gamma-Ray Logs

Uranium, thorium, and potassium are the main sources of natural gamma radiation. In the standard gamma-ray logging tool, gamma rays from all three elements and any other radioactive elements are combined. The energy, or wavelength, of gamma rays from these elements differ and can be separately sensed. Spectral gamma-ray sensors can distinguish radiation from uranium, thorium, and potassium. This information can be useful for analysis of some formations.

9.5.3 Dipmeter Logs

A dipmeter is designed to determine the direction and angle of dip of rock strata adjacent to the borehole. The spatial orientation of a plane requires measuring the elevation and geographical position of at least three points on the plane.

The dipmeter logging tool uses three or more microresistivity sensors mounted on caliper arms to make measurements on different sides of the borehole wall. Information recorded by the dipmeter logging tool is recombined by computer to provide planar orientation including bedding dip. Acoustic and optical borehole imaging tools can provide similar information about bedding dip by providing images of the borehole wall.

9.6 LOG CALIBRATION WITH FORMATION SAMPLES

Conclusions derived from logging data are subject to varying levels of error. To reduce that error and build confidence in interpretations of logs, it is useful to compare the logs to samples from the formation. The cuttings collected during the mud logging process provide one set of samples. Other samples can be collected with coring operations.

9.6.1 Mud Logs

The mud log is obtained during the drilling operation. It includes several tracks: rate of penetration, depth, composition of gas liberated at the surface from the drilling mud, plus word and strip chart descriptions of cutting samples.

The wellsite geologist (mud logger) usually collects samples of cuttings every 10 ft of drill progress. The geologist examines the cuttings to determine lithology and the presence of any oils for the formations in contact with the bit. To do this, the geologist must account for the travel or lag time from the bit to the shale shaker as the mud carries the cuttings to the surface. These shale shaker samples can also contain bits of rock from shallower formations. Consequently, the sample may not be entirely representative of the environment encountered by the drill bit at a particular location, but they are the first direct evidence of lithology obtained from the borehole. Part of the duty of the mud logger is to identify the portion of the sample that was most recently cut by the bit. These descriptions of samples, as well as the samples themselves, are routinely compared to open-hole logs.

The rate of penetration track shows the amount of time it takes to drill through a foot of rock and indicates the hardness of the formation. Rock hardness inferred from the rate of penetration gives information about the rock type, which can be compared to the lithology logs.

Oil and gas shows in the mud log can be compared to indications of hydrocarbon in the porosity and resistivity logs.

9.6.2 Whole Core

Core samples are obtained by replacing the drill bit with a ring-shaped coring bit. The coring bit drills into the rock and captures a cylindrical volume of rock through the hole in the bit. The rock sample is collected in a core barrel in the lower portion of the bottom-hole assembly (BHA). The need to replace the drill bit with a coring bit, obtain a core, and then replace the coring bit with a drill bit to resume drilling means that coring is a time-consuming and expensive process. Consequently, only a few wells are cored. Cored wells are usually cored over a limited section of the well where core samples can provide the most useful information.

Cores are especially useful for characterizing productive formations. Each core from the formation is a small sample of reservoir rock. The core can be used to determine important reservoir properties such as lithology, porosity, and permeability. Core properties can be detrimentally modified during the coring process because reduction in temperature and pressure as the core is lifted to the surface alters core fluid content. Core samples may be obtained using a process known as native-state coring that is designed to keep core samples at original *in situ* conditions. The more complicated native-state coring process increases cost and tends to limit native-state coring to special situations.

Cores and any results of core analysis are routinely compared to open-hole logs as a way to improve interpretation of the logs.

9.6.3 Sidewall Core

Samples of formations can also be collected by devices lowered on a wireline. Some of these devices use percussive means to drive a sampling cup into the formation. Other devices use rotary saws for cutting samples. While percussive sampling is faster, rotary cutting provides better samples.

9.7 MEASUREMENT WHILE DRILLING AND LOGGING WHILE DRILLING

Measurement while drilling (MWD) refers to real-time measurements of drilling progress with sensors in the BHA. MWD allows better control of the drilling operation for reaching targets efficiently. Originally, MWD sensors measured drilling direction only, but currently they can also measure weight on bit, torque, RPM, and other drilling-related parameters. Natural gamma-ray logging sensors are included in some MWD assemblies. All of this information is communicated to the surface with pressure pulses in the drilling mud. Opening and closing a valve in the BHA generates the pulses. Rates of 1–3 bits per second are common. A battery pack or a small turbine driven by drilling mud provides electrical power for the MWD equipment.

Over the years, more and more logging sensors have been included in BHAs, which led to the coinage of the acronym LWD for logging while drilling. Many of the logging tools available for open-hole logging are available for LWD. Logging information is communicated to the surface with pressure pulse systems in the MWD assemblies. In some cases, logging information is stored in memory devices for later retrieval. In place of mud pulse telemetry, some MWD and LWD operations use electromagnetic systems that communicate at about 10 bits per second. MWD and LWD depths are tied to driller depths. Drillers determine depths by summing the lengths of drill pipe joints in the hole relative to the kelly bushing.

9.8 RESERVOIR CHARACTERIZATION ISSUES

Well logs are a major source of data about the reservoir. Table 9.2 is a summary of the principal applications of some widely used well logs. The columns in Table 9.2 show that well logs provide information about rock type, fluid content, porosity, pressure prediction, and structural and sedimentary dip. This information is essential for characterizing the reservoir. Some common reservoir characterization issues are introduced below.

9.8.1 Well Log Legacy

Well logging technology has been changing for approximately a century. Well logs acquired during that period of time represent a data legacy that has a value as intellectual property because old well logs can still provide useful information about a reservoir at the time the well log was acquired. In many cases, however, changes in well logging tools required changes in analysis techniques.

9.8.2 Cutoffs

Well log measurements can provide information about reservoir rock over a range of properties that includes economically producible pay zones and zones that contribute very little oil and gas to the production stream. Reserves calculations are more accurate when the analysis of well logs includes only pay zones. A well log cutoff specifies the minimum value of a measured property so that unproductive rock volume does not get included in fluid in place and reserves calculations.

We illustrate the use of cutoffs in the calculation of reserves by recalling that the volume of reserves V_R is the product of original hydrocarbon in place and recovery factor R_F. Therefore the volume of reserves V_R is

$$V_R = \phi A h S_h R_F \tag{9.14}$$

TABLE 9.2 Principal Applications of Common Well Logs

Log Type	Lithology	Hydrocarbons	Porosity	Pressure	Dip
Electric					
SP	X				
Resistivity	X	X		X	
Radioactive					
Gamma ray	X				
Neutron		X	X		
Density		X	X		
Sonic	X	X	X	X	
Dipmeter					X

Source: After Selley and Sonnenberg (2015), page 86.

where ϕ is porosity, A is area, h is net thickness, and S_h is hydrocarbon saturation. The determination of porosity and saturation is often accompanied by the specification of porosity and saturation cutoffs. The thickness of the rock interval that has porosity or saturation values below the corresponding cutoffs is not included in the net thickness of the pay interval.

9.8.3 Cross-Plots

A cross-plot is a plot of one well log parameter against another. For example, Archie's equation for wetting-phase saturation can be written as

$$R_t = \phi^{-m} R_w S_w^{-n} \tag{9.15}$$

where we assume the coefficient $a = 1$ in the empirical relationship for formation resistivity factor $F(a,m)$. Archie's equation relates electrical conductivity of a sedimentary rock to its porosity and brine saturation. The logarithm of both sides of Archie's equation gives

$$\log R_t = -m \log \phi + \log R_w - n \log S_w \tag{9.16}$$

We obtain the Pickett cross-plot by plotting log R_t from a resistivity log versus log ϕ from the density log. The Pickett cross-plot is a visual representation of Archie's equation.

Another useful cross-plot is the plot of sonic travel time from a sonic log versus formation density from a density log. The resulting cross-plot indicates lithology.

Example 9.6 Porosity–permeability Cross-plot

The porosity–permeability ($\phi - K$) cross-plot relates porosity and permeability. The $\phi - K$ cross-plot for a well is obtained from core measurements to be $K = 0.0007 \exp(71.7\phi)$ where K is in md and ϕ is a fraction. Use the $\phi - K$ cross-plot to find cutoff porosity for cutoff permeability = 1 md.

Answer

Solve the $\phi - K$ cross-plot for porosity: $\phi = 0.139 \times \ln(1428/K)$. Substituting cutoff permeability in the $\phi - K$ cross-plot gives $\phi = 0.139 \times \ln(1428/1) = 0.10$. In this case, rock with less than 10% porosity would not be considered productive because the permeability would be too low.

9.8.4 Continuity of Formations between Wells

As shown in Figure 9.8, the continuity of formations between logged wells can be illustrated with a fence diagram that displays two-dimensional stratigraphic cross sections in three dimensions. The fence diagram is prepared by first positioning well logs in their correct spatial locations. Correlations are then drawn between

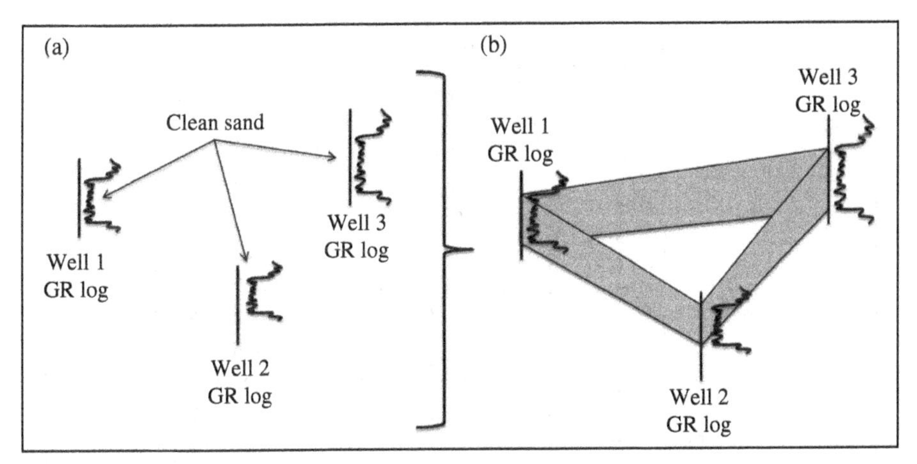

FIGURE 9.8 Illustration of a fence diagram showing correlation of a clean sand interval. (a) The clean sand interval is indicated by the GR logs. (b) Fence diagram displaying clean sand correlation.

neighboring well logs to form two-dimensional stratigraphic cross sections. The intersection of cross sections provides a three-dimensional representation of reservoir stratigraphy. Fence diagrams can show geologic discontinuities such as formation pinch-outs and unconformities. It must be remembered that stratigraphic sections drawn between well log locations are extrapolations and may be incorrect.

9.8.5 Log Suites

It is often desirable to combine different and complementary logging tools in the same logging run. An analysis of the combination of well logs makes it more likely that the characterization of the reservoir is correct. For example, the neutron–density logging tool combines a neutron logging tool and density logging tool. Both tools provide information about formation porosity but respond differently to the presence of hydrocarbon gas. The presence of hydrocarbon gas in the pore space increases the density log porosity and decreases neutron log porosity.

The gamma-ray logging tool can be included in a suite of well logging tools because the gamma-ray tool measures natural radioactivity. The gamma-ray logging tool can indicate shale content, identify lithology, correlate stratigraphic zones, and correct porosity log measurements in shale-bearing formations.

Figure 9.9 shows a combination of well logs for three wells: Wells 7, 3, and 9. Wells 7 and 9 are dry holes, and Well 3 is an oil producer. Well 3 is in the center of a line that extends from Well 7 to Well 9. The well logs in the figure are the SP log, seismic reflection coefficients (RC), and the resistivity log. The SP log and resistivity log are in arbitrary units. Depth is in feet.

Regional dip from Well 7 to Well 9 can be estimated by correlating the SP log and seismic RC at the top of the productive interval (at a depth of ~8450 ft). The resistivity in the upper part of the productive interval in Well 3 decreases at about 8550 ft.

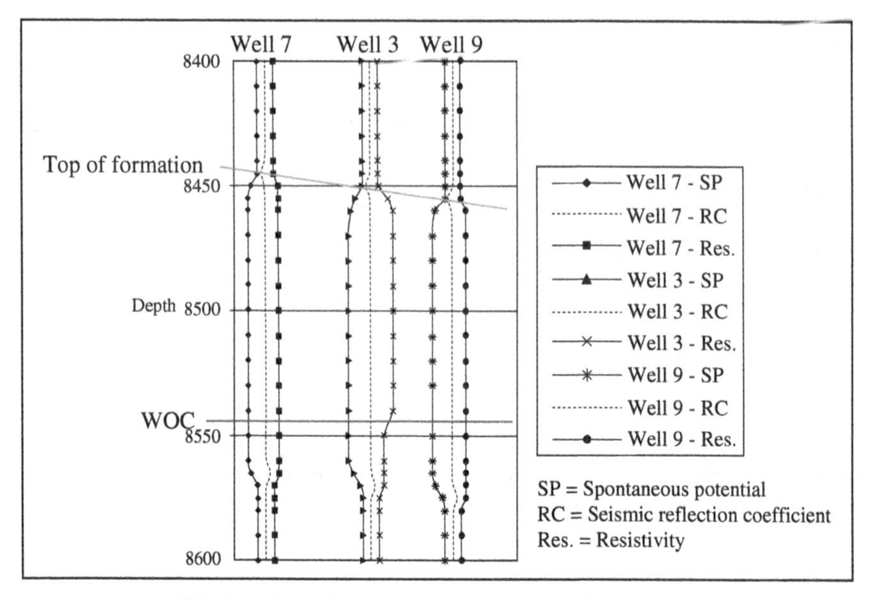

FIGURE 9.9 Combination of well logs (depth is in ft).

The decrease in resistivity in the lower part of the productive interval in Well 3 suggests that there is an increase in conductivity, which can be interpreted as an increase in water saturation. This indicates the presence of an oil–water contact (OWC).

Example 9.7 Resistivity Log and OWC

Estimate the OWC using the well logs for Well 3 in Figure 9.9.

Answer
The SP trace indicates a permeable formation from 8450 to 8570 ft. However the resistivity trace indicates a drop in resistivity at 8550 ft, suggesting the presence of formation water that contains solution salts (ions). Therefore the OWC in Well 3 appears to be between 8540 and 8550 ft.

9.8.6 Scales of Reservoir Information

Measured rock properties depend on the scale of the measuring technique. Porosity is routinely measured in rock cores. The measured porosity applies to a relatively small volume of the reservoir. By contrast, well logging tools can measure a few inches to a few feet into the formation. Therefore, well logs provide rock properties on a larger scale than rock cores. Figure 9.10 illustrates measurement techniques that vary from centimeter scale for cores to decameter scale for seismic measurements.

Seismic surveys sample a large region of the subsurface, but surface seismic data are considered "soft data" because seismic vibrations are detected at the surface after

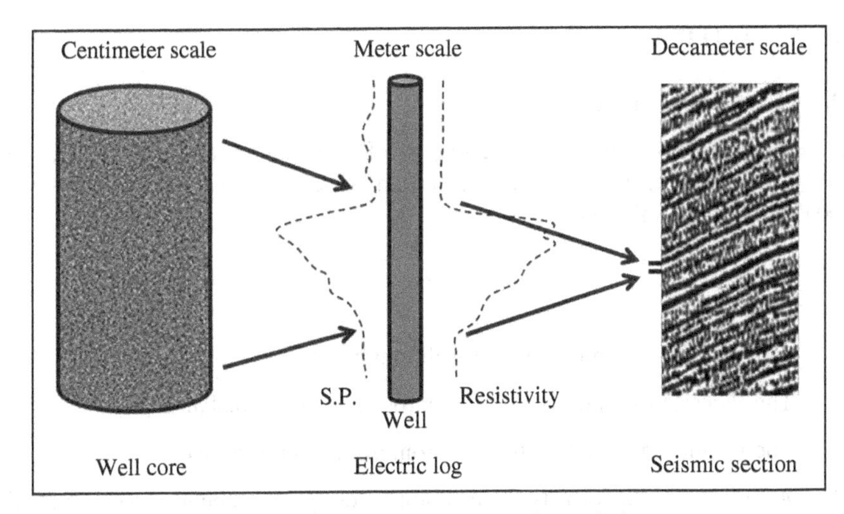

FIGURE 9.10 Range of data sampling techniques.

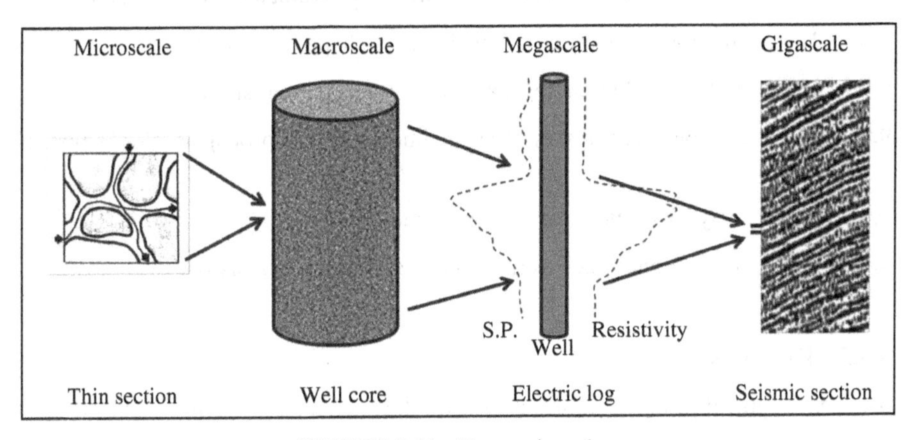

FIGURE 9.11 Reservoir scales.

propagating through many different media. Sonic logs can be used to calibrate surface seismic data at a particular well location and improve our confidence in the accuracy of the seismic measurements.

Figure 9.11 shows a scheme for classifying the scale of methods used in reservoir characterization. The four scales range from microscale to gigascale. The data sampling techniques shown in Figure 9.10 correspond to macroscale, megascale, and gigascale. Microscale is the smallest scale and is exemplified by an examination of a rock sample on the microscopic level. A thin section is a sliver of the rock surface placed on a microscope slide. It is often viewed with a microscope that uses light reflected from the surface of a rock rather than light that passes through the sliver of a rock. Scanning electron microscopes can also be used to examine a porous medium and can be placed in the microscale classification.

9.9 ACTIVITIES

9.9.1 Further Reading

For more information about well logging, see Selley and Sonnenberg (2015), Hyne (2012), Johnston (2010), Raymond and Leffler (2006), Asquith and Krygowski (2004), and Bassiouni (1994).

9.9.2 True/False

9.1 A log suite is a combination of well logging tools.

9.2 The Pickett cross-plot is a plot of porosity versus impedance.

9.3 The gamma-ray log detects shale from *in situ* radioactivity.

9.4 A well log and a porosity–permeability cross-plot can be used to estimate the Dykstra–Parsons coefficient.

9.5 A well log cross-plot is a plot of one well log parameter against another.

9.6 The caliper log gives information about the shape of the borehole.

9.7 Electrode logs, induction logs, and acoustic logs are resistivity logs.

9.8 The sonic log measures the transit time of sound propagating in a porous medium.

9.9 Gamma-ray logs detect gamma rays from NORM.

9.10 Archie's equation relates wetting-phase saturation to resistivity.

9.9.3 Exercises

9.1 **A.** Typical reservoir values for rock, oil, water, and gas compressibilities are

$c_r = 3\times10^{-6}$ / psia	$c_o = 10\times10^{-6}$ / psia
$c_w = 3\times10^{-6}$ / psia	$c_g = 500\times10^{-6}$ / psia

Suppose oil saturation is 0.8 in an oil–water system. Calculate total fluid compressibility using the aforementioned compressibilities. Hint: First, calculate water and gas saturation, and then calculate total fluid compressibility.

B. Calculate bulk modulus of fluid. Note that bulk modulus $K_f = 1/c_f$.

9.2 Plot hydrostatic pressure versus depth for a column of drilling mud in a well that is drilled to a depth of 5000 ft. The pressure gradient of drilling mud in the well is 0.48 psi/ft.

9.3 Suppose a density log is making measurements in a formation where the density of sandstone is 2.70 g/cc and the fluid density of brine is 1.03 g/cc. Use Equation 9.1 to plot porosity versus bulk density for the density log. Assume bulk density ranges from 2 to 2.70 g/cc.

9.4 Suppose an acoustic log is making measurements in a water-saturated limestone formation. The limestone matrix has a transit time per foot of $44\,\mu s/ft = 44 \times 10^{-6}\,s/ft$ corresponding to an acoustic velocity of 23 000 ft/s. The brine transit time per foot is $189\,\mu s/ft = 189 \times 10^{-6}\,s/ft$ corresponding to an acoustic velocity of 5300 ft/s. Use Equation 9.5 to plot porosity versus log transit time for the acoustic log. Assume log transit time per foot ranges from 40 to 100 μs/ft.

9.5 Fill in the following table using Equation 9.11 to calculate formation resistivity factor *F* for a resistivity log.

	Sand		Carbonate	
a	0.81	0.81	1	1
m	2	2	2	2
Porosity	0.1	0.2	0.1	0.2
F				

9.6 **A.** Fluid density for a volume with gas and water phases can be estimated using $\rho_f = S_g \times \rho_g + S_w \times \rho_w$ where ρ_g, ρ_w refer to gas density and water density, S_g, S_w refer to gas saturation and water saturation, and $S_g + S_w = 1$. Estimate fluid density when gas density is 0.00086 g/cc, water density is 1.03 g/cc, and water saturation is 30%. Hint: Convert water saturation to a fraction and calculate gas saturation from $S_g + S_w = 1$.

B. Suppose bulk density ρ_b is 2.20 g/cc from a density log, and density of rock matrix ρ_{ma} is 2.62 g/cc. Use the fluid density ρ_f from Part A and Equation 9.1 to estimate porosity.

9.7 Estimate the regional dip using the well logs in Figure 9.9. Hint: Fill in the table in the following. Use the RC trace to identify the top of the producing formation and then estimate the dip angle using tangent (dip angle) = (difference in depth between top of producing formation in Wells 7 and 9) divided by (distance between Wells 7 and 9).

Well Number	RC-Top (ft)	Distance from Well 7 (ft)
7		0
3		800
9		1400

9.8 The scale of reservoir sampling can be estimated by comparing the ratio of the area sampled to the drainage area of a well. The fraction of area sampled (f_{AS}) can be written as the ratio f_{AS} = area sampled/well drainage area.

A. What is the area of a vertical, circular wellbore that has a 6″ radius?

B. If the drainage area of a well is 40 acres, what fraction (f_{AS}) of this area is directly sampled by the wellbore?

C. Suppose a well log signal penetrates the formation up to 5 ft from the wellbore. What fraction of area is now sampled?

10

WELL COMPLETIONS

Well completion includes all the steps needed to prepare a newly drilled well for production. It frequently begins with placement of casing adjacent to the producing formation and ends with installation of production tubing and surface hardware. Nevertheless, the specifics of well completion are quite variable, being formation and operator dependent.

Completion of a formation occurs after open-hole logging and other tests of the drilled hole are finished. If analysis of the open-hole logs and other tests show that economic production of oil or gas is not possible, the hole will be abandoned by injecting cement into the well to isolate formations and to prevent contamination of surface water. Just beneath the surface, the casing system will be capped, and the well location will be restored to its predrilling condition. The abandonment procedure must be approved by the governmental agency that permitted the drilling process.

On the other hand, if analysis shows commercial promise, completion of the well will proceed. Costs of completion are included on the drilling AFE. We begin the following discussion of completions by introducing the concept of "skin" and showing how it relates to productivity of wells. Then we describe various options of completions: casing and liners, perforating practices, acidizing, and hydraulic fracturing. We conclude with descriptions of wellbore and surface hardware needed for production.

Introduction to Petroleum Engineering, First Edition. John R. Fanchi and Richard L. Christiansen.
© 2017 John Wiley & Sons, Inc. Published 2017 by John Wiley & Sons, Inc.
Companion website: www.wiley.com/go/Fanchi/IntroPetroleumEngineering

10.1 SKIN

Fluids and particulates in the drilling mud invade the formation immediately around the well during the drilling process. Drilling mud filtrate can change relative permeability or lead to mineral precipitates and scale buildup as the filtrate reacts with formation solids and native brine. Particulates in the drilling mud can plug pores in the formation. The extent of invasion of fluids and particulates varies, but the general result is reduced capacity for flow. Petroleum engineers refer to this as formation damage, and they quantitatively describe the extent of permeability damage with a dimensionless quantity "skin." An objective of well completion is to reduce skin.

To explore the concept of skin, consider radial flow through a damage-free cylindrical zone of radius r around a well of radius r_w. Darcy's law for state–state radial flow is

$$p - p_w = \frac{141.2 q \mu B_o}{kh} \ln\left(\frac{r}{r_w}\right) \tag{10.1}$$

The constant 141.2 in Equation 10.1 incorporates conversion factors so that field units can be used for the various parameters. Specifically, p_w is the pressure (psi) at wellbore radius r_w (ft), p is the pressure (psi) at radius r (ft), q is the liquid flow rate (STB/D), μ is the viscosity (cp), B_o is the formation volume factor (RB/STB), k is the permeability (md), and h is the formation thickness (ft). Imagine that this cylindrical zone is now damaged such that its permeability k_d is less than the undamaged permeability k. The pressure drop for the same flow rate q through the damaged zone is

$$p - p_{wd} = \frac{141.2 q \mu B_o}{k_d h} \ln\left(\frac{r}{r_w}\right) \tag{10.2}$$

Here, p_{wd} is wellbore pressure when the formation is damaged. The pressure drop $p-p_{wd}$ must be greater than $p-p_w$ when k_d is less than k. In other words, p_{wd} must be less than p_w. Solving the previous equations for the change in pressure as a result of the change in permeability gives

$$p_w - p_{wd} = \frac{141.2 q \mu B_o}{k_d h} \left(\frac{k}{k_d} - 1\right) \ln\left(\frac{r}{r_w}\right) \tag{10.3}$$

or

$$p_w - p_{wd} = \frac{141.2 q \mu B_o}{2\pi h} s \tag{10.4}$$

in which the skin s is a dimensionless pressure drop given by the following expression:

$$s = \left(\frac{k}{k_d} - 1\right) \ln\left(\frac{r_d}{r_w}\right) \tag{10.5}$$

The radius r_d is the radius of the damaged zone. Equation 10.5 is Hawkin's formula for skin. It shows that skin depends on the change of permeability as well as the size of the damaged zone relative to the well. Actual values of skin around wells are usually found from analysis of well tests.

Example 10.1 Skin from Hawkin's Formula

Use Hawkin's formula to estimate the skin of a damaged zone around a well with radius of 4 in., $k = 20$ md, and $k_d = 2$ md. The damaged zone extends 2 in. beyond the radius of the well.

Answer

Substitute the physical values into Equation 10.5:

$$s = \left(\frac{20\,\text{md}}{2\,\text{md}} - 1\right)\ln\left(\frac{6\,\text{in.}}{4\,\text{in.}}\right) = 3.6$$

Our next objective is to describe how skin affects well productivity. Consider a well in a large cylindrical zone of radius r_e and thickness h. Here, r_e is the drainage radius and p_{ave} is the average pressure in the zone. For example, r_e is 745 ft for a well in a 40-acre pattern. The difference in pressure between the average pressure of the drainage zone and the well pressure is

$$p_{ave} - p_w = \frac{141.2 q \mu B_o}{kh}\left(\ln\frac{r_e}{r_w} - \frac{1}{2} + s\right) \tag{10.6}$$

for steady-state flow. The units are field units: pressure (psi), flow rate (STB/D), viscosity (cp), formation volume factor (RB/STB), permeability (md), and formation thickness (ft). Petroleum engineers often rearrange this equation to obtain an expression for productivity index J:

$$J = \frac{q}{\left(p_{ave} - p_w\right)} = \frac{kh}{141.2 \mu B_o\left(\ln\frac{r_e}{r_w} - \frac{1}{2} + s\right)} \tag{10.7}$$

Productivity index depends on properties and dimensions of the formation, fluid properties, and skin. Productivity index is a handy term for comparing performances of any group of wells. It should be clear that increasing skin will decrease productivity index. The magnitude of skin depends on the extent of near-well damage. Severe damage could translate to skin of about 20. Skin for mild damage might be 5 or so. Skin can be negative as noted in the discussion of hydraulic fracturing. Negative skin corresponds to enhanced permeability in the near-well region.

Example 10.2 Productivity Index

Calculate J for the following values: well radius = 0.33 ft and r_e = 745 ft, formation thickness h = 100 ft, permeability = 8 md, oil viscosity = 1.5 cp, formation volume factor B_o = 1.05 RB/STB, and skin = 4.5.

Answer
Substitute the physical values into Equation 10.7:

$$J = \frac{(8\,\text{md})(100\,\text{ft})}{141.2(1.5\,\text{cp})(1.05\,\text{RB/STB})\left(\ln\frac{745\,\text{ft}}{0.33\,\text{ft}} - \frac{1}{2} + 4.5\right)} = 0.31\,\text{STB}/d/\text{psi}$$

10.2 PRODUCTION CASING AND LINERS

Usually, the first task of completion is to install production casing or other hardware in the well adjacent to the target formation. Casing is a pipe, usually steel, that extends from the producing formation to the surface. Figure 8.10 illustrates the placement of casing in a well. If the operator chooses to install casing, it is generally cemented in place with the cement covering any open formations and continuing upward to previously installed surface or intermediate casing. When cemented in place, casing mechanically supports the formation surrounding the hole. Furthermore, it allows for selective connection to the surrounding formation with perforations. After the cement has set, a cement-bond log is run on wireline to assess the quality and location of cement in the annulus. Along with the cement-bond sensor, the logging assembly has a gamma sensor and a casing-collar locator. The gamma response is needed to correlate the collar-depth chart to the formation intervals that will be perforated later. In addition to the cement-bond log, a cased and cemented well will be pressure tested to assess mechanical integrity. The nature of the pressure tests vary, but generally the wellhead equipment is tested to its maximum rated operating pressure, and the casing is tested to the maximum pressure that is expected during the remainder of completion operations.

If the producing formation is sufficiently strong, an operator may choose to leave the hole open, without casing. All fluid production is commingled in an open hole.

A liner is an alternative between cemented casing and open hole. It is a steel pipe that is perforated with slots or holes, and it is anchored to the bottom of surface casing or intermediate casing. The liner extends downward through the producing formation. A liner does not provide as much mechanical support as the support provided by cemented casing, but the liner can catch solids that slough from the wellbore. Undesirable portions of a formation can be isolated by placing a hole-free liner with packers. These packers have expandable sleeves that seal the annular gap between the liner and the formation. The elastomeric sleeve can be expanded by a variety of means, including rotation of the liner to actuate mechanical compression of a sleeve against the formation.

Some hydrocarbon formations consist of poorly consolidated sand. This sand can damage tubing or casing and can plug up surface equipment. To prevent production of sand, a variety of screened liners are used, sometimes combined with a "gravel pack" in the annular region between the liner and the formation. The "gravel" consists of sand that is about six times larger than the mean size of sand in the formation.

Many coalbed methane wells are cased, cemented, and then perforated. Some are completed with open holes. Some open holes are expanded by very high rate gas production that blows coal fragments and powder out of the well and forms a large cavity.

For oil and gas wells in shale formations, wells are usually drilled horizontally to maximize contact between the well and the very-low-permeability formation. Some operators case and cement the horizontals, and some case them without cementing.

10.3 PERFORATING

If a well is cased or lined and then cemented, the connection between the inside of the casing and the surrounding formation is established by shooting holes through the casing and cement and into the formation. In the 1930s and 1940s, perforations were shot with large caliber bullets. But developments in high explosives during World War II led to shaped charges for perforating. A few operators have used water jets with abrasives to perforate casing.

A shaped charge for perforation consists of a cup-shaped device with high explosive between an outer metallic cup and an inner tungsten layer. The inner tungsten layer becomes a high-speed metal jet when the explosive detonates. The metal jet punches a hole through the casing and the cement and into the formation. This tunnel in the formation is surrounded by a zone of crushed rock with the remains of the metal jet at the end of the tunnel.

To prepare for perforating, the shaped-charge cups are mounted on a steel strip or tube in a spiral pattern with a detonator cord running from the top to the bottom of the strip or tube. The detonator cord is wrapped to touch a primer on each cup. The strip or tube is then mounted in a steel pipe, or hollow carrier, to complete the "gun." At the top of a gun, a blasting cap is connected to the detonation cord. The blasting cap can be activated by an electric signal through wireline to the surface or by a pressure switch. The length of perforation guns varies from 1 ft to about 20 ft, but guns can be linked to form longer perforating assemblies. The diameter of guns ranges from about 2 to about 7 in. Larger diameter guns can carry larger shaped charges.

Perforating guns can be lowered to the target formation on wireline or tubing. Gravitational force is sufficient to pull the guns on wireline for vertical and modestly deviated wells. For horizontal wells, tractors are often used to pull the guns on wireline. When using wireline, the gun assembly will normally include a casing-collar locator. The collar-depth data from the locator on the gun is correlated to the collar depths observed during cement-bond logging, which are correlated with the gamma log from the cement-bond log to the open-hole logs. With the gun at the correct location, the wireline operator fires the shaped charges with an electric signal.

Before perforating a well, petroleum engineers are responsible for several design decisions. They first need to determine what zone or zones will be perforated. That decision depends on results from mud and open-hole logs plus experience in neighboring wells. They must also consider perforation hole size and length of penetration, number of shots per foot, angular phasing (e.g., 30, 60, 90, or 180°) between adjacent shots in the spiral of charges mounted in the gun, and gun length. Service companies have fired charges and measured penetration length and hole diameter for many rock samples. Service company personnel can provide support for estimating penetration length and hole diameter for a specific formation with its unique mechanical properties.

In 1997, Brooks published a method for predicting productivity of a perforated vertical well as a function of perforation penetration length L_p, perforation tunnel diameter d_p, formation damage length L_d, number of shots per foot n, the ratio of horizontal to vertical permeability k_h/k_v, and the skin s_p of the perforations. Just a portion of that method is described here. Brooks proposed a dimensionless description of perforations N_{pd}:

$$N_{pd} = \frac{N_{pd1}}{N_{pd2}} \tag{10.8}$$

with

$$N_{pd1} = \frac{\left(L_p - L_d\right) n^{3/2} d_p^{1/2}}{\left(k_h / k_v\right)^{5/8}} \tag{10.9}$$

$$N_{pd2} = \left(s_p + 1\right)\left(1 + \frac{\left(s_p + 1\right)}{N_{pd1}}\right) \tag{10.10}$$

Brooks concluded that $N_{pd} \geq 100$ is an efficiently perforated well. For $N_{pd} < 100$, efficiency of the perforation design can be increased by increasing penetration length and shot density. For $N_{pd} < 10$, the perforation design is very inefficient. Brooks' correlation for dimensionless productivity efficiency N_{pe} is

$$N_{pe} = 0.97\left(1 - \exp\left(-0.57\, N_{pd}^{\ 0.38}\right)\right) \tag{10.11}$$

This correlation is shown in Figure 10.1. Use of Brooks' method is demonstrated in the following examples.

Example 10.3 Brooks' Dimensionless Perforation Description, N_{pd}

Calculate N_{pd} for the following values: perforation penetration length $L_p = 16$ in.; perforation tunnel diameter $d_p = 0.4$ in.; formation damage length $L_d = 5$ in.; number of shots per foot $n = 4$ shots/ft; the ratio of horizontal to vertical permeability $k_h / k_v = 5$; and the perforation tunnel skin $s_p = 5$.

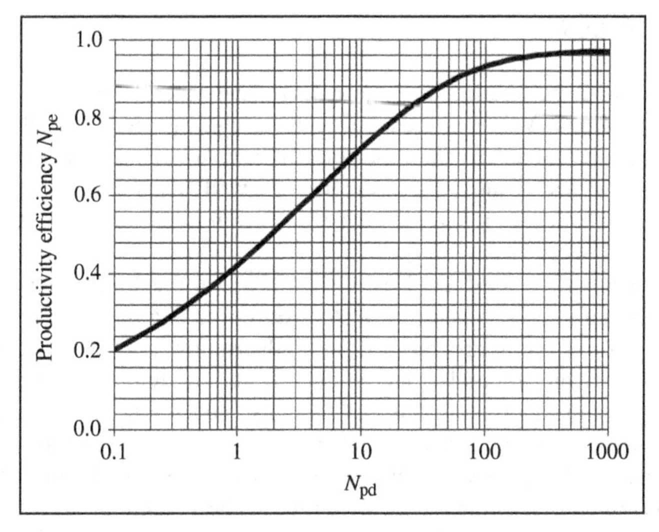

FIGURE 10.1 Brooks' correlation for productivity efficiency of perforations.

Answer

Substitute the physical values into Equations 10.9 and 10.10 and then Equation 10.8:

$$N_{pd1} = \frac{(16\,\text{in.} - 5\,\text{in.})(\text{four shots/ft})^{3/2}(0.4\,\text{in.})^{1/2}}{5^{5/8}} = 0.49$$

$$N_{pd2} = (5+1)\left(1 + \frac{(5+1)}{0.49}\right) = 79.5$$

$$N_{pd} = \frac{0.49}{79.5} = 0.006.$$

Brooks' dimensionless perforation description in the previous example is now used to estimate productivity efficiency.

Example 10.4 Brooks' Productivity Efficiency, N_{pe}

Calculate N_{pe} for $N_{pd} = 0.006$.

Answer

Substitute $N_{pd} = 0.006$ into Equation 10.11:

$$N_{pe} = 0.97\left(1 - \exp\left(-0.57(0.006)^{0.38}\right)\right) = 0.08$$

According to Brooks, this perforation plan is very inefficient. It can be improved by increasing penetration length and shot density and cleaning the perforations to reduce tunnel skin.

10.4 ACIDIZING

The next step after perforating is to clear debris and other damage for a cased and cemented hole. If reservoir pressure is sufficiently high, opening the well and starting production may be enough to clear the perforations. Acidizing is another option for cleaning perforations. At any time in the life of a well, acids can be used to remove mineral deposits in the well and the near-well region in the formation that are limiting production rate.

Four acids are often used. Hydrochloric (HCl), formic, and acetic acids attack carbonate minerals such as calcite ($CaCO_3$), dolomite ($CaMg(CO_3)_2$), and siderite ($FeCO_3$). Mixtures of hydrochloric and hydrofluoric (HF) acids are used in treatments of silicate minerals such as quartz, various feldspars, and clays. The volumetric amount of mineral that can be consumed by these acids depends on reaction stoichiometry and can be calculated. Tables 10.1 and 10.2 show the results of these calculations. Volumetric dissolving power is the ratio of mineral volume consumed per volume of acid solution. The volume of acid needed for a treatment equals the volume of mineral to be removed divided by the volumetric dissolving power for that acid and its concentration as demonstrated in the following example.

TABLE 10.1 Dissolving Power (Volume of Mineral per Volume of Acid Solution) for Carbonate Minerals

		Volumetric Dissolving Power			
Mineral	Acid	5 wt% Acid	10 wt% Acid	15 wt% Acid	30 wt% Acid
Calcium carbonate	HCl	0.026	0.053	0.082	0.175
	Formic	0.020	0.041	0.062	0.129
	Acetic	0.016	0.031	0.047	0.096
Dolomite	HCl	0.023	0.046	0.071	0.152
	Formic	0.018	0.036	0.054	0.112
	Acetic	0.014	0.027	0.041	0.083

Source: Adapted from Schecter (1992).

TABLE 10.2 Dissolving Power (Volume of Mineral per Volume of Acid Solution) for Minerals in Sandstone

		Volumetric Dissolving Power				
Mineral	Acid	2 wt% Acid	3 wt% Acid	4 wt% Acid	6 wt% Acid	8 wt% Acid
Quartz	HF	0.006	0.010	0.018	0.019	0.025
Albite	HF	0.008	0.011	0.015	0.023	0.030

Source: Adapted from Schecter (1992).

Example 10.5 HCl Volume for Removing Calcium Carbonate

Estimate the volume of 15% HCl needed to remove calcium carbonate in the perforated zone around a vertical well in sandstone formation with the following properties: 18% porosity and 12% of grain volume is $CaCO_3$. The treatment is to extend 1 ft beyond well radius (0.33 ft).

Answer

First, use volumetric relations to find the volume of $CaCO_3$ within the treatment radius per foot of formation:

$$\text{Volume CaCO}_3\text{/ft} = \pi\left(r_t^2 - r_w^2\right)(1-\phi)\, f_{CaCO_3}$$
$$= \pi\left(1.33^2 - 0.33^2\right)\text{ft}^2\,(1-0.18)(0.12) = 0.51\,\text{ft}^3\text{/ft}$$

Second, use the dissolving power of calcium carbonate from Table 10.1 for 15% HCl to find the acid treating rate:

$$\text{Volume HCl/ft} = \frac{0.51\,\text{ft}^3\text{/ft}}{0.082} = 6.26\,\text{ft}^3\text{/ft} = 47\,\text{gal/ft}$$

There are several complicating factors that should be considered in design of acid treatments, especially in sandstone formations. For example, dissolution of iron rust in the casing can lead to ferric hydroxide $Fe(OH)_3$ precipitate in the formation. Acids can interact with asphaltenes in the oil of the formation to produce sludges. Tests with formation oil samples before an acid treatment are essential. If sludges form, then injection of an aromatic solvent to displace oil away from the near-well region is an option. In sandstone formations, precipitation of calcium fluoride can be avoided by a preflush of hydrochloric acid to dissolve and displace any calcium carbonate in the near-well region. Precipitation of colloidal silica $Si(OH)_4$ is another complication of sandstone acidizing. To prevent significant precipitation of colloidal silica, some engineers recommend injection of HF at a high rate followed by rapid production of the spent acid.

Historically, a typical acid treatment for sandstones was as follows: 50 gal/ft 15% HCl preflush, 50–200 gal/ft of 3/12% HF/HCl ("mud acid"), and 200 gal/ft brine postflush; then immediate flow back. More recently, lower acid concentrations are being used. It is believed that lower concentrations will lead to less precipitates and reduced risk of unconsolidating the formation. In principle, testing of acids with reservoir rock samples should help with design; but that probably does not happen very often.

10.5 HYDRAULIC FRACTURING

The goal of hydraulic fracturing or fracking is to increase the number and capacity of flow paths between the wellbore and the surrounding formation. Fracking increases oil or gas production to a well in the same way that a freeway increases traffic flow

FIGURE 10.2 Hydraulic fracturing operation in Mansfield, Texas.

to or from a city. The flow capacity of the frack is related to the width of the fracture and the permeability of the proppants that hold the fracture open. The permeability of the proppants increases with size of the proppants and is roughly analogous to speed on a freeway. The width of the fracture is analogous to the number of freeway lanes. The desired flow capacity of a frack equals or slightly exceeds the flow capacity of the surrounding formation, just as the needed freeway size relates to the traffic capacity of the surrounding community.

In simple terms, fracking consists of pumping a slurry of liquid (usually water) and solid proppant (sand or ceramic) into a well at sufficient rate that the bottom-hole pressure rises to the point of splitting and then propagating a fracture into the formation. A fracking operation is shown in Figure 10.2. Storage containers shown in the figure are needed to store injection liquids and proppant. Trucks are needed to blend the mixture to form the slurry and pump the slurry into the well. The trucks shown in the figure are backed up to the wellheads. The slurry must be injected with enough pressure to work against stresses and strengths of the formation. Thus, describing stresses in formations is a good starting point for understanding fracking.

Stress in the vertical direction σ_v in a formation is a consequence of the weight of overlying formations and can be estimated as follows:

$$\sigma_v = \rho g h \tag{10.12}$$

where ρ is the density of overburden rock, g is the gravitational acceleration, and h is the depth to the formation. If a porous formation contains pressurized fluids, we can think of an effective vertical stress as follows:

$$\sigma_{ve} = \sigma_v - \alpha p \tag{10.13}$$

in which p is the fluid pressure in the pores of the formation and α is Biot's constant with values ranging from 0.7 to 1.0. Biot's constant approaches 1.0 for modest to

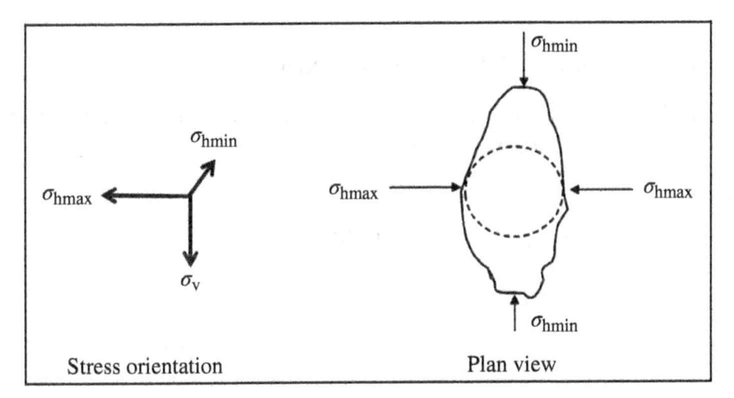

FIGURE 10.3 Orientation of three principal stresses and plan view of borehole breakout.

high-permeability rock. (This Biot refers to Maurice Anthony Biot who died in 1985 and is not the same as the namesake of the Biot number in heat transfer—that was Jean-Baptiste Biot who died in 1862.)

In addition to vertical stress, we need to describe stresses in the horizontal directions. In general, there are two horizontal stresses: the smaller horizontal stress is minimum horizontal stress σ_{hmin}, and the larger horizontal stress is maximum horizontal stress σ_{hmax}. Figure 10.3 shows the orientation of the three principal stresses and a plan view of wellbore failure known as borehole breakout for a vertical well. Borehole breakout occurs in the direction of minimum principal stress and can be detected using a caliper log. The orientation of borehole breakout can be used to indicate the orientation of principal horizontal stresses.

The effective horizontal stress can be found from the effective vertical stress using Poisson's ratio ν:

$$\sigma_{he} = \frac{\nu \sigma_{ve}}{1-\nu} \tag{10.14}$$

Poisson's ratio is the ratio of radial to axial strain for a cylindrical sample in tension. For an incompressible material, Poisson's ratio is 0.5. Poisson's ratio is about 0.3 for carbonates, about 0.2 for sandstones, and greater than 0.3 for shales. The actual horizontal stress can be expressed using Biot's constant:

$$\sigma_h = \sigma_{he} + \alpha p \tag{10.15}$$

This horizontal stress is the minimum horizontal stress, σ_{hmin}. To find the maximum horizontal stress, we need to include the additional component of tectonic stress:

$$\sigma_{hmax} = \sigma_{hmin} + \sigma_{tectonic} \tag{10.16}$$

The "plane" of a hydraulic fracture will be roughly perpendicular to the direction of the smallest of these three stresses (σ_v, σ_{hmin}, and σ_{hmax}), which is usually σ_{hmin}. The above relationships for stress provide first-order estimates, but they are usually on the low side. Applications of these relationships are demonstrated using examples. The first two examples show how to calculate the overburden stress gradient.

Example 10.6 Density of Brine-saturated Rock

Find the density of rock saturated with saline water. The water density is $65\,\text{lb}_\text{m}/\text{ft}^3$, specific gravity of rock grains is 2.65, and rock porosity is 0.22.

Answer
Rock density is the weighted average of the water and grain densities:

$$\rho_{\text{Rock}} = (0.22)\left(65\,\text{lb}_\text{m}/\text{ft}^3\right) + (1-0.22)(2.65)\left(62.4\,\text{lb}_\text{m}/\text{ft}^3\right) = 143\,\text{lb}_\text{m}/\text{ft}^3$$

Example 10.7 Overburden Stress Gradient

Find the overburden stress gradient if the overburden density is $160\,\text{lb}_\text{m}/\text{ft}^3$. The overburden stress gradient is $\sigma_\text{v}/h = \rho g$. Express your result in psi/ft.

Answer
Substitute overburden density into $\sigma_\text{v}/h = \rho g$ and convert units:

$$\frac{\sigma_\text{v}}{h} = \left[\frac{\left(160\,\text{lb}_\text{m}/\text{ft}^3\right)\left(32\,\text{ft}/\text{s}^2\right)}{\left(\dfrac{32\,\text{lb}_\text{m}\,\text{ft}/\text{s}^2}{\text{lb}_\text{f}}\right)}\right]\left(\frac{1\,\text{ft}^2}{144\,\text{in.}^2}\right) = 1.11\,\text{psi/ft}$$

The following example shows how to calculate horizontal stresses.

Example 10.8 Suite of Stresses

Find the stresses in a sandstone layer at depth of 10000 ft. Assume overburden stress gradient is 1.0 psi/ft, formation pressure is 4500 psi, Biot's constant is 0.95, Poisson's ratio for the sandstone is 0.25, and the tectonic stress is 2000 psi.

Answer
Apply Equations 10.12–10.16 with the physical values provided:

$$\sigma_\text{v} = 10000\,\text{psi}$$

$$\sigma_\text{ve} = 10000\,\text{psi} - 0.95(4500\,\text{psi}) = 5725\,\text{psi}$$

$$\sigma_\text{he} = \frac{0.25(5725\,\text{psi})}{1-0.25} = 1908\,\text{psi}$$

$$\sigma_{hmin} = \sigma_h = 1908 \text{ psi} + 0.95(4500 \text{ psi}) = 6183 \text{ psi}$$

$$\sigma_{hmax} = 6183 \text{ psi} + 2000 \text{ psi} = 8183 \text{ psi}$$

During the fracking process, pressure measurements at the wellhead can provide feedback on the above estimates of stress. One pressure value that may be observed during preliminary stages of fracking is the frack closure pressure. It appears on the pressure–time plot as a quick change of slope. The frack closure pressure is key to selecting the type of proppant for a fracture. If the closure pressure exceeds the strength of the proppant, the proppant will be crushed, and the benefit of the fracture can be entirely lost. And if the proppant is sufficiently strong to prevent crushing, embedment of proppant particles in the walls of the formation will reduce the fracture benefit. Table 10.3 lists proppant categories and their approximate closure stress limits. As there are many types of sand and ceramic proppants, the specific limitations of a proppant should be known before using it in a frack.

Next, we consider fracture performance, its relation to conductivity of the fracture and the formation to be fracked, and the length of the fracture. In 1960, McGuire and Sikora published a short paper on this topic. They used an "electric analyzer" in their predigital-computer-age effort to study productivity for a quadrant of a 40-acre (1320 ft on a side) drainage area with a fracture on one side. The length of a side of the 40-acre quadrant is 1320 ft/2 = 660 ft. Their Figure 2, which is reproduced in Figure 10.4, shows the productivity benefits of fracturing on the vertical axis as a function of relative conductivity of the fracture on the horizontal axis and the ratio R that equals fracture length L_f divided by quadrant side length L_q, which is 660 ft for a 40-acre drainage area. The axes and the parameter R in Figure 10.4 are defined by the following equations:

Vertical axis (left hand side):

$$\text{Modified productivity index ratio} = \frac{J}{J_o}\left(\frac{7.13}{\ln\left(0.472\left(L_q/r_w\right)\right)}\right) \tag{10.17}$$

Horizontal axis:

$$\text{Relative conductivity} = \frac{wk_f}{k}\sqrt{\frac{40}{A}} \tag{10.18}$$

TABLE 10.3 Approximate Closure Pressure Limits for Proppant Categories

Proppant	Closure Pressure Limit (psi)	Temperature Limit (°F)
Sand	4000	NA
Resin-coated sand	8000	250
Intermediate-strength ceramic	10000	NA
High-strength ceramic	>12000	NA

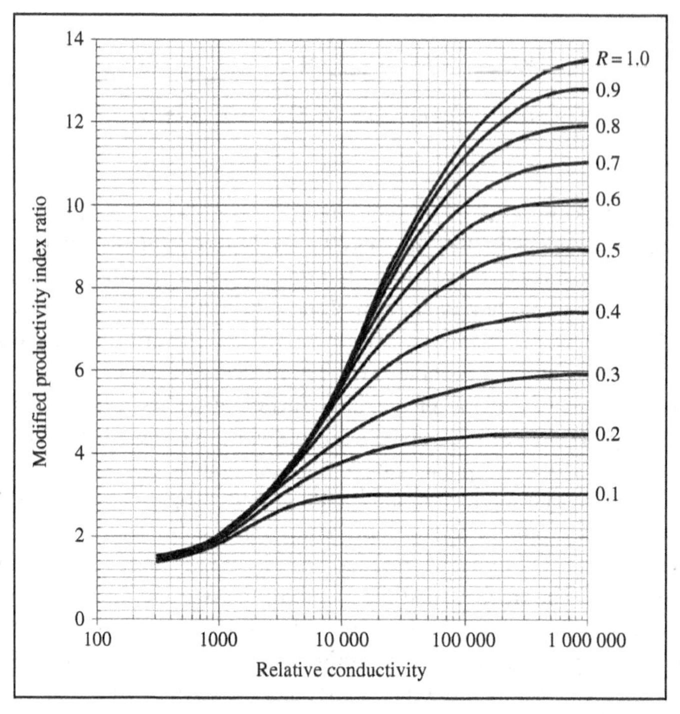

FIGURE 10.4 Relationship between modified productivity index ratio and relative conductivity. (Source: Adapted from Figure 2 of McGuire and Sikora (1960).)

Value of R (right-hand side of figure):

$$R = \frac{L_f}{L_q} \tag{10.19}$$

The terms in the previous equations and associated units are defined as follows:

J = productivity index after fracking
J_o = productivity index before fracking
L_q = length of the side of the drainage quadrant, ft
r_w = wellbore radius, ft
w = propped width of fracture, in.
k_f = permeability of proppant, md
k = average formation permeability, md
A = well spacing, acres
L_f = fracture length from wellbore, ft

The vertical axis combines productivity index ratio with a geometric factor (shown in brackets) that the authors added to scale the results to different drainage areas and well diameters. According to the authors (McGuire and Sikora, 1960, page 2), the modified productivity index ratio is "the ratio of generalized productivity indexes for fractured to unfractured cases multiplied by a scaling factor." In addition, relative conductivity "expresses the ability of the fracture to conduct fluid relative to that of the formation. It is the ratio of two products—fracture permeability times fracture width divided by formation permeability times the width of the drained area (drainage radius)." Relative conductivity should be dimensionless. For convenience, the authors replaced drainage radius with $\sqrt{A}$ where A is the spacing in acres. The constant 40 in Equation 10.18 is a scaling factor. The magnitude of the scaling factor $\sqrt{40/A}$ is 1 when A is 40 acres.

A key to using Figure 10.4 is strict compliance with the units specified earlier. Use of Figure 10.4 is illustrated with the following examples. We begin by calculating relative conductivity for a specified well pattern and formation.

Example 10.9 Relative Conductivity

Find the relative conductivity for a fracture in a 40-acre pattern. The formation permeability is 2 md, the fracture permeability is 150 darcies, and the fracture width is 0.3 in.

Answer

Substitute the given values into the definition of relative conductivity using the specified units:

$$\text{Relative conductivity} = \frac{wk_\text{f}}{k}\sqrt{\frac{40}{A}} = \frac{(0.3\,\text{in.})(150\,000\,\text{md})}{(2\,\text{md})}\sqrt{\frac{40}{40\,\text{acres}}} = 22\,500$$

The productivity index ratios for two different well radii are now compared.

Example 10.10 Productivity Index Ratio

A. Find the ratio of productivity indices for a reservoir with a 120-ft fracture half-length using the relative conductivity and other input from the previous example. The well radius r_w is 3 in.

B. Repeat the preceding example with well radius r_w equal to 4 in.

Answer

A. The ratio $R = L_\text{f}/L_\text{q} = 120/660 = 0.18$. For relative conductivity of 22 500, the reading on the vertical axis of Figure 10.4 is about 4.0, so the ratio of productivity indices is 4.0. In other words, productivity after fracking is four times greater than the productivity of the unfracked formation. Next, the productivity scaling term $7.13/\ln(0.472\,L_\text{q}/r_\text{w}) = 7.13/\ln\left((0.472)(660\,\text{ft})/(0.25\,\text{ft})\right) = 1.00$.

B. For $L_\text{f}/L_\text{q} = 0.18$ and relative conductivity of 22 500, the reading on the vertical axis of Figure 10.4 is still about 4.0. However, the productivity scaling term is slightly different: $7.13/\ln((0.472)(660\,\text{ft})/(0.33\,\text{ft})) = 1.04$. So the ratio of productivity indices is $4.0/1.04 = 3.8$.

The procedure for estimating fracture length is illustrated by the following example.

Example 10.11 Fracture Length

If the relative conductivity of a fracture is 1000 in., what length should be specified for the fracture?

Answer

For this low relative conductivity, the ratio of productivity indices is about 1.8 for $L_f/L_q = 0.1$, and it is very insensitive to increasing fracture length. Rather than worry about fracture length, it would be better to find how to increase relative conductivity—either by making a wider fracture or providing for higher fracture permeability. If relative conductivity could be increased to 10 000 in., then L_f/L_q as high as 0.4 could make sense, allowing for a ratio of productivity indices as high as about 5.6.

The relative conductivity of a fracture depends on two things that a petroleum engineer can control or influence during frack design: the permeability of the proppant pack and the width of the fracture, which relates to the amount of proppant placed in the fracture. The permeability of the proppant pack varies with size of the proppant particles. The size range of proppant is expressed by mesh range. Table 10.4 gives opening sizes for a short list of mesh numbers. For a proppant in the 30–50 US mesh range, its particles fall through a 30 US mesh sieve and are caught on a 50 US mesh sieve; as a result its particles are smaller than 0.060 cm and larger than 0.025 cm. Proppant pack porosities usually fall between 35 and 40%.

An engineer can estimate the permeability k(cm^2) of a clean proppant pack using the average diameter d (cm) of the proppants and the porosity of the pack:

$$k = \frac{1}{150}\frac{\phi^3 d^2}{(1-\phi)^2} \tag{10.20}$$

TABLE 10.4 Sizes of Openings for a Range of US and Tyler Mesh Numbers

US Mesh Standard	Tyler Mesh Standard	Opening Size (cm)
12	10	0.170
14	12	0.140
16	14	0.118
18	16	0.100
20	20	0.085
30	28	0.060
40	35	0.043
50	48	0.030
60	60	0.025
70	65	0.021
80	80	0.018
100	100	0.015

In a fracture, the proppant pack may include contaminating particulates of various sizes plus any remaining fracturing fluids that will decrease permeability below that of the previous equation. Permeability will also decrease if proppants break under stress, if proppant is embedded in fracture walls, and if fracture walls spall in the presence of liquids. As a result, the actual permeability of proppant in a fracture will be lower than the estimate from the previous equation.

Example 10.12 Proppant Permeability

Find the permeability of a proppant pack with porosity of 0.38 and average particle diameter of 0.063 cm.

Answer

$$k = \frac{1}{150}\frac{(0.38)^3(0.063\,\text{cm})^2}{(1-0.38)^2} = 3.78\times10^{-6}\,\text{cm}^2$$

Using the unit conversion 1 darcy $= 9.87\times10^{-9}\,\text{cm}^2$, we obtain proppant permeability $k = 383$ darcies.

The mass of proppant in a fracture typically ranges from 50 000 to 500 000 pounds. This mass m_p relates to the size of the fracture and properties of the proppant pack:

$$m_p = \rho_p Aw(1-\phi) = \rho_p L_{ee} wh(1-\phi) \tag{10.21}$$

where ρ_p is the density of proppant particles, A is the fracture area, w is the fracture width, ϕ is the porosity of the pack, L_{ee} is the end-to-end length of the fracture, and h is the fracture height. The units for Equation 10.21 can be any consistent set of units.

10.5.1 Horizontal Wells

Horizontal wells in very-low-permeability formations such as shales are typically cased, cemented, and fracked with 10 or more stages starting at the "toe" of the well and working back to the "heel" where the well bends up to the surface. In each stage, the casing must be perforated, then the formation is fracked, and finally a plug is placed in the casing on the heel side of the perforations to isolate the fracked interval from the next stage. To efficiently finish all these stages, service companies have developed hardware and methods to cycle quickly from perforating to fracking and to plugging. In some methods, the casing or liner is not cemented. In other methods, ball-drop plugs are combined with the casing to isolate one frack stage from another. No doubt, fracking technology will continue to evolve.

The permeability of shale formations is in the range of microdarcies and lower. With such low permeability, the relative conductivity (defined in Eq. 10.18) of the propped fracture can be increased with smaller fracture permeability.

10.6 WELLBORE AND SURFACE HARDWARE

All of the previous sections deal with connecting the formation to the wellbore. In addition, wellbore and surface hardware are needed to complete the well and then produce oil, gas, and associated water. Wellbore hardware includes production tubing, nipples, subsurface safety valves, packers, and pumping equipment. Surface hardware includes the wellhead, the Christmas tree, a pump driver, a separator, storage tanks, and pipelines. In the following discussion, we briefly describe production tubing systems and then pumping or artificial lift systems and introduce surface facilities.

Production tubing consists of many 30 to 40-foot lengths of pipe joined together. Production tubing extends from the wellhead at the surface to the producing zone. The weight of the tubing is supported by the wellhead. A short piece of pipe, or landing nipple, is placed at or near the lower end of the tubing. The inside dimensions of the landing nipple are machined to fit with tools and other hardware used during workovers and other operations later in the life of the well. A packer may be installed at the lower end of the tubing to seal the annular gap between the tubing and the casing. In some wells, subsurface safety valves are installed in the tubing near the surface so that fluid flow can be stopped in the event of damage to surface valves and other equipment.

Pumping, or "artificial lift," equipment is needed in many wells to lift liquids to the surface because reservoir pressure is not sufficient on its own. Common methods for artificial lift include sucker rod pumps, electric submersible pumps (ESP), gas lift, or progressive cavity pumps (PCPs). These four methods are described here.

The rocking motion of a pump jack (also known as horsehead, nodding donkey, grasshopper, etc.) is often encountered in oil country. The pump jack raises and lowers the sucker rod that drives a piston pump near the bottom of the tubing.

In ESP, multiple impellers are mounted on a shaft driven by an electric motor. Power for the motor is provided by an electric cable that runs along the side of the tubing to the surface. Submersible pumps can be used in oil or gas wells to pump liquid volumes at high rates. They are also common in coal gas production, offshore production, and environmentally sensitive areas where the footprint of surface facilities needs to be minimized.

In some wells, gas is injected at the surface into the tubing–casing annulus. The gas flows through "gas-lift" valves into the tubing to help lift the liquid to the surface. The gas mixes with the liquid (oil or water) and reduces the density of the gas–liquid mixture. If the density is low enough, the reservoir pressure may be able to push the mixture to the surface.

Invented by Rene Moineau in 1930, PCPs consist of a helical steel shaft or rotor that fits inside a helical rubber stator. When the rotor turns, cavities between the rotor and stator advance along their axis. Liquid inside the cavities is forced toward the surface. A PCP mounted near the end of production tubing is typically driven by a motor mounted on the wellhead and connected to the PCP rotor by a steel shaft. Although PCPs are used to lift liquids to the surface, they also function as downhole motors in drilling operations.

As noted previously, surface facilities of a well consist of the wellhead, the Christmas tree, a pump driver, a separator, storage tanks, and pipelines. Pump drivers were described earlier. The wellhead provides mechanical support for the casing and tubing and access through valves to annular spaces between successive casing strings and tubing. The Christmas tree is bolted to the top of the wellhead and is connected to the tubing. It is used to control fluids produced from the tubing. The Christmas tree usually splits into two or more branches adorned with valves and pressure gauges.

Oil and gas wells produce oil, water, and gas in varying quantities and ratios. For example, some gas wells produce gas with a little condensate and some water, while other gas wells produce a lot of water. Connected to the Christmas tree, separators must cope with the challenges of separating these fluids. Most separators operate at 100–200 psi and depend on the differences in density among phases to separate the fluids by gravity segregation. To facilitate separation of oil and water and to prevent formation of ice and gas hydrates, most separators are heated, especially in cold weather. Effluent gas from the separator passes through a backpressure regulator that keeps pressure constant in the separator. Fluid levels in a separator are maintained with level-control valves. At least two flow lines leave a separator: one carries gas to a central gas plant, the other carries liquids. For small rates of liquid flow, the liquids line goes to storage tanks at the well site. For high liquid rates, the liquids line goes to a central processing facility. As needed, trucks can unload liquid from storage tanks on location.

10.7 ACTIVITIES

10.7.1 Further Reading

For more information about completions, see Economides et al. (2013), Hyne (2012), Denehy (2011), van Dyke (1997), Brooks (1997), Schecter (1992), and McGuire and Sikora (1960).

10.7.2 True/False

10.1 Skin can be negative or positive with units of feet.

10.2 Skin depends on the depth of penetration of formation damage.

10.3 Production tubing is routinely cemented to the borehole wall.

10.4 Liners extend from the surface down to the depth of the producing formation.

10.5 The number of shots per foot equals the number of shaped charges per foot.

10.6 If Brooks' N_{pd} is 40 for a perforation plan, the design can be improved by selecting shaped charges that will give more penetration.

10.7 Perforating guns are commonly used to punch holes in tubing.

10.8 Acetic acid is used for treatments of silicate minerals.

10.9 The permeability of a propped fracture increases with size of the proppant.

10.10 A drilling AFE does not include costs for completion.

10.7.3 Exercises

10.1 Estimate the skin for a well if the damaged zone extends 4 in. beyond the radius of the well, which is 3 in. The native formation permeability is 10 md, and the permeability of the damaged zone is 3 md.

10.2 Find the productivity index for a well with the following properties: well radius = 4 in. and r_e = 550 ft; formation thickness h = 20 ft; permeability = 15 md; oil viscosity = 0.95 cp; formation volume factor B_o = 1.55 RB/STB; and skin = 6.

10.3 Use the following values to find Brooks' N_{pd}: perforation penetration length L_p = 20 in.; perforation tunnel diameter d_p = 0.3 in.; formation damage length L_d = 4 in.; number of shots per foot n = 4/ft; the ratio of horizontal to vertical permeability k_h/k_v = 10; and the perforation tunnel skin s_p = 2.

10.4 For a particular perforation plan, Brooks' N_{pd} = 150. What should be done to improve the plan?

10.5 Find the McGuire–Sikora relative conductivity for a fracture in an 80-acre pattern. The formation permeability is 0.01 md, the fracture permeability is 50*d*, and the fracture width is 0.2 in.

10.6 Find the parameter *R* used in Figure 10.4 for a fracture in an 80-acre pattern. The length of the fracture from the wellbore is 220 ft.

10.7 Referring to Figure 10.4, if *R* is 0.3 and the relative conductivity is 50 000, find the ratio of productivity indices J/J_o. The well drains a 40-acre pattern and the radius of the well is 4 in.

10.8 Estimate the permeability of a frack filled with 100 US mesh proppant. The porosity is 36%.

11

UPSTREAM FACILITIES

The oil and gas industry can be divided into upstream, midstream, and downstream sectors. The upstream sector includes operations intended to find, discover, and produce oil and gas. The downstream sector consists of crude oil refining, natural gas processing, and marketing and distributing products derived from crude oil and natural gas. The midstream sector connects the upstream and downstream sectors and includes transportation, storage, and wholesale marketing of hydrocarbons. Upstream facilities are needed for managing fluid production and preparing it for transportation. These facilities are described here.

11.1 ONSHORE FACILITIES

Fluid flow through the top of the well is controlled by a Christmas tree, which is installed on top of the wellhead. The Christmas tree is a collection of valves and fittings to control fluid flow. The wellhead is used to control pressure and support casing and tubing. It consists of the casing head and tubing head. Figure 11.1 illustrates a Christmas tree and wellhead.

The flowline connects the wellhead assembly to the separator. The flowline can be buried to reduce seasonal temperature variations and protect the line from weather and wildlife.

Introduction to Petroleum Engineering, First Edition. John R. Fanchi and Richard L. Christiansen.
© 2017 John Wiley & Sons, Inc. Published 2017 by John Wiley & Sons, Inc.
Companion website: www.wiley.com/go/Fanchi/IntroPetroleumEngineering

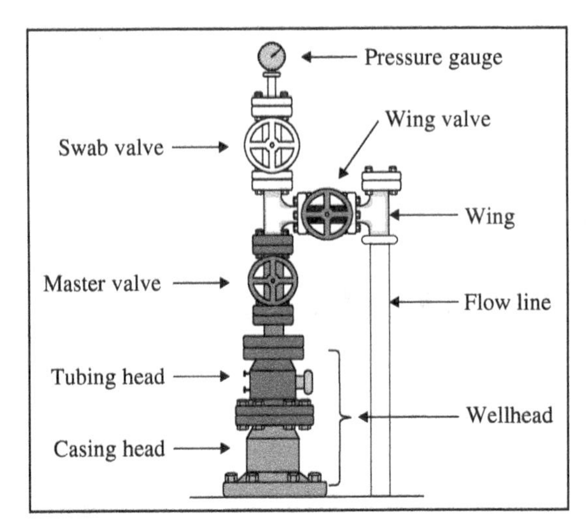

FIGURE 11.1 Christmas tree and wellhead.

Fluids flow from the wellhead assembly to separation and storage facilities. Separators are used to separate gas, oil, and water phases based on fluid density. If gas, oil, and water phases are in a vertical column, the gas phase will be at the top of the column, the water phase will be at the bottom of the column, and the oil phase is between the gas phase and water phase. Separators take advantage of gravity segregation to separate the fluids after production.

A two-phase separator will be used if one liquid phase and one gas phase are being produced. The liquid phase can be oil or water. If water, oil, and gas are being produced simultaneously, a three-phase separator is needed. The three-phase separator has separate outlets for oil, water, and gas. The gas outlet is near the top of the separator, while the water outlet is near the bottom of the separator.

In some instances oil and water will mix and create an emulsion. A chemical emulsion breaker can be used to separate oil and water. A chemical analysis of produced water can determine the compounds dissolved in the water phase. A heater treater is a separator that uses heat to separate oil and water.

The process of treating produced fluids starts at the wellhead where they are produced (Figure 11.2). The fluids flow through pipes to a separator where the different phases are separated. Each fluid phase is then moved to its own treatment equipment where it is measured, tested, treated, and/or gathered for transport to another facility.

If chemical flooding is used in the system, then the produced fluid will typically be an oil–chemical emulsion that must be broken. Once broken, the oil and chemicals are separated out into tanks where they can be tested, the oil can be collected for the next step in the production process, and the chemicals can be reused. Chemical flooding is complex and expensive.

Other equipment includes dehydrators (for removing water vapor from gas), oil and water storage tanks, flowlines, wellheads, compressors, and automation equipment. Automation equipment is used to monitor and, in some cases, control wells. On-site storage tanks store produced oil and water until the liquids can be transported away

FIGURE 11.2 Oilfield production equipment.

from the field. A tank battery is a collection of storage tanks at a field. Gas from the separator is usually routed to a flowline or compressor. The compressor boosts the pressure of the gas so that it can be injected into a flowline.

A central processing unit, or gathering center, is a location for collecting fluids from multiple wells. The fluids produced from all connected wells flow through gathering center separator(s) and into commingled storage tanks. The gathering center can save money on processing fluids, but it can reduce the operator's ability to analyze production from each individual well.

Surface facilities such as drilling rigs, storage tanks, and compressor stations are needed to drill, complete, and operate wells. The surface area required for installing all of the facilities needed to develop a resource is called the footprint. The size of the footprint has an impact on project economics and environmental impact. As a rule, it is desirable to minimize the size of the footprint.

Drilling rigs may be moved from one location to another on trucks, ships, or offshore platforms; or drilling rigs may be permanently installed at specified locations. The facilities may be located in desert climates in the Middle East, stormy offshore environments in the North Sea, arctic climates in Alaska and Siberia, and deepwater environments in the Gulf of Mexico and off the coast of West Africa.

Example 11.1 Pipeline Capacity

A. A gathering center receives oil from 16 wells. Each well can produce up to 5000 bbl liquid per day per well. Maximum liquid flow rate is the flow rate when all wells are producing at capacity. What is the maximum liquid flow rate?

B. The pipeline from the gathering center to a processing facility can carry 50 000 bbl liquid per day. Can all of the wells produce at maximum liquid flow capacity?

Answer

A. $16\,\text{wells} \times (5000\,\text{bbl/day/well}) = 80\,000\,\text{bbl/day}$.

B. No. The pipeline would have to be expanded or production from the wells must be limited to 50 000 bbl liquid per day.

11.2 FLASH CALCULATION FOR SEPARATORS

The fluid stream produced from a well enters a separator where the fluid phases separate. The focus of this section is the separation of the gas phase and the liquid hydrocarbon phase. The compositions and volumes of the gas and liquid phases can be estimated with a flash calculation. Flash calculations are also used in models of many gas plant and refinery operations, as well as in compositional reservoir models.

To understand a flash calculation, consider adding F total moles of a mixture of hydrocarbons to a vessel operating at temperature T and pressure P in Figure 11.3. The mole fraction for each component in F is z_i where the subscript i denotes component i. In the vessel, the mixture, or feed, equilibrates to yield a gas of molar amount G with mole fractions y_i and a liquid of molar amount L with mole fractions x_i.

The number of moles in the feed must equal the sum of the moles in the gas and liquid phases. The total mole balance is

$$F = G + L \tag{11.1}$$

Similarly, the moles of component i in the feed must equal the sum of the moles of component i in the gas and liquid phases. The component balance for component i is

$$z_i F = y_i G + x_i L \tag{11.2}$$

At equilibrium, the ratio of y_i to x_i is the k-value:

$$k_i = \frac{y_i}{x_i} \tag{11.3}$$

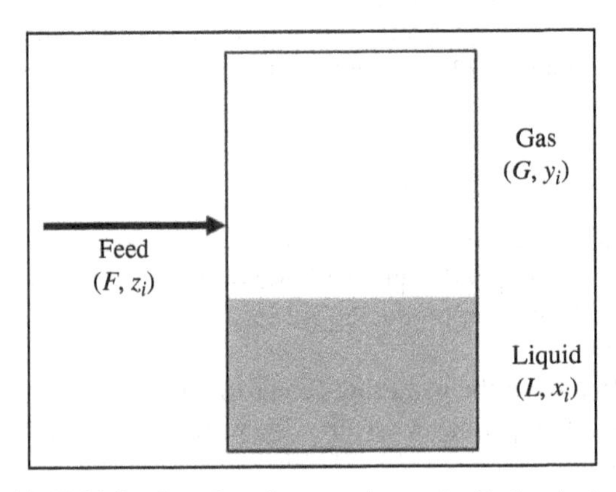

FIGURE 11.3 Sketch and nomenclature for flash calculation.

The k-value depends on temperature, pressure, and composition. For low pressures (<100 psi), k-values are equal to the ratio of vapor pressure to total pressure: $k_i = P_{\text{vapi}} / P$. Components with high vapor pressures will have higher k-values, while components with lower vapor pressure have lower k-values. For higher pressures, equations of state can be used for accurate estimates of k-values. The ideal gas law is an example of an equation of state. Charts and tables of k-values are found in some reference books. The Wilson equation is often used for qualitative estimates of k-values at pressure P and temperature T:

$$k_i = \frac{P_{\text{ci}}}{P} \exp\left[5.37(1+\omega_i)\left(1-\frac{T_{\text{ci}}}{T}\right)\right] \tag{11.4}$$

with critical pressure P_{ci}, critical temperature T_{ci}, and acentric factor ω_i for component i. Temperatures for Equation 11.4 must be absolute temperatures.

Example 11.2 Estimate k-value for Methane

Estimate the k-value for methane at 100°F and 150 psia. Use $P_c = 666$ psia, $T_c = -117$°F, and $\omega = 0.010$.

Answer
First convert from Fahrenheit to absolute temperatures in degrees Rankine:

$$T = 100°\text{F} + 460 = 560°\text{R}$$
$$T_{\text{ci}} = -117°\text{F} + 460 = 343°\text{R}$$

Now, find k_{methane}:

$$k_{\text{methane}} = \left(\frac{666\,\text{psia}}{150\,\text{psia}}\right)\exp\left[5.37(1+0.010)\left(1-\frac{343°\text{R}}{560°\text{R}}\right)\right] = 36.3$$

Rearranging Equation 11.3, the mole fraction y_i of component i in the gas phase can be written as

$$y_i = k_i x_i \tag{11.5}$$

Using this expression in Equation 11.2 gives

$$z_i F = k_i x_i G + x_i L = (k_i G + L)x_i \tag{11.6}$$

The mole fraction x_i of component i in the liquid phase is

$$x_i = \frac{z_i F}{k_i G + L} \tag{11.7}$$

The sum of x_i for all n components in the liquid phase must equal unity; therefore

$$\sum_{i=1}^{n} \frac{z_i F}{k_i G + L} = 1 \tag{11.8}$$

For convenience, we set $F = 1$ so that G and L become fractions of total feed moles in the gas and the liquid phases. As such, $G + L = 1$. Equation 11.8 becomes

$$\sum_{i=1}^{n} \frac{z_i}{1 + (k_i - 1)G} = 1 \tag{11.9}$$

A flash calculation consists of finding the value for G that satisfies the previous equation for a given set of feed compositions z_i and k-values. In solving Equation 11.9, we are finding the composition of the liquid phase, because each member of the sum is x_i. We calculate the composition of the gas phase y_i using Equation 11.5.

When a mixture of hydrocarbons is placed in a vessel at low pressure, the mixture could be entirely in the vapor phase. In this case, the flash calculation would fail. Similarly, at a high pressure, the mixture could be entirely liquid and the flash calculation would fail. If the following two conditions are met, the mixture exists as two phases and the flash calculation will not fail:

$$\sum_{i=1}^{n} z_i k_i > 1 \tag{11.10}$$

and

$$\sum_{i=1}^{n} \frac{z_i}{k_i} > 1 \tag{11.11}$$

Example 11.3 Two-phase Check for Flash Calculation

A mixture of methane, propane, and normal pentane equilibrates in a vessel at 100°F and 150psia. With feed mole fractions and k-values from the Wilson equation in the following table, will the mixture be one phase or separate into two phases?

Component	Mole Fraction	k-Value
Methane	0.55	36.3
Propane	0.30	1.3
Normal pentane	0.15	0.1

Answer

Substituting mole fractions and k-values from the previous table into Equations 11.10 and 11.11 yields 20.4 and 1.8, respectively. Hence, the mixture will separate into two phases, gas and liquid.

Example 11.4 Flash Calculation

Complete the flash calculation for the mixture in the previous example. Verify that $G = 0.89$.

Answer
Use the kernel of Equation 11.9 to find the mole fractions x_i in the liquid phase. Then use Equation 11.4 to find mole fractions y_i in the gas phase. Vary G from 0.88 to 0.90 until the sums of x_i and y_i are both equal to 1.000.

Component	Mole Fraction z_i	k-Value	$x_i = \frac{z_i}{1+(k_i-1)G}$	$y_i = k_i x_i$
Methane	0.55	36.3	0.02	0.62
Propane	0.30	1.3	0.24	0.31
Normal pentane	0.15	0.1	0.74	0.07

11.3 PRESSURE RATING FOR SEPARATORS

Separators and storage tanks on well locations are usually cylindrical in shape. Storage tanks at refineries are mostly cylindrical, but some are spherical. These tanks may be considered thin-walled pressure vessels (Hibbeler, 2011) when the thickness t of the wall is small relative to the inner radius of the cylinder r_{ic} or sphere r_{is}. The thin-walled criterion may be written as

$$\frac{t}{r_{ic}} < 0.1 \text{ or } \frac{t}{r_{is}} < 0.1 \tag{11.12}$$

The principal stresses in thin-walled pressure vessels are hoop (circumferential) stress and axial (longitudinal) stress when the external pressure is small relative to the internal pressure from a filled container (see Figure 11.4). The hoop stress σ_h of a

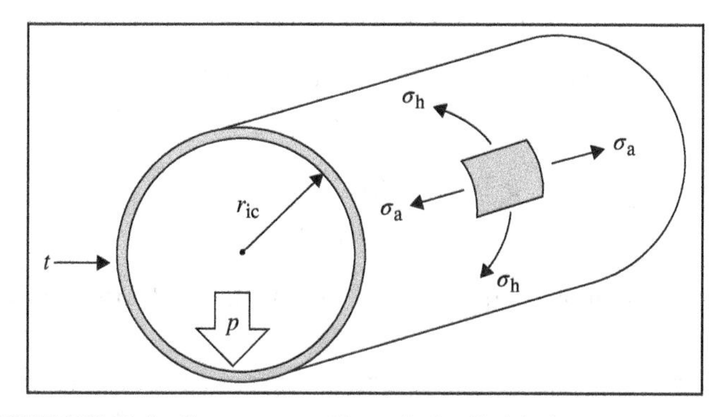

FIGURE 11.4 Stresses on a thin-walled cylindrical pressure vessel.

thin-walled cylindrical pressure vessel with internal pressure p, thickness t, and inner radius r_{ic} is

$$\sigma_h = \frac{(p - p_a) r_{ic}}{t} \tag{11.13}$$

with p_a equal to local atmospheric pressure. The axial force on the ends of a closed cylindrical tank produces axial stress, which is directed along the longitudinal axis of the cylinder. The axial stress σ_a of a thin-walled cylindrical pressure vessel with internal pressure p, thickness t, and inner radius r_{ic} is

$$\sigma_a = \frac{(p - p_a) r_{ic}}{2t} \tag{11.14}$$

The axial stress is half of the hoop stress.

The surface (tangential) stress of a spherical pressure vessel is the same in all directions because of the spherical symmetry of the tank. The surface stress σ_s of a thin-walled spherical pressure vessel with internal pressure p, thickness t, and inner radius r_{is} is

$$\sigma_s = \frac{(p - p_a) r_{is}}{2t} \tag{11.15}$$

Equations 11.13–11.15 can be used to estimate the maximum operating pressure for thin-walled pressure vessels. The maximum stress should equal the maximum safe operating stress for the vessel wall. Steel pressure vessels are often designed for a maximum stress of 20000 psi. That specification can vary with the alloy and the manufacturer. The maximum working pressure for a separator should be on the nameplate for the vessel.

Example 11.5 Separator Tank

A separator tank with inner radius of 20 in. is fabricated from 0.25-in.-thick steel. For an internal pressure of 250 psig, determine the hoop and axial stress in psi.

Answer

Solve using Equation 11.13. Note that 250 psig equals the difference between internal pressure and ambient, or local atmospheric, pressure:

$$\sigma_h = \frac{(p - p_a) r_{ic}}{t} = \frac{(250\,\text{psi})(20\,\text{in.})}{0.25\,\text{in.}} = 20000\,\text{psi}$$

Depending on the alloy, 250 psig may be the maximum safe operating pressure for the vessel.

11.4 SINGLE-PHASE FLOW IN PIPE

Pipes are used extensively to move oil, gas, and water on the well location and from the location to gathering facilities. We introduce the factors that influence fluid flow in pipe by considering the relatively simple case of single-phase flow in pipe. Flow downstream from the separator should be single-phase flow.

The movement of fluid in pipe can be laminar or turbulent flow. In laminar fluid flow, fluid moves parallel to the direction of bulk flow. By contrast, fluid moves in all directions relative to the direction of bulk flow when fluid flow is turbulent. For a fluid with a given density and dynamic viscosity flowing in a tube of fixed diameter, the flow regime is laminar at low flow velocities and turbulent at high flow velocities. One parameter that is often used to characterize fluid flow is Reynolds number N_{Re}.

Reynolds number expresses the ratio of inertial (or momentum) forces to viscous forces. For fluid flow in a conduit, the Reynolds number is

$$N_{Re} = \frac{\rho v D}{\mu} \tag{11.16}$$

where ρ is the fluid density, v is the bulk flow velocity, D is the tube inner diameter for flow in a tube, and μ is the dynamic viscosity of the fluid. The choice of units must yield a dimensionless Reynolds number. In *Système Internationale* (SI) units, a dimensionless Reynolds number is obtained if fluid density is in kg/m³, flow velocity is in m/s, tube diameter is in m, and dynamic viscosity is in Pa·s. Note that $1\,\text{cp} = 1\,\text{mPa}\cdot\text{s} = 10^{-3}\,\text{Pa}\cdot\text{s}$.

The Reynolds number for flow in a cylindrical pipe can be written in terms of volumetric flow rate q. The bulk flow velocity v of a single-phase fluid flowing in the cylindrical pipe is related to volumetric flow rate q by

$$v = \frac{q}{A} = \frac{4q}{\pi D^2} \tag{11.17}$$

Substituting Equation 11.17 into 11.16, then

$$N_{Re} = \frac{\rho v D}{\mu} = \frac{4\rho q}{\pi \mu D} \tag{11.18}$$

Fluid flow in cylindrical pipes is laminar if $N_{Re} < 2000$, and it is considered turbulent at larger values of the Reynolds number.

Example 11.6 Reynolds Number

Suppose oil is flowing through a circular pipe with volumetric flow rate $q = 1000$ barrels/day. The oil density is $\rho = 0.9\,\text{g/cc} = 900\,\text{kg/m}^3$ and the dynamic viscosity of oil is $\mu = 2\,\text{cp} = 0.002$ Pa·s. The pipe has a 3 in. inner diameter. Calculate Reynolds number.

Answer

First express all variables in SI units. The inner diameter in SI units is

$$D = 3\,\text{in.}\left(0.0254\,\text{m / in.}\right) = 0.0762\,\text{m}$$

Volumetric flow rate in SI units is

$$q = 1000\ \text{barrels/day}\left(0.159\ \text{m}^3/\text{barrel}\right)\left(1\ \text{day}/86400\ \text{sec}\right) = 0.00184\ \text{m}^3/\text{s}$$

Reynolds number is

$$N_{\text{Re}} = \frac{4\rho q}{\pi \mu D} = \frac{4\left(900\,\text{kg/m}^3\right)\left(0.00184\,\text{m}^3/\text{s}\right)}{\pi\left(0.002\,\text{Pa}\cdot\text{s}\right)\left(0.0762\,\text{m}\right)} = 13800$$

The relationship between fluid flow velocity and pressure change along the longitudinal axis of the cylindrical pipe is obtained by performing an energy balance calculation. Figure 11.5 shows the geometry of an inclined cylindrical pipe with length L along the longitudinal axis and angle of inclination θ. The single-phase fluid has density ρ and dynamic viscosity μ. It is flowing in a gravity field with acceleration g.

We make two simplifying assumptions in our analysis that allow us to minimize external factors and consider only mechanical energy terms. We assume that no thermal energy is added to the fluid and that no work is done on the system by its surroundings, for example, no mechanical devices such as pumps or compressors are adding energy to the system. An energy balance with these assumptions yields the pressure gradient equation

$$\frac{\Delta p}{\Delta L} = \left[\frac{\Delta p}{\Delta L}\right]_{PE} + \left[\frac{\Delta p}{\Delta L}\right]_{KE} + \left[\frac{\Delta p}{\Delta L}\right]_{\text{fric}} \tag{11.19}$$

where p is the pressure. We have written the pressure gradient along the longitudinal axis of the pipe as the sum of a potential energy term

$$\left[\frac{\Delta p}{\Delta L}\right]_{\text{PE}} = \rho g \sin\theta, \tag{11.20}$$

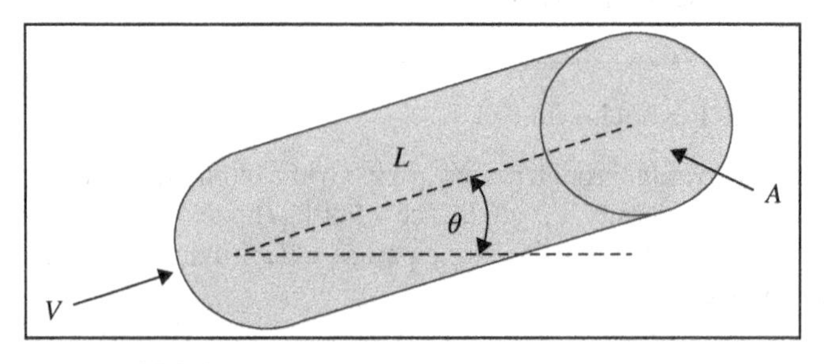

FIGURE 11.5 Flow in an inclined cylindrical pipe.

a kinetic energy term

$$\left[\frac{\Delta p}{\Delta L}\right]_{\text{KE}} = \rho v \frac{\Delta v}{\Delta L}, \tag{11.21}$$

and a friction term

$$\left[\frac{\Delta p}{\Delta L}\right]_{\text{fric}} = f \frac{\rho v^2}{2D} \tag{11.22}$$

that depends on a dimensionless friction factor f. If the flow velocity of the fluid does not change appreciably in the pipe, the kinetic energy term can be neglected and the pressure gradient equation reduces to the simpler form

$$\frac{\Delta p}{\Delta L} = \rho g \sin\theta + f \frac{\rho v^2}{2D} \tag{11.23}$$

Equation 11.23 is valid for single-phase incompressible fluid flow. If we assume the right-hand side is constant over the length L of the pipe, the pressure change from one end of the pipe to the other is

$$\Delta p = \rho g L \sin\theta + f \frac{\rho v^2}{2D} L \tag{11.24}$$

The friction factor f depends on flow regime. For laminar flow with Reynolds number $N_{\text{Re}} < 2000$, the friction factor is inversely proportional to Reynolds number:

$$f = \frac{64}{N_{\text{Re}}} \tag{11.25}$$

For turbulent flow, the friction factor depends on Reynolds number and pipe roughness. Pipe roughness can be quantified in terms of relative roughness ξ which is a fraction defined relative to the inner diameter of the pipe as

$$\xi = d_{\text{p}}/D < 1 \tag{11.26}$$

in which d_{p} is the distance of a protrusion from the pipe wall. Typical values of pipe relative roughness ξ range from 0.0001 (smooth) to 0.05 (rough). The length of protrusions inside the pipe may change during the period that the pipe is in service. For example, buildup of scale or pipe wall corrosion can change the relative roughness of the pipe. One correlation for friction factor for turbulent flow is (Beggs, 1991, page 61)

$$\frac{1}{\sqrt{f}} = 1.14 - 2\log\left(\xi + \frac{21.25}{N_{\text{Re}}^{0.9}}\right) \tag{11.27}$$

Example 11.7 Friction Factor

Calculate friction factor for fluid flowing through a pipe. Assume the relative roughness of the pipe wall is 0.00014 and Reynolds number = 13 800.

Answer

Friction factor is calculated from Equation 11.27:

$$\frac{1}{\sqrt{f}} = 1.14 - 2\log\left(0.00014 + \frac{21.25}{(13800)^{0.9}}\right)$$

Solving gives $f = 0.0287$.

11.5 MULTIPHASE FLOW IN PIPE

The description of single-phase fluid flow in pipes presented earlier is relatively simple compared to multiphase flow. In particular, two-phase flow is characterized by the presence of flow regimes or flow patterns. The flow pattern represents the physical distribution of gas and liquid phases in the flow conduit. Forces that influence the distribution of phases include buoyancy, turbulence, inertia, and surface tension. The relative magnitude of these forces depends on flow rate, the diameter of the conduit, its inclination, and the fluid properties of the flowing phases.

Flow regimes for vertical flow are usually represented by four flow regimes (Brill, 1987; Brill and Mukherjee, 1999): bubble flow, slug flow, churn flow, and annular flow. Figure 11.6 illustrates the four flow regimes. Bubble flow is the movement of gas bubbles in a continuous liquid phase. Slug flow is the movement of slug units; each slug unit consists of a gas pocket, a film of liquid surrounding the gas pocket that is moving downward relative to the gas pocket, and a liquid slug with distributed gas bubbles between two gas pockets. Churn flow is the chaotic movement of

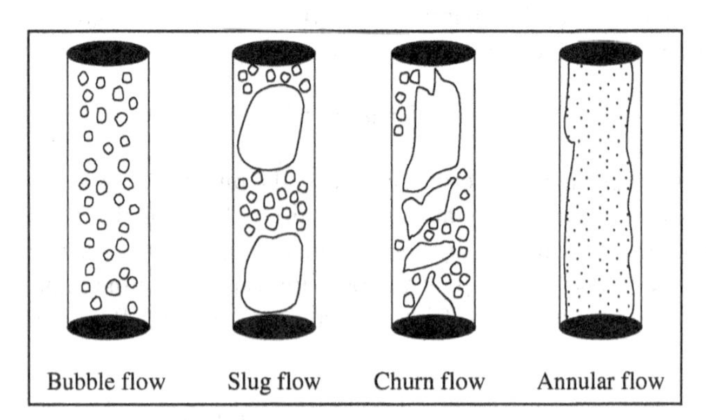

FIGURE 11.6 Flow regimes for vertical two-phase flow. (Source: Brill and Mukherjee (1999).)

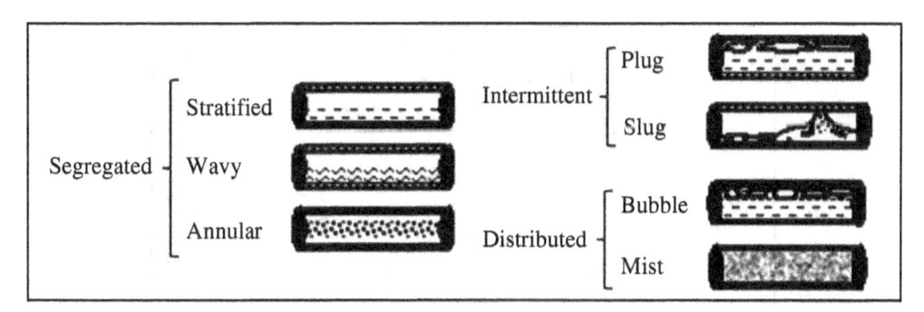

FIGURE 11.7 Flow regimes for horizontal two-phase flow. (Source: Brill and Mukherjee (1999).)

distorted gas pockets and liquid slugs. Annular flow is the upward movement of a continuous gas phase in the center of the conduit with an annular film of liquid flowing upward between the central gas phase and the wall of the conduit and with dispersed liquid droplets being lifted by the gas phase.

Following Beggs and Brill (1973), Brill and Mukherjee (1999) represent multiphase flow in horizontal conduits using the seven flow regimes shown in Figure 11.7. These flow regimes are not universally accepted. For example, the set of flow regimes used by Petalas and Aziz (2000) to model multiphase flow in pipes includes dispersed bubble flow, stratified flow, annular-mist flow, bubble flow, intermittent flow, and froth flow. Froth flow was described as a transition zone between dispersed bubble flow and annular-mist flow and between annular-mist flow and slug flow.

11.5.1 Modeling Multiphase Flow in Pipes

The identification of qualitative flow regimes discussed earlier influences the structure of analytical and numerical models used to quantify multiphase flow in pipes. The flow regimes are used to construct flow regime maps, also called flow pattern maps, which are log-log plots of superficial gas velocity versus superficial liquid velocity. Figure 11.8 illustrates a flow pattern map.

Historically, predictions of multiphase flow in pipes began in the 1950s when investigators used data from laboratory test facilities and, to a lesser extent, field data to prepare empirical flow pattern maps (Brill, 1987; Brill and Arirachakaran, 1992). Early models of multiphase flow were extrapolations of single-phase flow models. Single-phase terms in the pressure gradient equation introduced earlier were replaced with mixture variables. Thus, the terms in the pressure gradient equation for single-phase flow given by Equation 11.19 become

$$\left[\frac{\Delta p}{\Delta L}\right]_{\mathrm{PE}} = \rho_{\mathrm{m}} g \sin\theta \tag{11.28}$$

for potential energy,

$$\left[\frac{\Delta p}{\Delta L}\right]_{\mathrm{KE}} = \rho_{\mathrm{m}} v_{\mathrm{m}} \frac{\Delta v_{\mathrm{m}}}{\Delta L} \tag{11.29}$$

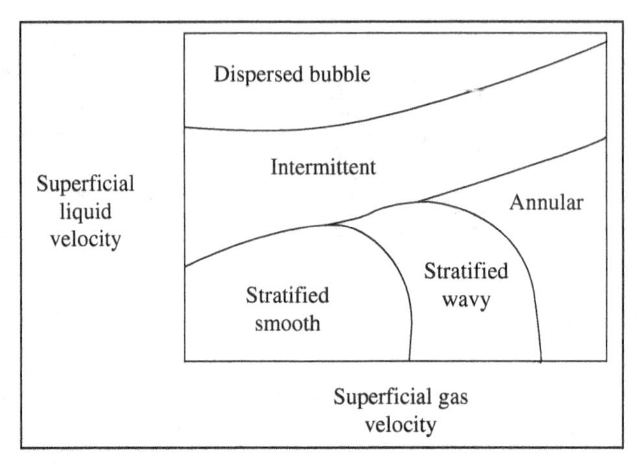

FIGURE 11.8 Illustration of a flow pattern map. (Source: Brill and Arirachakaran (1992).)

for kinetic energy, and

$$\left[\frac{\Delta p}{\Delta L}\right]_{\text{fric}} = f\frac{\rho_{\text{m}} v_{\text{m}}^2}{2D} \tag{11.30}$$

for friction. The subscript *m* attached to variables on the right-hand side of Equations 11.28 through 11.30 denotes that the associated variable is calculated for a mixture.

Models based on mixture variables are called homogeneous models. In addition to homogeneous models, two other approaches are often used: empirical correlations and mechanistic models. Empirical correlations depend on fitting experimental data and field data to models that contain groups of physical parameters. The empirical correlations approach can yield useful and accurate results quickly, but does not provide a scientific basis for extrapolation to significantly different systems. By contrast, mechanistic models are based on physical mechanisms that describe all significant flow mechanisms. Modern mechanistic modeling still requires some empiricism to determine poorly known or difficult to measure parameters.

Homogeneous models do not account for slip between fluid phases. Drift-flux models are designed to model slip between fluid phases flowing in a pipe, as well as model countercurrent flow. Countercurrent flow is the movement of heavy and light phases in opposite directions when fluid flow is slow or there is no net fluid flow in the pipe.

11.6 WELL PATTERNS

The effectiveness of a hydrocarbon recovery process depends on many factors, including factors that are beyond our control, such as depth, structure, and fluid type. Other factors that influence efficiency can be controlled, however. They include the number and type of wells, well rates, and well locations. The distribution of wells is known as the well pattern. The selection of a development plan depends on a comparison of the economics of alternative development concepts. Reservoir flow models are especially useful tools for performing these studies.

In many reservoirs, an injection program is used to push oil from injection wells to producing wells. In many reservoirs, the injected fluid is water, which is immiscible with the oil. Immiscible fluid displacement between one injection well and one production well is the simplest pattern involving injection and production wells. A variety of other patterns may be defined. Some examples are shown in Figure 11.9. A representative pattern element for the five-spot pattern is identified using shaded wells.

The ratio of the number of producing wells to the number of injection wells is shown in Table 11.1 for common well patterns. The patterns in Table 11.1 and Figure 11.9 are symmetric patterns that are especially effective for reservoirs with relatively small dip and large areal extent. The injectors and producers are generally interspersed. Other patterns in which injectors and producers are grouped together may be needed for reservoirs with significant dip. For example, a peripheral or flank injection pattern may be needed to effectively flood an anticlinal reservoir.

In addition to reservoir geometry and the displacement process, the well pattern depends on the distribution and orientation of existing production wells and the desired spacing of wells. Wells may be oriented vertically, horizontally, or at some deviation angle between horizontal and vertical. The orientation of a well depends on such reservoir features as formation orientation and, if fractures are present, fracture orientation. For example, if a reservoir contains many fractures that are oriented in a particular direction, recovery is often optimized by drilling a horizontal well in a direction that intersects as many fractures as possible. Recovery is optimized because recovery from fractured reservoirs usually occurs by producing fluid that flows from the matrix into the fractures and then to the wellbore.

Well spacing depends on the area being drained by a production well. A reduction in well spacing requires an increase in the density of production wells. The density of production wells is the number of production wells in a specified area. Well density can be increased by drilling additional wells in the space between wells in a process called infill drilling.

11.6.1 Intelligent Wells and Intelligent Fields

It is often necessary in the management of a modern reservoir to alter the completion interval in a well. These adjustments are needed to modify producing well fluid ratios such as water-oil ratio or gas-oil ratio. One way to minimize the cost associated with completion interval adjustments is to design a well that can change the completion interval automatically. This is an example of an "intelligent well."

Intelligent wells are designed to give an operator remote control of subsurface well characteristics such as completion interval. In addition, intelligent wells are being designed to provide information to the operator using downhole measurements

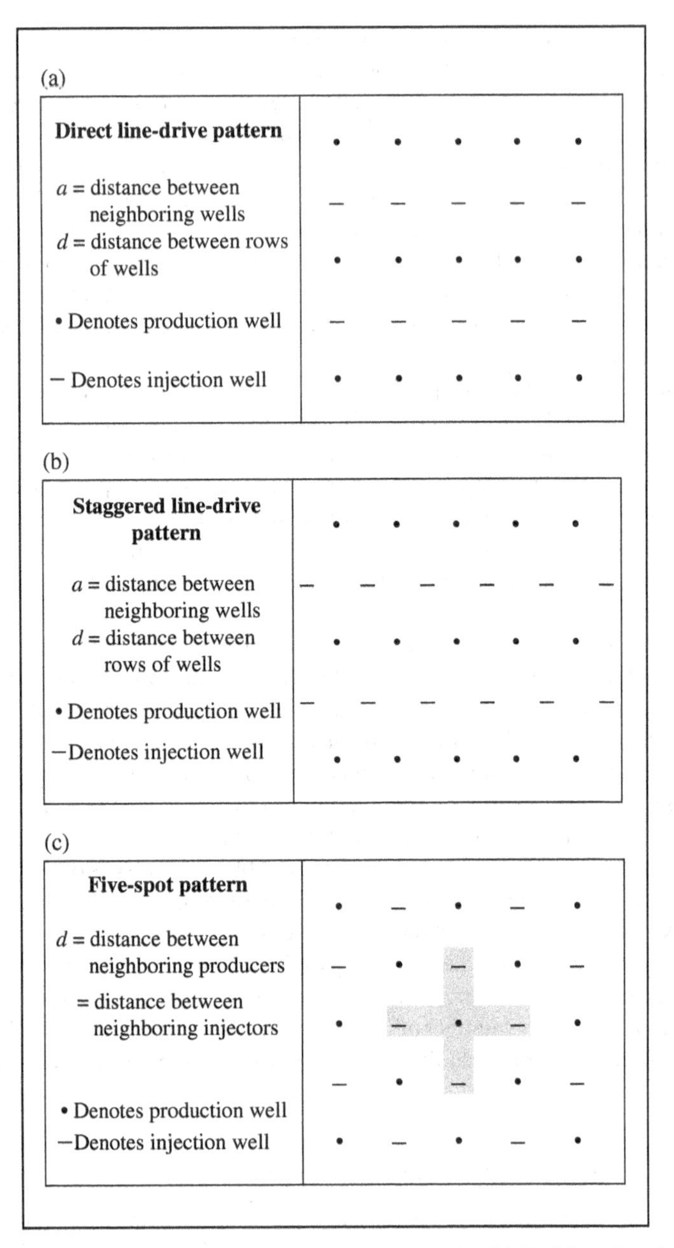

FIGURE 11.9 (a) Well locations in direct line-drive pattern. (b) Well locations in staggered line-drive pattern. (c) Well locations in five-spot pattern.

of physical properties such as pressure, temperature, and seismic vibrations. One goal of intelligent well technology is to convey a stream of continuous and real-time information to the operator who can monitor and make adjustments as needed to achieve reservoir management objectives.

TABLE 11.1 Producer-to-Injector Ratios for Common Well Patterns

Well Pattern	Producer-to-Injector Ratio
Four-spot	2
Five-spot	1
Direct line drive	1
Staggered line drive	1
Seven-spot	1/2
Nine-spot	1/3

Intelligent wells can be found in an intelligent field (i-field). An i-field is an integrated operation that uses improved information and computer technology to operate a field. The operation uses supervisory control and data acquisition (SCADA), which is computer technology designed to monitor the field. Operators are able to use the technology to respond to changing conditions in real time. Capital investment in i-field technology can be larger than in fields without i-field technology, but a good design can result in a decrease in operating expenses by increasing automation of a field and allow remote control of a field in a difficult environment. The desired result is less direct human intervention.

11.7 OFFSHORE FACILITIES

Hydrocarbon production using offshore drilling rigs in shallow water began in the early twentieth century. Today there are many types of offshore platforms as shown in Figure 11.10. Drilling jack-up rigs can be installed in varying depths of water. Jack-up rigs can be floated to drilling locations where the legs are lowered to the seabed. Fixed platforms are set on steel jackets that sit on the seabed. Pilings hammered into rock beneath the seafloor can be used to support the jacket and platform. A compliant tower has a platform set atop a scaffold. A bottom-setting platform in dry dock is shown in Figure 11.11. The platform has a derrick and must be used in relatively shallow water.

Wellheads can be installed on the seabed. Fluid produced through subsea wellheads is routed to either a platform or through flowlines in a riser to a floating rig or ship. If well depths are beyond the reach of human divers, remotely operated vehicles (ROVs) must be used to install and maintain equipment.

The definition of deep water for classifying production of oil and gas in deepwater conditions depends on the operator. Seitz and Yanosek (2015) defined deepwater production as production in water depths greater than 450 m (≈1500 ft). Subsea completions can be used in deep water. Subsea facilities in deep water are maintained using ROVs.

A key factor in selecting an offshore platform is water depth. Gravity-based platforms can be used in up to 1000 ft of water. Steel jacket platforms are typically used in up to 1500 ft of water depth. Compliant towers can be used at up to 3000 ft. Tension leg platforms can be used at depths of up to 5000 ft. Production spars can be used at

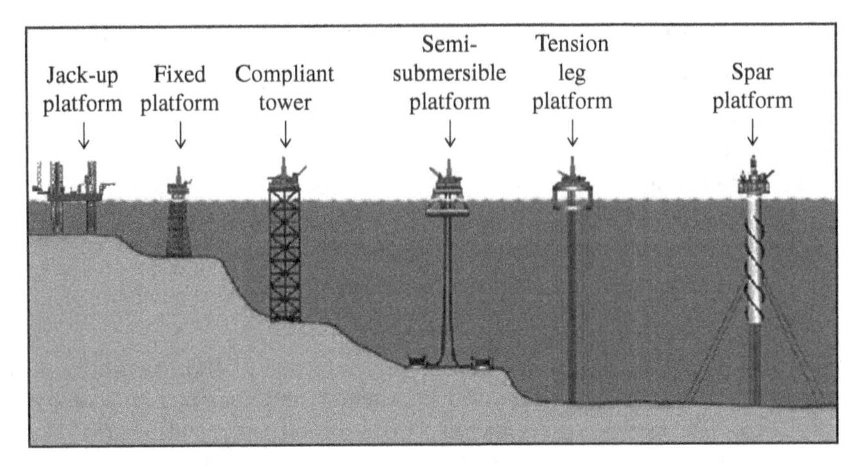

FIGURE 11.10 Examples of offshore platforms.

FIGURE 11.11 Offshore platform in dry dock, Galveston, Texas. Source: energy.fanchi.com (2003).

depths of up to 7500 ft, and floating production offloading (FPO) vessels and subsea completions can be used at arbitrary depths. ROV technology imposes a practical limit on depth since ROVs may be the only way to access equipment on the seabed.

Offshore drilling has many similarities to onshore drilling. Both use rotary drilling technology, and a top drive can be used on an offshore rig. The top drive rotates the drill string and is attached to the traveling block so it can move up and down as

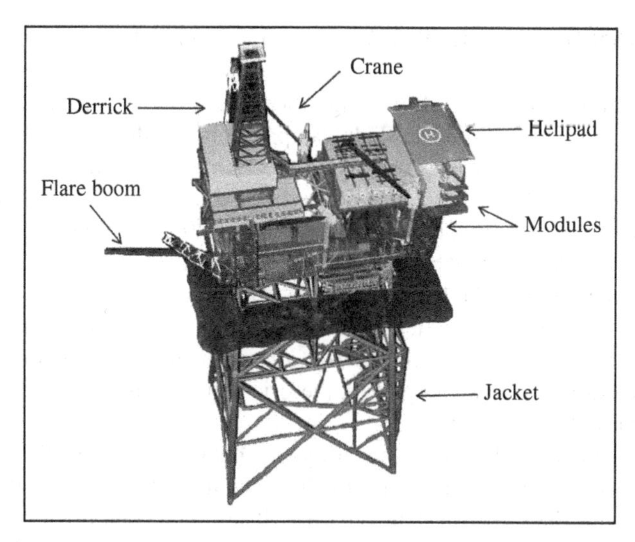

FIGURE 11.12 Key components of an offshore platform.

pipe is raised and lowered. A key difference between onshore and offshore drilling is the limited amount of space offshore for operations, such as storing pipe, installing wellheads with BOP stacks, and storing drilling mud in tanks. Offshore rigs are often supported by service vessels and helicopters. Helicopters can transport personnel, and service vessels are typically used to transport supplies and equipment to and from the platform.

Platforms often include a derrick, crane, deck, modules, and sometimes a helipad (Figure 11.12). Prefabricated modules contain offices, crew quarters, control rooms, and heavy equipment. They can be added and removed as needed. Drillships have derricks aligned over moon pools in the center of the ship. The moon pool is an opening in the hull that provides access to the water beneath the ship from the drill floor. Drillships are mobile drilling platforms and are especially useful for exploratory drilling.

Example 11.8 Platform Extent

An oil reservoir is 30 km long and 20 km wide. Can a single platform be used to develop a field this size? Assume that extended reach drilling is limited to 10 km.

Answer
A single platform cannot develop the field. The areal reach of the platform is a circular area with a 10 km radius. Other alternatives must be considered to reach the undeveloped parts of the field. For example, a second platform, a satellite platform, or subsea templates connected back to the platform with pipelines are options to consider.

11.8 URBAN OPERATIONS: THE BARNETT SHALE

The combination of hydraulic fracturing, horizontal drilling, and a reasonable gas price catalyzed the growth of the shale gas production industry. In 2002, Mitchell Energy implemented horizontal drilling in the Barnett Shale in North Texas. The Barnett Shale in the United States is up to 450 ft thick and covers an area of over 5000 mi². A vertical well, which may pass through 400 ft of shale, can extract gas from a much smaller volume than a horizontal well that could run through two miles or more of shale. Figure 11.13 shows that a horizontal well can intersect more of the reservoir when the areal extent of the reservoir is substantially larger than the thickness of the reservoir. The orientation of man-made fractures associated with hydraulic fracturing depends in part on the orientation of the well and also on the rock properties of the reservoir.

Hydraulic fracturing had been used for years, but the results had limited success. Horizontal drilling made it possible for horizontal wellbores to contact a much larger volume of the reservoir than vertical drilling. The combination of hydraulic fracturing and horizontal drilling increased the rate that gas could be produced from targets like gas-rich, low-permeability shale. Multiple wells drilled from a single well pad decrease the need for large numbers of potentially noisy and aesthetically unappealing surface well sites. With multiwell drilling sites, the number of locations that must have pipelines for transporting fluids and that must be operated and maintained is minimized, saving time and money for operators. Horizontal drilling makes it possible to locate a well site outside of a highly populated area and drill underneath the population without being seen or heard. This development has further increased interest in shale gas because a large volume of shale gas is found under large cities. A small operational footprint increases public support for urban drilling.

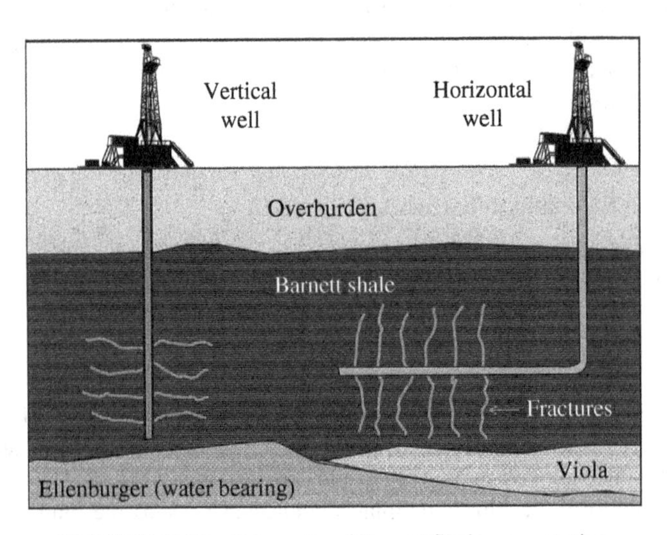

FIGURE 11.13 Schematic of Barnett Shale cross section.

11.9 ACTIVITIES

11.9.1 Further Reading

For more information about surface facilities, see Hyne (2012), Raymond and Leffler (2006), and van Dyke (1997).

11.9.2 True/False

11.1 A semisubmersible is a deepwater drilling platform that floats on pontoons.

11.2 Dynamic positioning is used to keep an offshore drilling rig on station by computer.

11.3 A barge, jack-up, or semisubmersible is often used to work over deepwater wells.

11.4 In a two-phase separator, gas rises to the top of the separator while the liquid falls to the bottom.

11.5 The wellhead includes the casing head, tubing head, and conductor pipe.

11.6 Helipads are required on all offshore platforms.

11.7 Oil and gas storage tanks are considered pressure vessels.

11.8 Crude oil is converted into transportation fuel at the separator.

11.9 Fluid flow in pipe depends on the friction of the pipe.

11.10 Reynolds number is the dimensionless ratio of inertial forces to viscous forces.

11.9.3 Exercises

11.1 Estimate the k-value for propane at 100°F and 150 psia. Use $P_c = 616$ psia, $T_c = 206$°F, and $\omega = 0.15$.

11.2 Perform the two-phase check on a mixture with 80 mol% methane, 15 mol% propane, and 5 mol% n-pentane using the k-values in Example 11.3 for 100°F and 150 psia.

11.3 Perform the two-phase check on a mixture with 1 mol% methane, 9 mol% propane, and 90 mol% n-pentane using the k-values in Example 11.3 for 100°F and 150 psia.

11.4 A spherical gas tank is fabricated by bolting two 0.5-in.-thick steel hemispheres together. If the gas is at a pressure of 500 psi, determine the inner and outer radii of the spherical tank if the maximum allowed surface stress is 10 ksi (1 ksi = 1000 psi).

11.5 Calculate the volume in a section of cylindrical pipe with inner diameter ID = 6 in. and length $L = 30$ ft. Express your answer in ft^3.

11.6 **A.** A platform can drill up to 32 wells with a production capacity of 12 000 bbl liquid/day/well. What is the maximum liquid flow rate possible if all wells are producing at full capacity?

B. The pipeline from the platform to a gathering center onshore can carry 400 000 bbl liquid/day. Can the platform produce at full liquid flow capacity?

11.7 Suppose water is flowing through a circular pipe with volumetric flow rate $q = 1000$ barrels/day. The water density is $\rho = 1\,\text{g/cc} = 1000\,\text{kg/m}^3$ and the dynamic viscosity of water is $\mu = 1\,\text{cp} = 0.001\,\text{Pa·s}$. The pipe length is 8000 ft and has a 5 in. inner diameter. Calculate Reynolds number.

11.8 Calculate friction factor for fluid flowing through a pipe. Assume the relative roughness of the pipe wall is 0.000144 and Reynolds number = 18 400.

11.9 **A.** A cylindrical steel pipe with 48 in. inner diameter is 0.25 in. thick. The pipe carries water under an internal pressure $p = 50\,\text{psi}$. The circumferential stress in the steel is $S = pd/2t$ where p is internal pressure, d is inner diameter of pipe, and t is pipe thickness. Calculate circumferential stress in the steel S.

B. If the pressure is raised to 100 psi in the 48-in. inner-diameter steel pipe, what thickness of steel would be required for an allowable circumferential stress $S = 10000\,\text{psi}$? Hint: Solve $S = pd/2t$ for t.

C. Do you need to replace the steel pipe?

11.10 **A.** Suppose 100 000 gallons of oil are spilled at sea. How many barrels of oil were spilled?

B. If the oil has a specific gravity of 0.9, determine the mass of oil spilled. Express your answer in kg and lbm. Hint: Find density of oil ρ_o from oil specific gravity γ_o and water density ρ_w (1g/cc $= 1000\,\text{kg/m}^3$), where $\rho_o = \rho_w \gamma_o$. Mass of oil spilled = volume spilled × density.

C. Estimate the area covered by the spill if the thickness of the spill is 1 mm. Express your answer in m², km², mi², and acres. Use area covered by spill = volume spilled/thickness of spill.

12

TRANSIENT WELL TESTING

Transient well tests measure changes in reservoir pressure associated with changes in well rates. Measurement and analysis of the time-dependent pressure response can provide information about reservoir structure and the expected flow performance of the reservoir. Transient well tests often motivate changes in the way the well is operated or the field is managed. This chapter discusses a selection of transient well tests for oil wells and gas wells and illustrates how transient well test information is used.

12.1 PRESSURE TRANSIENT TESTING

Pressure transient testing (PTT) is a widely used method for obtaining information about the reservoir far from the well. The PTT method relies on measuring the changes in pressure at the wellbore as a function of time that accompany fluid flow rate changes in the wellbore. Changes in pressure are known as pressure transients. For example, the flow rate at a production or injection well can be increased or decreased. Pressure gauges measure the resulting pressure changes as a function of time. Plots of pressure and a time derivative of pressure as functions of time are prepared. The interpretation of these plots gives information about fluid flow through different parts of the reservoir.

A pressure transient test can be used to estimate flow capacity (permeability times thickness) at the well, average reservoir pressure in the drainage area of the well, the distance to reservoir boundaries and faults from the well, and formation damage in

Introduction to Petroleum Engineering, First Edition. John R. Fanchi and Richard L. Christiansen.
© 2017 John Wiley & Sons, Inc. Published 2017 by John Wiley & Sons, Inc.
Companion website: www.wiley.com/go/Fanchi/IntroPetroleumEngineering

the vicinity of the well. Flow capacity and formation damage measurements indicate the quality of a completion. Some pressure transient tests provide fluid samples that are suitable for measuring fluid properties.

The integration of information from PTT with data from other sources can be used to help characterize the reservoir. For example, flow rate changes at one well can lead to pressure changes that can be observed at another well if the reservoir does not have an impermeable barrier between the two wells. This gives an idea about the continuity of the flow path between the two wells. Another example of integrating data is the combination of PTT flow capacity with formation thickness. In this case, an estimate of formation permeability in the region of the well investigated by the pressure transient test is estimated by dividing flow capacity from the pressure transient test by the net thickness of the formation from a well log.

12.1.1 Flow Regimes

Figure 12.1 displays different flow regimes at the well resulting from a rate change. The circle on the well pressure axis is initial pressure p_i. The early time response is dominated by wellbore and near-wellbore effects. Wellbore effects are associated with the volume of the wellbore, while near-wellbore effects are associated with the quality of the borehole wall. The wellbore volume is known as wellbore storage, and the quality of the borehole wall is quantified by a variable known as skin.

The transition region is the period of time between wellbore-dominated flow and the infinite-acting region. The infinite-acting region is the period of time when the pressure response behaves as if it is not affected by either the wellbore (inner boundary) or outer boundary. A departure from the trend set during the infinite-acting region indicates that the pressure transient is being affected by the reservoir boundary.

12.1.2 Types of Pressure Transient Tests

Well flow rate can be either increased or decreased in both production wells and injection wells. Consequently, the four possible types of pressure transient tests

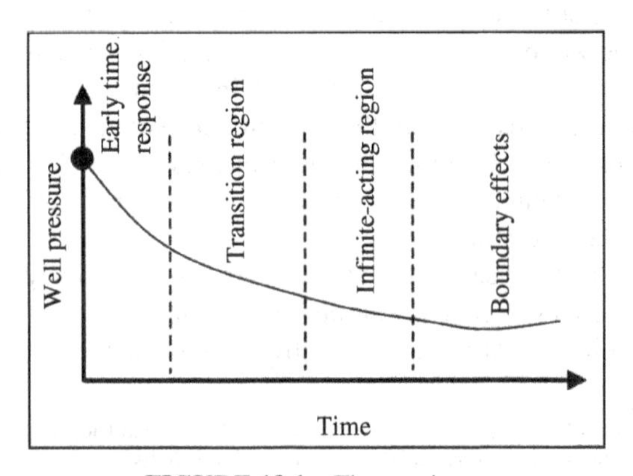

FIGURE 12.1 Flow regimes.

TABLE 12.1 Types of Pressure Transient Test

Well Type	Change in Flow Rate	Pressure Transient Test
Production well	Decrease	Pressure buildup
	Increase	Pressure drawdown
Injection well	Decrease	Pressure falloff
	Increase	Injectivity

associated with rate changes in production and injection wells are summarized in Table 12.1. Wellbore pressure is lower than formation pressure in production wells. If the production well is shut in, the pressure in the wellbore will increase. The corresponding PTT is called a pressure buildup (PBU) test. If the production flow rate is increased, the wellbore pressure will decrease. In this case, the corresponding PTT is called a pressure drawdown test.

In the case of injection wells, the wellbore pressure is higher than formation pressure. If the injection well is shut in, the pressure in the wellbore will decrease, and the corresponding PTT is called a pressure falloff test. By contrast, an injectivity test is run by increasing the injection rate and monitoring the associated increase in wellbore pressure. Injectivity tests can be run to confirm that an injection well can achieve a desired injection rate.

Many transient well tests are distinguished by the way flow rate is changed. Flow rate can be changed incrementally or in abrupt steps. A decrease from the existing flow rate to shut-in is an abrupt change in rate and occurs in PBU and pressure falloff tests. On the other hand, incremental changes from one rate to another are used in gas well deliverability tests. Deliverability tests in gas wells provide information about the flow capacity of gas wells and are discussed later.

Tests designed to measure the change in pressure at one well as a result of an incremental change in rate at another well are called pressure pulse tests or interference tests. In this case, a change in flow rate at one well can induce a change in pressure at another well if the wells are in pressure communication. Pressure communication between wells means that a pressure change at one well can be detected at another well.

12.2 OIL WELL PRESSURE TRANSIENT TESTING

The behavior of a pressure transient induced in the formation by a change in flow rate is modeled using the diffusivity equation for a single-phase liquid in a homogeneous medium. The diffusivity equation in radial coordinates is the partial differential equation

$$\frac{\partial^2 p}{\partial r_{\mathrm{D}}^2}+\frac{1}{r_{\mathrm{D}}}\frac{\partial p}{\partial r_{\mathrm{D}}}=\frac{\partial p}{\partial t_{\mathrm{D}}} \tag{12.1}$$

where p is the fluid pressure, r_D is the dimensionless radius, and t_D is the dimensionless time. Dimensionless radius is defined as

$$r_D \equiv \frac{r}{r_w} \tag{12.2}$$

where r is the radial distance from the well (ft) and r_w is the well radius (ft). Dimensionless time is defined in terms of a group of parameters, namely,

$$t_D \equiv 0.000264 \frac{kt}{\phi(\mu c_T)_i r_w^2} \tag{12.3}$$

where k is the permeability (md), t is the time (hr), ϕ is the porosity (fraction), μ is the viscosity of the mobile, single-phase liquid (cp), and c_T is the total compressibility (psia^{-1}). The group of parameters $k/\phi\mu c_T$ is called the diffusivity coefficient. The subscript i is used to show that liquid viscosity μ and total compressibility c_T are evaluated at initial pressure. Dimensionless radius increases as radial distance increases, and dimensionless time increases as time increases.

The diffusivity equation is based on several assumptions: the equation applies to single-phase flow of a slightly compressible liquid; the liquid flows in the horizontal, radial direction only; and changes in formation and liquid properties are negligible. In practice, formation and liquid properties do depend on pressure and can experience slight changes as pressure changes. As a matter of consistency, initial values are used to analyze pressure transient tests. The horizontal flow assumption implies that gravity effects are neglected. In addition, it is assumed that flow rate is proportional to pressure gradient so that Darcy's law applies. A consequence of the assumptions is that the well is able to produce or inject liquid throughout the thickness of the formation so that flow is only along the radial direction. Total system compressibility is

$$c_T = c_r + c_o S_o + c_w S_w + c_g S_g \tag{12.4}$$

where c_r is the rock compressibility and fluid-phase compressibility is the saturation-weighted average of oil-, water-, and gas-phase compressibilities.

The following discussion applies to any well satisfying the previous assumptions. In our case, we focus on PTT of oil wells, but similar techniques can be applied to PTT of water wells. Gas wells are discussed in later sections.

The diffusivity equation is solved by treating the well as a line source with constant flow rate. The line source assumption implies that the volume of a real well does not have a significant impact on our model of the pressure transient test. In general, this is a reasonable assumption because wells are usually only a few inches in diameter and the pressure transient test can probe the reservoir to a radial distance of hundreds of feet. The volume of the well does impact early time pressure response, as noted previously, and must be considered when interpreting PTT results.

Solutions of the diffusivity equation depend on the specified boundary conditions. During the infinite-acting time, flow in a reservoir behaves as if it does not have an outer boundary and the reservoir can be treated like an infinite reservoir. The solution of the diffusivity equation for flow of oil in a porous medium with outer boundary at $r_D \to \infty$ is

$$p_D(r_D, t_D) = -\frac{1}{2}\left[\text{Ei}\left(\frac{r_D^2}{4t_D}\right)\right] \quad (12.5)$$

The term Ei(···) is the exponential integral

$$-\text{Ei}(-x) = \int_x^\infty \frac{e^{-u}}{u}\,du \quad (12.6)$$

Dimensionless pressure p_D in oil field units is

$$p_D = \frac{kh}{141.2qB\mu}(p_i - p_{wf}) \quad (12.7)$$

where k is the permeability (md), h is the thickness (ft), q is the flow rate (STB/d), B is the formation volume factor (RB/STB), μ is the viscosity (cp), p_i is the initial reservoir pressure (psia), and p_{wf} is the well flowing pressure (psia).

Equation 12.5 is valid from the outer diameter of the well at dimensionless radius $r_D = 1$ to the outer boundary of the reservoir at $r_D \to \infty$. If we use an approximation of the exponential integral solution that is valid when $t_D / r_D^2 > 10$ at $r_D = 1$ and express variables in oil field units, we obtain

$$p_{wf} = p_i - 162.6\frac{qB\mu}{kh}\left(\log t + \log\frac{k}{\phi\mu c_T r_w^2} + 0.87S - 3.23\right) \quad (12.8)$$

Skin typically ranges from $-5 < S < 50$. A positive value of skin S indicates well damage. If skin S is negative, it suggests that the well is stimulated.

Example 12.1 Dimensionless Time

Calculate dimensionless time given the following data: wellbore radius $r_w = 0.25$ ft, permeability $k = 150$ md, time $t = 2$ hr, porosity $\phi = 0.14$, initial liquid viscosity $\mu = 0.9$ cp, and initial total compressibility $c_T = 8 \times 10^{-6}$ psia^{-1}.

Answer

Solve $t_D \equiv 0.000264\dfrac{kt}{\phi(\mu c_T)_i r_w^2}$ using the previous data:

$$t_D = 0.000264\frac{150\,\text{md} \times 2\,\text{hr}}{0.14 \times 0.9\,\text{cp} \times \left(8 \times 10^{-6}/\text{psia}\right) \times (0.25\,\text{ft})^2} = 1.26 \times 10^6$$

12.2.1 Pressure Buildup Test

The concepts of pressure transient test analysis can be exhibited by outlining the analysis for the PBU test. The PBU test measures the pressure response at the well when the well is shut in after the well has been producing. It is necessary to have an average flow rate for the production period. Flow rates in real wells tend to fluctuate. An average flow rate q is determined by producing the well for a period t_F; thus

$$q(\mathrm{STB/D}) = \frac{\text{cumulative production}(\mathrm{STB})}{t_F} \tag{12.9}$$

After the well is shut in, the elapsed time measured from shut-in is denoted Δt. Bottom-hole pressure is recorded as a function of time during the shut-in period.

The PBU test is analyzed using the superposition principle. The superposition principle says that the total pressure change at the location of the shut-in well is equivalent to a linear sum of the changes in pressure at two wells: the actual well and an imaginary well. Both the actual well and imaginary well are located at the same location. The pressure change at the real well is due to the production rate q for the entire period of the test plus the production rate $-q$ from the imaginary, or image, well for the shut-in period of the test. The rates for the PBU test are shown in Figure 12.2. The solid line in the figure is the total rate obtained by adding the rates of the real well and the image well.

Dimensionless pressure is a linear function of the logarithm of time during the infinite-acting time period as in Equation 12.8; thus

$$p_D = \frac{1}{2}\left[\ln(t_D) + \text{constant}\right] \tag{12.10}$$

Applying the superposition principle and using oil field units give the shut-in pressure at the well p_{ws} (psia) as

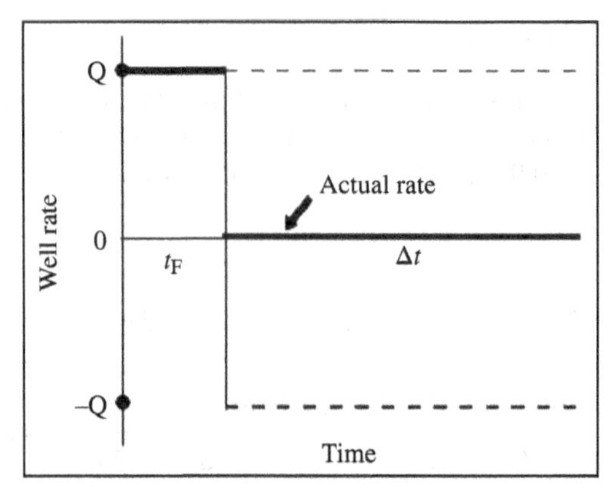

FIGURE 12.2 Rates for the PBU test.

$$p_{ws} = p_i - 141.2\frac{qB\mu}{kh}\left[p_D\left(\left[t_F + \Delta t\right]_D\right) - p_D\left(\Delta t_D\right)\right] \tag{12.11}$$

where p_i is the initial reservoir pressure (psia), q is the stabilized flow rate (STB/D), B is the formation volume factor (RB/STB), μ is the viscosity (cp), k is the permeability (md), and h is the net thickness (ft). The term $p_D([t_F + \Delta t]_D)$ is dimensionless pressure evaluated at dimensionless time $[t_F + \Delta t]_D$, and the term $p_D(\Delta t_D)$ is dimensionless pressure evaluated at dimensionless time Δt_D. Dimensionless pressure in Equation 12.11 is replaced by Equation 12.10 to obtain

$$p_{ws} = p_i - m\log\left(t_H\right) \tag{12.12}$$

where t_H is the dimensionless Horner time:

$$t_H = \frac{t_F + \Delta t}{\Delta t} \tag{12.13}$$

The hour is the unit of both times t_F, Δt to be consistent with the unit of time used in the dimensionless time calculation. The variable m is given by

$$m \equiv 162.6\frac{qB\mu}{kh} \tag{12.14}$$

The unit of m is psia per log cycle when pressure is in psia, q is stabilized flow rate (STB/D), B is formation volume factor (RB/STB), μ is viscosity (cp), k is permeability (md), and h is net thickness (ft). The concept of log cycle is clarified later.

Example 12.2 Horner Time

A well flows for 8 hr with a stabilized rate before being shut in. Calculate Horner time at a shut-in time of 12 hr.

Answer

The previous data gives Horner time $t_H = \frac{t_F + \Delta t}{\Delta t} = \frac{8+12}{12} = 1.667$.

Equation 12.12 is the equation of a straight line if we plot p_{ws} versus the logarithm of Horner time on a semilogarithmic plot. The slope of the straight line is $-m$. An example of the semilogarithmic plot for a PBU test is shown in Figure 12.3. The plot is called a Horner plot.

The early time behavior of a Horner plot is displayed at large Horner times on the right-hand side of the plot. Later times correspond to smaller Horner times on the left-hand side of the plot. The wellbore storage effect appears as the rapid increase of shut-in pressure at early times in Figure 12.3 and does not end until Horner time ≈ 25. The infinite-acting period appears as the straight line in the Horner time

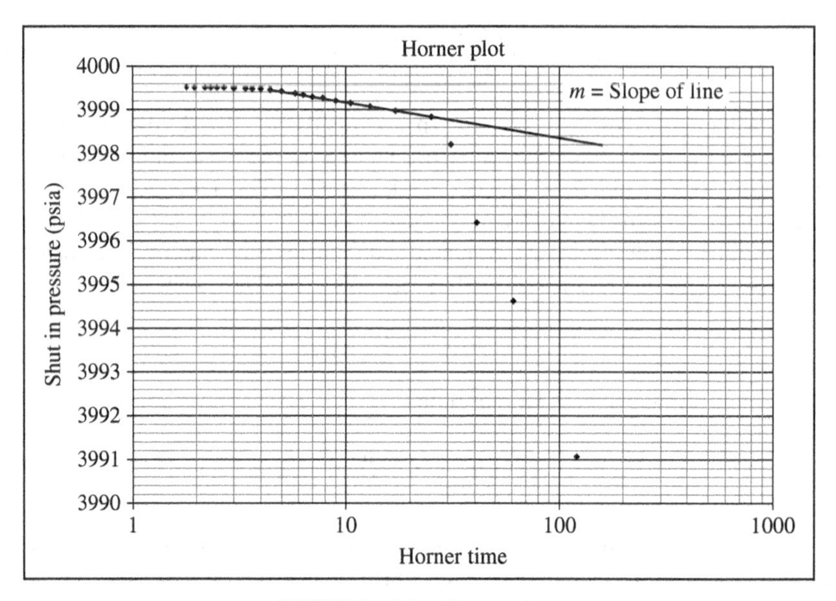

FIGURE 12.3 Horner plot.

ranges from ~25 to ~4. The slope of the straight line drawn through the infinite-acting period is the slope m where m has units of psia/log cycle. The value of m is the difference in pressure over a log cycle. In Figure 12.3, the slope m is approximately 0.8 psia/log cycle for the log cycle between Horner time = 10 and 100. In this PBU example, a reservoir boundary effect is observed for Horner times less than 4 when the slope of the infinite-acting line changes.

The flow capacity in the region of investigation of the transient well test is obtained by rearranging Equation 12.14 to obtain

$$kh \equiv 162.6 \frac{qB\mu}{m} \tag{12.15}$$

The slope m is obtained from the Horner plot. Net formation thickness h obtained from a well log can be used with the flow capacity kh from Equation 12.15 to estimate permeability in the region of investigation. The permeability estimate from transient well testing represents a larger sample volume than laboratory measurements of permeability using cores.

Dimensionless skin factor S can be estimated from the Horner plot as

$$S = 1.151 \left[\frac{p_{1\,\text{hr}} - p_{\text{wf}}(t_{\text{F}})}{m} - \log\left(\frac{k}{\phi \mu c_{\text{T}} r_{\text{w}}^2} \right) + 3.23 \right] \tag{12.16}$$

where $p_{\text{wf}}(t_{\text{F}})$ is the well flowing pressure at the end of the stabilized flow period t_{F} and $p_{1\,\text{hr}}$ is the pressure of the infinite-acting period at shut-in time $\Delta t = 1$ hr. It is often necessary to extrapolate the straight line drawn through the infinite-acting period to the Horner time at shut-in time $\Delta t = 1$ to determine $p_{1\,\text{hr}}$.

Example 12.3 Permeability from the Horner Plot

Calculate the permeability of a drainage area based on a pressure buildup test in an oil well with the following data: flow rate $q = 78$ STB/day, oil formation volume factor $B = 1.13$ RB/STB, oil viscosity $\mu = 0.7$ cp, formation thickness $h = 24$ ft, and Horner plot slope $m = 10$ psia/log cycle.

Answer

Solve $kh \equiv 162.6 \frac{qB\mu}{m}$ for permeability; thus

$$k = 162.6 \frac{qB\mu}{mh} = 162.6 \frac{78 \times 1.13 \times 0.7}{10 \times 24} \approx 42 \text{ md}$$

12.2.2 Interpreting Pressure Transient Tests

Pressure transient tests can give insight into flow patterns within the reservoir. Some common flow patterns are illustrated in Figure 12.4: radial flow in a horizontal plane, linear flow in a vertical plane, and spherical flow. Flow rate depends on the difference in pressure between the well and the reservoir, and the arrows in the figure indicate the direction of flow.

Flow patterns can be inferred from a log–log diagnostic plot. This plot is prepared by plotting pressure versus time on a log–log scale. It also presents a plot of a function known as the pressure derivative as a function of time on a log–log scale. The "pressure derivative" function is the product of shut-in time Δt and the derivative of pressure with respect to shut-in time, namely, $\Delta t \cdot dp/d(\Delta t)$ or $dp/d \ln(\Delta t)$. The pressure derivative is a more sensitive indicator of reservoir and flow characteristics than the pressure response.

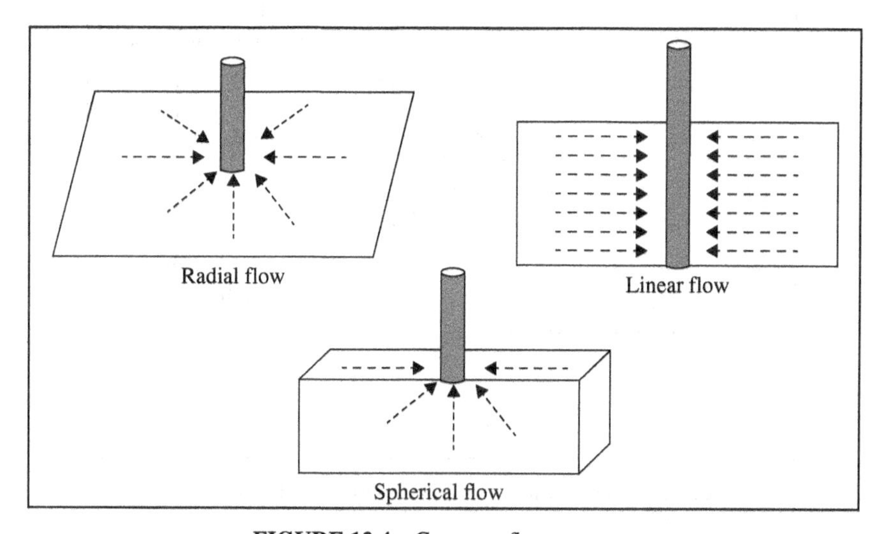

FIGURE 12.4 Common flow patterns.

An example of a log–log diagnostic plot is presented in Figure 12.5 for a partially completed well. The flow pattern for a partial completion is presented as the spherical flow pattern in Figure 12.4. The spherical flow pattern appears after the wellbore storage pressure response. The pressure response does not show as much structure as the pressure derivative response labeled $dp/d\ \ln(\Delta t)$. The spherical flow pattern does not continue until the end of the test. At some point in time spherical flow changes to radial flow as fluid flow encounters the upper and lower boundaries of the reservoir. The radial flow pattern appears as the pressure derivative response with a zero slope.

Figure 12.6 illustrates the slope of the pressure derivative curve on a log–log diagnostic plot for wellbore storage, linear flow, radial flow, and spherical flow. PTT analysts look for these slopes when they are interpreting PTT results.

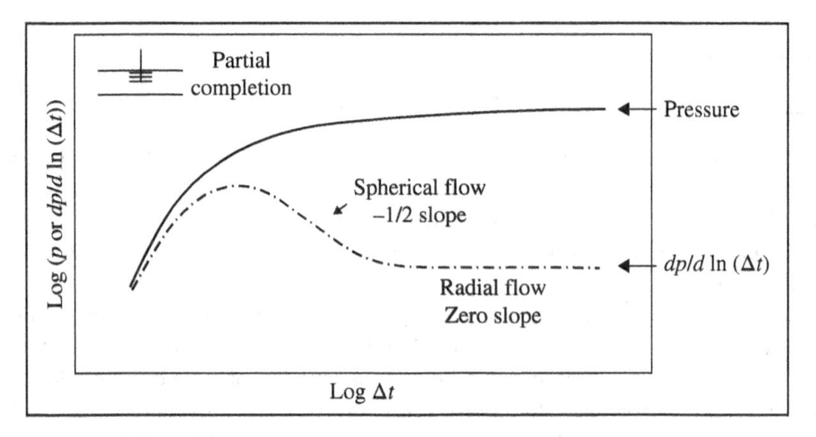

FIGURE 12.5 Log–log diagnostic plot of partially completed well.

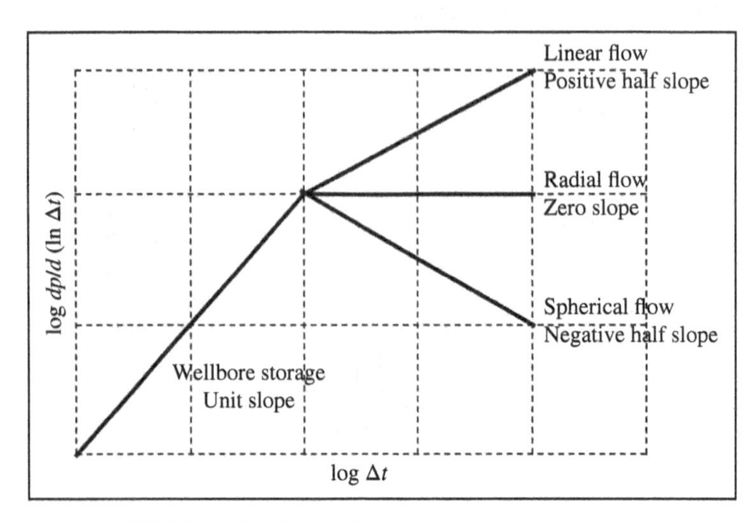

FIGURE 12.6 Slopes of flow patterns on a log–log plot.

12.2.3 Radius of Investigation of a Liquid Well

The change in pressure induced by a change in flow rate at the well travels through the reservoir at a rate that is a function of the properties of reservoir rock and fluid. The distance traveled by the pressure transient away from the wellbore depends on the elapsed time following the change in flow rate. This distance is known as radius of investigation. The Horner analysis of a PBU test gives the following equation for radius of investigation:

$$r_{\text{inv}} = 0.029\sqrt{\frac{k\Delta t}{\phi\mu c_{\text{T}}}} \tag{12.17}$$

where radius of investigation r_{inv} is in feet, Δt is the shut-in time (hr), k is the permeability (md), ϕ is the porosity (fraction), μ is the viscosity (cp) of the mobile liquid phase, and c_{T} is the total compressibility (1/psia).

Radius of investigation is an approximation of the distance traveled by the pressure transient. This can be useful in reservoir characterization because it can be used to estimate the distance to a feature in the reservoir that causes a change in the slope of the pressure measured at the well. The shut-in time corresponding to the change in slope can be used to estimate the distance to the feature that caused the change in slope. Features that can change the slope of the pressure response include no-flow barriers such as sealing faults or permeability pinch-outs. The radius of investigation gives information about the distance to the feature. The interpretation of the feature should be consistent with all available data.

Example 12.4 Oil Well Radius of Investigation

Calculate radius of investigation of a pressure buildup (PBU) test in an oil well when the shut-in time (Δt) is 24 hr. Use the following data: permeability $k = 150$ md, porosity $\phi = 0.14$, initial liquid viscosity $\mu = 0.9$ cp, and initial total compressibility $c_{\text{T}} = 8 \times 10^{-6}\,\text{psia}^{-1}$.

Answer

$$r_{\text{inv}} = 0.029\sqrt{\frac{k\Delta t}{\phi\mu c_{\text{T}}}} = 0.029\sqrt{\frac{150\,\text{md} \times 24\,\text{hr}}{0.14 \times 0.90\,\text{cp} \times 8.0 \times 10^{-16}\,/\text{psia}}} \approx 1733\,\text{ft}$$

12.3 GAS WELL PRESSURE TRANSIENT TESTING

PTT of a gas well relies on the same general principles as PTT of an oil well. Gas flow rate is changed, and the resulting change in pressure at the well is recorded as a function of time. Results of gas well PTT can be used for two purposes: to determine reservoir characteristics and to determine gas well deliverability.

Gas well deliverability tests are discussed in the next section, and PTT for characterizing gas flow in the reservoir is discussed here. Conversion factors correspond to standard temperature 60°F = 520°R and standard pressure 14.7 psia.

12.3.1 Diffusivity Equation

The starting point of gas well PTT is a diffusivity equation. Pressure was the dependent function in the diffusivity equation for oil well PTT. By contrast, the diffusivity equation for gas well PTT must account for nonlinear gas properties. This can be done by expressing the diffusivity equation for gas flow in terms of real gas pseudopressure $m(p)$ (psia²/cp), which is defined as

$$m(p) = 2\int_{p_{\mathrm{ref}}}^{p} \frac{p'}{\mu Z} dp' \tag{12.18}$$

where p_{ref} is the reference pressure (psia), p' is the dummy variable (psia), Z is the real gas compressibility factor (fraction), and μ is the gas viscosity (cp). Real gas pseudopressure $m(p)$ accounts for the pressure dependence of Z and μ.

The diffusivity equation for single-phase gas flow in the radial direction is

$$\frac{\partial^2 m(p)}{\partial r_{\mathrm{D}}^2} + \frac{1}{r_{\mathrm{D}}} \frac{\partial m(p)}{\partial r_{\mathrm{D}}} = \frac{\partial m(p)}{\partial t_{\mathrm{D}}} \tag{12.19}$$

Dimensionless radius and dimensionless time are defined as

$$r_{\mathrm{D}} \equiv \frac{r}{r_{\mathrm{w}}};\ t_{\mathrm{D}} \equiv 0.000264 \frac{kt}{\phi(\mu c_{\mathrm{T}})_{\mathrm{i}} r_{\mathrm{w}}^2} \tag{12.20}$$

where r is the radial distance from the well (ft), r_{w} is the well radius (ft), k is the permeability (md), t is the time (hr), ϕ is the porosity (fraction), μ is the gas viscosity (cp), and c_{T} is the total compressibility (psia^{-1}). The subscript i shows that gas viscosity μ and total compressibility c_{T} are evaluated at initial pressure. Total system compressibility is

$$c_{\mathrm{T}} = c_{\mathrm{r}} + c_{\mathrm{o}} S_{\mathrm{o}} + c_{\mathrm{w}} S_{\mathrm{w}} + c_{\mathrm{g}} S_{\mathrm{g}} \tag{12.21}$$

where c_{r} is the rock compressibility and fluid-phase compressibility is the saturation-weighted average of oil-, water-, and gas-phase compressibilities. If all three phases are present, the diffusivity equation for single-phase gas flow is based on the assumption that gas is the only mobile phase.

12.3.2 Pressure Buildup Test in a Gas Well

We illustrate analysis of gas well PTT by analyzing a PBU test in a gas well. A gas flow rate is maintained at a stabilized rate for a duration t_{F}. The well is shut in after the stabilized flow period, and the pressure response at the well is recorded as a function of shut-in time Δt. The superposition principle is combined with the observation that dimensionless real gas pseudopressure increases linearly with the logarithm of time during the infinite-acting period to find the solution of the real gas diffusivity equation as

$$m(p_{ws}) = m(p_i) + \frac{1637qT}{kh}\left[\log\left(\frac{kt}{\phi(\mu c_T)r_w^2}\right) - 3.23 + 0.869S'\right] \tag{12.22}$$

$$S' \equiv S + D|q|$$

where $m(p_{ws})$ is the dimensionless real gas pseudopressure evaluated at well shut-in pressure p_{ws}, $m(p_i)$ is the dimensionless real gas pseudopressure evaluated at initial pressure p_i, q is the surface flow rate of gas (MSCF/D), T is the reservoir temperature (°R), k is the permeability (md), h is the formation thickness (ft), S is the skin (dimensionless), and D is the non-Darcy flow coefficient ((MSCF/D)$^{-1}$). The skin factor S is the skin factor S characterizing formation damage plus a factor proportional to gas rate q that accounts for turbulent gas flow at high flow rates. Application of these equations to the PBU case gives

$$m(p_{ws}) = m(p_i) + 1637\frac{qT}{kh}\log(t_H) \tag{12.23}$$

where t_H is the dimensionless Horner time:

$$t_H = \frac{t_F + \Delta t}{\Delta t} \tag{12.24}$$

Results of the buildup test are analyzed by first plotting $m(p_{ws})$ versus Horner time $(t_F + \Delta t)/\Delta t$ on a semilogarithmic Horner plot. Flow capacity

$$kh = 1637\frac{qT}{m} \tag{12.25}$$

is estimated from the slope m of the Horner plot. The slope m should not be confused with real gas pseudopressure $m(p)$. The skin factor S is given by

$$S' = 1.151\left[\frac{m(p_{1hr}) - m(p_{wf})}{m} - \log\left(\frac{k}{\phi\mu c_T r_w^2}\right) + 3.23\right] \tag{12.26}$$

where $m(p_{1\,hr})$ is real gas pseudopressure extrapolated to a shut-in pressure of 1 hr.

12.3.3 Radius of Investigation

The radius of investigation r_{inv} for a PBU test in a gas well is an estimate of the distance the pressure transient moves away from the well during shut-in time Δt. It is estimated using

$$r_{inv} = 0.0325\sqrt{\frac{k\Delta t}{\phi\mu c_T}} \tag{12.27}$$

for $r_{inv} \le r_e$ where r_{inv} is the radius of investigation (ft), r_e is the drainage radius (ft), k is the permeability (md), Δt is the shut-in time (hr), ϕ is the porosity (fraction), μ is the gas viscosity (cp), and c_T is the total compressibility (1/psia). The radius of investigation for gas wells has the same functional dependence as the radius of investigation for oil wells, but the coefficient is larger for a gas well than an oil well.

Example 12.5 Gas Well Radius of Investigation

Calculate radius of investigation of a pressure buildup (PBU) test in a shale gas well when the elapsed time (t) is 1, 7, 30, and 365 days. Use the following data: permeability $k=0.0001$ md, porosity $\phi=0.05$, initial gas viscosity $\mu=0.02$ cp, and initial total compressibility $c_T=1\times10^{-4}$ per psia.

Answer

Radius of investigation at 1 day = 24 hr:

$$r_{inv} = 0.0325\times\sqrt{\frac{k\Delta t}{\phi\mu c_t}} = 0.0325\sqrt{\frac{0.0001\,\text{md}\times 24\,\text{hr}}{0.05\times0.02\,\text{cp}\times1.0\times10^{-4}\,/\,\text{psia}}} \approx 5\,\text{ft}$$

Radius of investigation at 7 days = 168 hr is 13 ft.
Radius of investigation at 30 days = 720 hr is 28 ft.
Radius of investigation at 365 days = 8760 hr is 96 ft.
Notice that radius investigated in shale by a PBU test is relatively short even after a relatively long elapsed time.

12.3.4 Pressure Drawdown Test and the Reservoir Limit Test

A procedure that is analogous to the PBU test analysis is used to analyze pressure drawdown tests in gas wells. The drawdown test measures pressure response to a gas flow rate q for a gas well that was shut in long enough to achieve a static pressure, that is, a pressure that does not change. The change in real gas pseudopressure is

$$\begin{aligned}\Delta m(p) &= m(p_i) - m(p_{wf}) \\ &= 1637\frac{qT}{kh}\left[\log\left(\frac{kt}{\phi\mu c_T r_w^2}\right) - 3.23 + 0.869S'\right] \\ S' &\equiv S + D|q|\end{aligned} \tag{12.28}$$

where t is the time the well is flowing at rate q. Results of the drawdown test are analyzed by first plotting $\Delta m(p)$ versus log t. Flow capacity

$$kh = 1637\frac{qT}{m} \tag{12.29}$$

is estimated from the slope m.

The pressure drawdown test can be used to estimate the limits of a gas reservoir. The reservoir limit test is a pressure drawdown test that continues until pseudosteady-state (PSS) gas flow is achieved. The beginning of PSS flow begins at time

$$t_s \approx 380 \frac{\phi \mu c_T A}{k} \tag{12.30}$$

where t_s is the stabilization time in hr, ϕ is the porosity (fraction), μ is the gas viscosity (cp), c_T is the total compressibility (psia^{-1}), k is the permeability (md), and A is the drainage area of the well in ft^2. The drainage area for a radial system may be approximated as a circular area $A = \pi r_e^2$ where r_e is the drainage radius. Stabilization time t_s is an approximation because drainage radius r_e is not well known.

Example 12.6 Gas Well Stabilization Time

A. Stabilization time is inversely proportional to permeability. Assume stabilization time in a conventional 10 md gas reservoir is 1 hr. How long is stabilization time in a tight gas reservoir with 1 microdarcy = 0.001 md permeability?

B. How long is stabilization time in a shale gas reservoir with 1 nanodarcy = 1×10^{-6} md permeability?

Answer

A. If we assume all other factors are equal, we have the relationship

$$(t_s k)_{\text{conventional}} = (t_s k)_{\text{tight gas}} = 10\,\text{md} \cdot \text{hr}$$

Stabilization time in a tight gas reservoir is 10 000 hr.

B. If we assume all other factors are equal, we have the relationship

$$(t_s k)_{\text{conventional}} = (t_s k)_{\text{shale gas}} = 10\,\text{md} \cdot \text{hr}$$

Stabilization time in the shale gas reservoir is 10 million hr.

These results show that stabilization time is most practical in conventional gas reservoirs.

12.3.5 Rate Transient Analysis

Rate transient analysis (RTA) uses flow rate and flowing wellbore pressure to forecast production rate for very low-permeability reservoirs. RTA is a mechanistic method that combines Darcy's law, an equation of state, and material balance to derive solutions that depend on different boundary conditions. The boundary conditions represent different well orientations and reservoir geometries. The solution that best matches production data is used to calculate reservoir characteristics such as permeability and original gas in place. The quality of the match depends on the quality of data, which can be affected by noise in the data and the frequency of data sampling. Data noise is introduced by mechanisms that are not included in RTA, such as multiphase flow, interference from other wells, and the allocation of rates to different geologic layers.

Stabilization time and radius of investigation depend on permeability. Tight gas reservoirs and gas shales have permeabilities that are orders of magnitude smaller than conventional gas reservoirs. It is often too expensive to conduct a pressure transient test in a tight gas well or shale gas well until the reservoir boundary is reached because the stabilization time is too long or the radius of investigation is too short for a realistic well test duration.

12.3.6 Two-Rate Test

The turbulence factor D is determined by conducting two transient well tests at two different flow rates $\{q_1, q_2\}$. Each test yields a skin value $\{S'_1, S'_2\}$. The two equations for skin are solved for the two unknowns S and D in the expression $S' = S + D|q|$ given flow rates $\{q_1, q_2\}$ and skins $\{S'_1, S'_2\}$.

12.4 GAS WELL DELIVERABILITY

The gas well deliverability test is used to predict gas flow rate as reservoir pressure declines (Canadian Energy Resources Conservation Board, 1975; Beggs, 1984; Ahmed, 2000; Lee, 2007). The result of a deliverability test is a relationship between pressure measurements and corresponding flow rates.

Changes in gas flow rate cause changes in pressure measured at the well. The choice of a deliverability test depends on the length of time needed to stabilize pressure changes. Stabilized reservoir pressure p_r is obtained by shutting in the well until reservoir pressure stops changing. Stabilization time is estimated as

$$t_s = 1000 \frac{\phi \mu r_e^2}{k p_r} \tag{12.31}$$

where the variables are stabilization time t_s (hr), stabilized reservoir pressure p_r (psia), porosity ϕ (fraction), gas viscosity μ at p_r (cp), outer radius of drainage area r_e (ft), and effective permeability k (md). Stabilization time t_s increases as reservoir permeability decreases.

Gas deliverability tests for a single well include the conventional backpressure test, the isochronal test, and the modified isochronal test. The conventional backpressure test consists of a few equal duration flow periods. The test begins at reservoir pressure p_{res}. A reasonable number of flow periods is four, as shown in Figure 12.7. The flow rate is increased from one production period to the next. The flow rate is maintained in each flow period until the flowing pressure at the well stabilizes. The pressure at the well does not return to p_{res} until the test is completed.

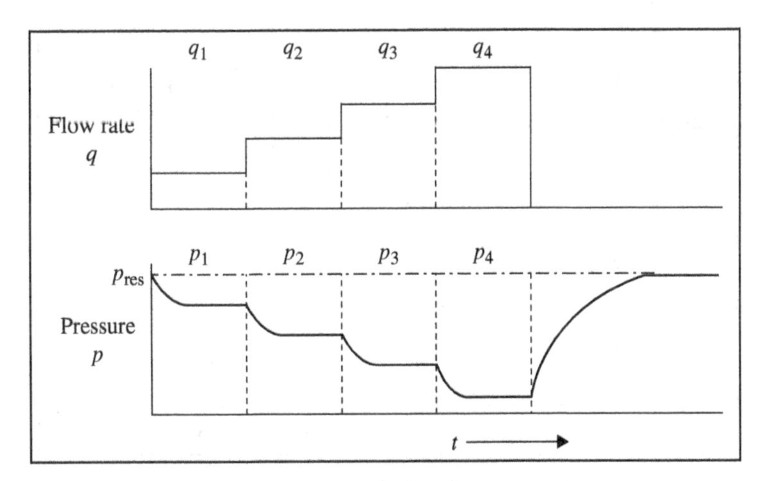

FIGURE 12.7 Conventional backpressure test.

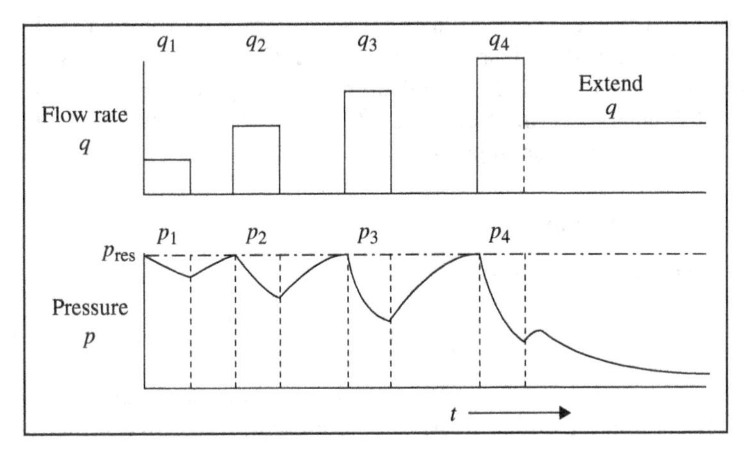

FIGURE 12.8 Isochronal test.

The isochronal test can be conducted more quickly than the conventional backpressure test and is especially useful in low-permeability reservoirs with long stabilization times. The isochronal test also uses equal duration flow periods, and the flow rate is increased from one production period to the next. The difference between the isochronal test and the conventional backpressure test is the shut-in period that follows each flow period shown in Figure 12.8. The shut-in period provides time for the pressure in the well to return to p_{res} observed at the beginning of the test. An extended flow rate is maintained after the last equal duration flow period.

Production time is lost during the isochronal test while the well is shut in. Another approach, the modified isochronal test, is faster but less accurate than the isochronal test. In this case, the flow periods and shut-in periods are the same length of time, as shown in Figure 12.9. The shut-in periods are too short in the modified isochronal test to allow shut-in pressures to return to the pressure p_{res} at the beginning of the test.

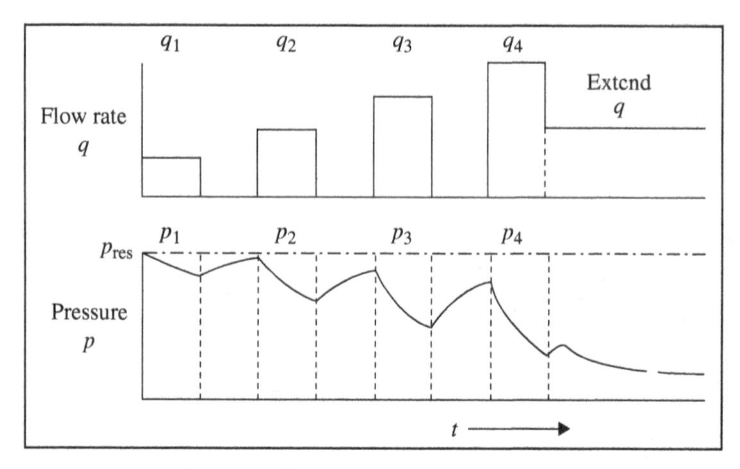

FIGURE 12.9 Modified isochronal test.

Two methods can be used to evaluate gas deliverability tests: simplified backpressure analysis (SBA) and laminar–inertial–turbulent (LIT) analysis. The SBA method uses pressures, while the LIT method uses real gas pseudopressures. According to Lee (2007, page 758), the SBA method is valid at low pressures (<2000 psia). The LIT method is valid for all pressures.

12.4.1 The SBA Method

The SBA method uses the backpressure equation

$$q_{sc} = C\left(p_r^2 - p_{wf}^2\right)^n \equiv C\left(\Delta p^2\right)^n \tag{12.32}$$

where C and n are empirical parameters, q_{sc} is the gas flow rate at standard conditions (MMSCFD), p_r is the stabilized reservoir pressure (psia), and p_{wf} is the flowing well pressure (psia). The logarithm of the backpressure equation for measurement i yields

$$\log q_{sci} = n \log\left(\Delta p^2\right)_i + \log C \tag{12.33}$$

A plot of $\log q_{sci}$ versus $\log(\Delta p^2)_i$ yields a straight line. The values of C and n are determined by using least squares analysis to fit Equation 12.33 to deliverability test data. The slope of the line is n and the intercept is log C. The absolute open flow (AOF) of a gas well is the rate at flowing well pressure $p_{wf} = 0$.

Example 12.7 SBA Method

Application of the SBA method to data from a gas well deliverability test gives $C = 5.89 \times 10^{-2}$ and $n = 0.3477$. Estimate the AOF for the well at stabilized reservoir pressure equal to 1815 psia. The deliverability test uses well rates in MMSCFG/day.

Answer
The AOF is the rate when $p_{wf} = 0$ psia so that

$$q_{sc} = C\left(p_r^2 - p_{wf}^2\right)^n = C\left(p_r^2\right)^n = 5.89\times10^{-2}\left(1815\times1815\right)^{0.3477} = 10.9\ \text{MMSCFG/day}$$

12.4.2 The LIT Method

Any gas well deliverability test can be analyzed using the LIT method. Deliverability test data are fit to the LIT equation

$$m(p_r) - m(p_{wf}) \equiv \Delta m = aq_{sc} + bq_{sc}^2 \tag{12.34}$$

where q_{sc} is the gas flow rate at standard conditions (MMSCFD), $m(p_r)$ is the real gas pseudopressure corresponding to p_r (psia²/cp), and $m(p_{wf})$ is the real gas pseudopressure corresponding to p_{wf} (psia²/cp). The term aq_{sc} represents uniform laminar flow. More chaotic inertial and turbulent flow is represented by the term bq_{sc}^2. Equation 12.34 can be rearranged to give

$$\frac{\Delta m_i}{q_{sci}} = bq_{sci} + a \tag{12.35}$$

for each measurement i. This is a straight line with slope b and intercept a when we plot $\Delta m_i/q_{sci}$ versus q_{sci}. Least squares analysis of deliverability test data yields values of a and b. The quadratic formula is used to rearrange Equation 12.34 to express flow rate q_{sc} as

$$q_{sc} = \frac{-a + \sqrt{a^2 + 4b\Delta m}}{2b} \tag{12.36}$$

in terms of real gas pseudopressure and empirical parameters a, b. The condition for determining AOF using the LIT method is $m(p_{wf}) = 0$.

Example 12.8 LIT Method

Application of the LIT method to data from a gas well deliverability test gives $a = 3.19\times10^{-6}$ psia²/cp/MMSCFD and $b = 1.64\times10^{-6}$ psia²/cp/(MMSCFD²). Estimate the AOF for the well at $p_r = 1948$ psia with pseudopressure $m(p_r) = 3.13\times10^8$ psia²/cp. The deliverability test uses well rates in MMSCFG/day.

Answer
The AOF is the rate when $m(p_{wf}) = 0$ psia²/cp so that

$$q_{sc} = \frac{-a + \sqrt{a^2 + 4bm(p_r)}}{2b} = 12.9\ \text{MMSCFG/day}$$

12.5 SUMMARY OF TRANSIENT WELL TESTING

Transient well tests can provide information about individual well performance, wellbore damage, reservoir pressure, reservoir fluid flow capacity, and forecasts of fluid production. PTT can be used to estimate reservoir permeability, well skin, the distance to reservoir boundaries, structural discontinuities, and communication between wells. Deliverability tests provide a relationship between flow rate and flowing wellbore pressure and can be used to estimate AOF rate.

Many transient well tests are performed on a single well, while others require changing rates or monitoring pressures in two or more wells. The wells can have different orientations and can be in many different geologic environments including both conventional and unconventional reservoirs. The number of transient well tests is limited by factors such as cost and the desire to continue production without interruption.

Transient well test validity depends on the quality of measured production data as well as data used in the calculations, such as fluid property data, rock compressibility, formation thickness, and porosity. One way to improve the interpretation of transient well tests is to compare transient well test results with geological and geophysical models of the reservoir. There should be consistency in the interpretations.

12.6 ACTIVITIES

12.6.1 Further Reading

For more information about transient well testing, see Economides et al. (2013), Satter et al. (2008), Horne (1995), Earlougher (1977), and Matthews and Russell (1967).

12.6.2 True/False

12.1 The LIT test is the laminar–inertial–turbulent gas deliverability test.

12.2 The falloff test measures pressure increase after an injection well is shut in.

12.3 The buildup test measures pressure increases after a producing well has been shut in.

12.4 Pressure communication between wells means that a pressure change at one well can be detected at another well.

12.5 Pressure buildup and pressure drawdown tests are conducted in production wells.

12.6 The diagnostic analysis of a transient well test begins by plotting pressure versus time on a Cartesian plot.

12.7 Horner time has the unit of time.

12.8 A positive skin represents stimulation.

12.9 The wellbore storage effect is due to the volume of the wellbore.

12.10 Radius of investigation is proportional to permeability.

12.6.3 Exercises

12.1 Calculate Horner time for each of the shut-in times shown in the following table. The well is allowed to flow for 8 hr before being shut in.

Shut-In Time (hr)	Horner Time
1	
12	
24	
36	

12.2 What is the radius of investigation of a pressure buildup (PBU) test in an oil well when elapsed time (Δt) is 25 hr? Use the pressure buildup data in the following table.

$$r_{\text{inv}} = 0.029\sqrt{\frac{k\Delta t}{\phi\mu c_t}}$$

Pressure Buildup Test	
Production time (t_P)	100.0 hr
Initial pressure (p_i)	2300 psi
Initial time	0.0 hr
Reservoir temperature (T)	180.0 °F
Porosity (ϕ)	0.18
Thickness (h)	24.0 ft
Total compressibility (c_t)	1.0×10^{-5} psia^{-1}
Viscosity (μ)	0.90 cp
Formation volume factor (B)	1.30 RB/STB
Well flow rate (q)	180.0 STB/D
Fluid type	Liquid
Permeability (k)	150 md

12.3 **A.** The mapped boundary closest to a well is approximately 1500 ft away from the well. Use the information in Exercise 12.2 to calculate the minimum time (in hr) needed to investigate this distance using a PBU test.

B. Is 25 hr of shut-in time enough time to see the boundary effect?

12.4 **A.** Use the Horner plot in Figure 12.3 to find the slope m.

B. Use the pressure buildup data in the following table and slope m to calculate flow capacity kh = permeability × thickness.

C. Use kh to calculate k.

Pressure Buildup Test	
Production time (t_F)	24.0 hr
Initial pressure (p_i)	4000 psi
Initial time	0.0 hr
Reservoir temperature (T)	160.0 °F
Porosity (ϕ)	0.22
Thickness (h)	120.0 ft
Total compressibility (c_t)	6.0×10^{-6}/psia
Viscosity (μ)	0.695cp
Formation volume factor (B)	1.32 RB/STB
Well flow rate (q)	100.0 STB/D
Fluid type	Liquid

12.5 **A.** Well PI = well productivity index = flow rate divided by pressure drawdown. Suppose the well originally produces 10000 STBO/day at a pressure drawdown of 10 psia. What is well PI?

B. The well PI declined 4% a year for the first three years of production. What is the well PI at the end of year 3?

C. The well PI at the beginning of year 4 is 750 STBO/day/psia. A well test showed an increase in skin and the skin is positive. Does the well need a workover (Y or N)?

13

PRODUCTION PERFORMANCE

Production can begin as soon as wells are completed and surface facilities are installed. Production performance is evaluated using a variety of techniques. It is first necessary to acquire field performance data. Data is used to estimate reserves and forecast reservoir performance. Some production performance evaluation techniques include bubble mapping, decline curve analysis, and material balance analysis. They are discussed here.

13.1 FIELD PERFORMANCE DATA

A database should include as much data as possible about field performance. Sources of field performance data include seismic surveys, core analysis (geological and petrophysical), well logging, pressure transient analysis, production and injection monitoring, and tracer surveys. Production and injection data include fluid rates and volumes for all wells. Data can be collected in a room full of boxes on shelves or digitized in a computer-based system. The database needs to be accessible and can exist in many forms.

Data are needed to describe and manage the reservoir. Seismic data is a source of data about reservoir structure. It can be calibrated by well data, which includes well logs and core data. The framework of the reservoir can be inferred from well data that has been correlated between wells and takes into account the prevailing

Introduction to Petroleum Engineering, First Edition. John R. Fanchi and Richard L. Christiansen.
© 2017 John Wiley & Sons, Inc. Published 2017 by John Wiley & Sons, Inc.
Companion website: www.wiley.com/go/Fanchi/IntroPetroleumEngineering

depositional model. The reservoir framework and structure should be consistent. Flow tests on cores give information about permeability. Well logs are a primary source of information about net pay, fluid contacts, porosity, and water saturation. In some cases, data from different scales provide information about the same feature.

Data from multiple sources must be consistent from one source to another. For example, well log and seismic data provide information about the depths of fluid contacts. If there is a difference between the two sources of data, they must be reconciled. Similarly, differences between permeability from flow tests in cores and permeability from transient well tests may have to be reconciled.

Field performance data is acquired from a variety of sources. Seismic data is acquired from seismic surveys by geophysicists. Well logs are typically acquired and analyzed by geoscientists and engineers. The database should include well pressure measurements and data characterizing the performance of surface and subsurface equipment. Subsurface equipment includes casing, tubing, subsea templates, and electric submersible pumps (ESPs). Surface equipment includes well heads, flow lines, separators, compressors, and other facilities like offshore platforms.

Data needs to be trustworthy. In some cases it is possible to ascertain the uncertainty associated with data. For example, multiple measurements of properties such as porosity and permeability can be used to conduct a statistical analysis if the data is available. It is a significant challenge to maintain a complete, up-to-date, and accurate database.

Example 13.1 Well PI and Workover Analysis

A. Suppose a well originally produces 10 000 STBO/day at a pressure drawdown of 10 psia. What is the well PI?

B. The well PI declined 5% a year for the first 2 years of production. What is the well PI at the end of year 2?

C. The well PI at the beginning of year 3 is 750 STBO/day/psia. A well test showed an increase in skin and the skin is positive. Does the well need a workover?

Answer

A. $\text{PI} = \dfrac{10000(\text{STBO/day})}{10\,\text{psia}} = 1000\dfrac{(\text{STBO/day})}{\text{psia}}$

B. $1000\dfrac{(\text{STBO/day})}{\text{psia}} \times (1-0.05) \times (1-0.05) = 902.5\dfrac{(\text{STBO/day})}{\text{psia}}$

C. Yes. $S > 0$ implies that well stimulation could improve flow.

13.1.1 Bubble Mapping

Bubble mapping can be applied to spatially distributed variables such as flow rates, reservoir pressures, fluid and rock properties, and cumulative production. It is a procedure that provides a visual comparison of spatially distributed variables in an

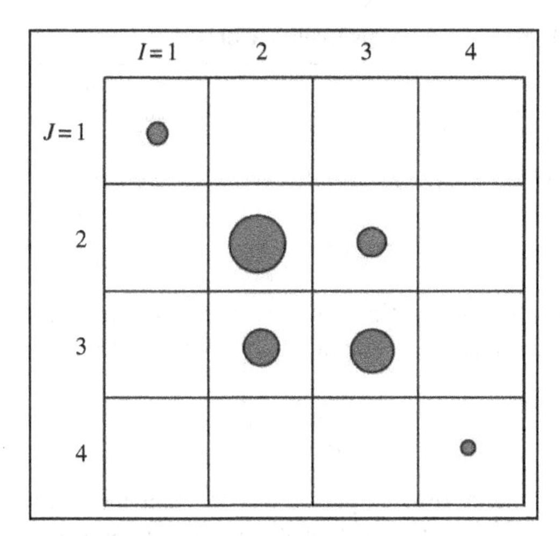

FIGURE 13.1 Illustration of a bubble map.

area of interest. For example, cumulative oil production for a well at a particular point in time can be plotted as a circle (or bubble) centered on the well location displayed on a map. If this is done for several wells as in Figure 13.1, the resulting map will provide a graphic comparison of the relative amount of cumulative oil production at each well. The location of wells in the figure is specified in terms of the I, J indices. The radius of the circle indicates the magnitude of the variable. In our example, a large circle indicates large cumulative oil production relative to other wells with smaller circles.

Bubble maps can be used to look for trends in the distribution of a variable. For example, if a small bubble representing the gas production rate at a well is surrounded by a set of large bubbles in an area, the small bubble may represent an anomalous measurement or a well that is damaged. In this case, a workover could increase the gas production rate at the well.

13.2 DECLINE CURVE ANALYSIS

Decline curve analysis (DCA) is an empirical technique for predicting oil or gas well production (Arps, 1945; Towler, 2002; Economides et al., 2013). The technique fits a curve to measurements of flow rate as a function of time. Some reservoirs, such as oil reservoirs with strong water influx, have enough energy to sustain relatively constant oil production rate for an extended period of time. As a rule, production flow rate declines with time once a well is completed and production begins.

An exponential equation has been used to predict future production by fitting the exponential equation to historical decline rates for many production wells. Although the exponential equation provides a good fit of production rate as a function of time for some wells, a hyperbolic equation provides a better fit of the decline in

TABLE 13.1 Arps Decline Curves

Decline Curve	n	Index
Exponential	0	$q = q_i e^{-at}$
Hyperbolic	$0<n<1$	$q^{-n} = nat + q_i^{-n}$
Parabolic	1	$q^{-1} = at + q_i^{-1}$

production rate as a function of time for other wells. The following equation includes both exponential and hyperbolic relationships (Arps, 1945):

$$\frac{dq}{dt} = -aq^{n+1} \tag{13.1}$$

The equation assumes that flowing pressure is constant and factors a and n are empirically determined constants. The empirical constant n ranges from 0 to 1. The process of fitting Equation 13.1 to production rate data is called DCA.

The shape of the decline curve depends on the value of n as shown in Table 13.1. The term q_i is initial rate.

The unknown parameters in the decline curves are determined by fitting the decline curves to historical data. For example, the natural logarithm of the exponential decline curve is

$$\ln q = \ln q_i - at \tag{13.2}$$

Equation 13.2 is the equation of a straight line $y = mx + b$ with slope m and intercept b if we define the independent variable x as time t and the dependent variable y as $\ln q$. In this case, the term $\ln q_i$ is the intercept b, and the slope m of the straight line is $-a$.

Total production in a time interval is called cumulative production for that time interval. Cumulative production for each of the decline curves in Table 13.1 can be calculated analytically. Rate is integrated with respect to time from initial rate q_i at time $t=0$ to rate q at time t. As an illustration, cumulative production N_p for the exponential decline equation is

$$N_p = \int_0^t q\,dt = \frac{q_i - q}{a} \tag{13.3}$$

Rearranging Equation 13.2 gives the decline factor a for the exponential decline case:

$$a = -\frac{1}{t}\ln\frac{q}{q_i} \tag{13.4}$$

Future production is estimated by extrapolating the decline curve to a specified final rate. The final rate is usually determined as the lowest rate that is still economically viable. For this reason, the specified final rate is called economic rate or abandonment rate. Reserves are the difference between cumulative production at abandonment and current cumulative production.

Example 13.2 Exponential Decline of Oil Rate

A. The initial oil rate of an oil well is 4800 STB/day. The rate declines to 3700 STB/day after two years of continuous production. Assume the decline is exponential. What is the decline factor?

B. When will the oil rate decline to 100 STB/day? Express your answer in years after the beginning of production.

Answer

A. Rearrange $q = q_i e^{-at}$ to estimate decline factor a:

$$a = -\frac{1}{t}\ln\left(\frac{q}{q_i}\right) = -\frac{1}{2}\ln\left(\frac{3700}{4800}\right) \approx 0.130/\text{yr}$$

B. Find the time when rate declines to 100 STB/day from the initial rate of 4800 STB/day:

$$t = -\frac{1}{a}\ln\left(\frac{q}{q_i}\right) = -\frac{1}{0.130}\ln\left(\frac{100}{4800}\right) \approx 29.7\text{yr}$$

13.2.1 Alternative DCA Models

The Arps exponential model does not always adequately model the decline rate of unconventional reservoir production. Valkó and Lee (2010) introduced the stretched exponential decline model (SEDM) into DCA as a generalization of the Arps exponential model. The SEDM is based on the idea that several decaying systems comprise a single decaying system (Phillips, 1996; Johnston, 2006). If we think of production from a reservoir as a collection of decaying systems in a single decaying system, such as declining production from multiple zones, then SEDM can be viewed as a model of the decline in flow rate. The SEDM has three parameters q_i, τ, n (or a, b, c):

$$q = q_i \exp\left[-\frac{t^n}{\tau}\right] = a\exp\left[-\left(\frac{t}{b}\right)^c\right] \tag{13.5}$$

Parameter q_i is flow rate at initial time t. The Arps exponential decline model is the special case with $n=1$.

A second decline curve model is based on the logarithmic relationship between pressure and time in a radial flow system. The logarithmic decline model with parameters a and b is

$$q = a\ln(t) + b \tag{13.6}$$

It is referred to as the LNDM model.

Example 13.3 SEDM Model of Shale Gas Decline

The SEDM model was used to model gas production rate for a shale gas well. The SEDM parameters for decline in shale gas rate for this well are $q_i = 8 \times 10^4$ MSCF/month, $\tau = 0.3$month, and $n = 0.2$. Estimate the gas rate after 3 years of production. Express time in months since the SEDM model parameters were calculated using monthly gas production.

Answer

The gas rate at $t=3$ years = 36 months is

$$q = q_i \exp\left[-\left(\frac{t}{\tau}\right)^n\right] = 8 \times 10^4 \exp\left[-\left(\frac{36}{0.3}\right)^{0.2}\right] = 5911 \text{ MSCF/mo}$$

13.3 PROBABILISTIC DCA

The probabilistic estimate of reserves is a Monte Carlo procedure that uses the workflow outlined in Figure 13.2 (Fanchi et al., 2013). Each step of the probabilistic DCA method in Figure 13.2 is briefly described in the following text.

Step 1: Gather rate-time data

Acquire production rate as a function of time. Remove significant shut-in periods so rate-time data represents continuous production.

Step 2: Select a DCA model and specify input parameter distributions

The number of input parameters depends on the DCA model chosen. The SEDM model requires three parameters, and the LNDM model requires two parameters. Parameter distributions may be either uniform or triangle distributions.

Step 3: Specify constraints

Available rate-time production history is used to decide which DCA trials are acceptable. Every DCA model run that uses a complete set of model input parameters constitutes a trial. The results of each trial are then compared to user-specified criteria. Criteria options include an objective function, rate at the end of history, and cumulative production at the end of history. The objective function quantifies the quality of the match by comparing the difference between model rates and observed rates. Objective functions with smaller values are considered better matches than objective functions with larger values because uncertainty has been reduced and forecasts are more closely grouped together.

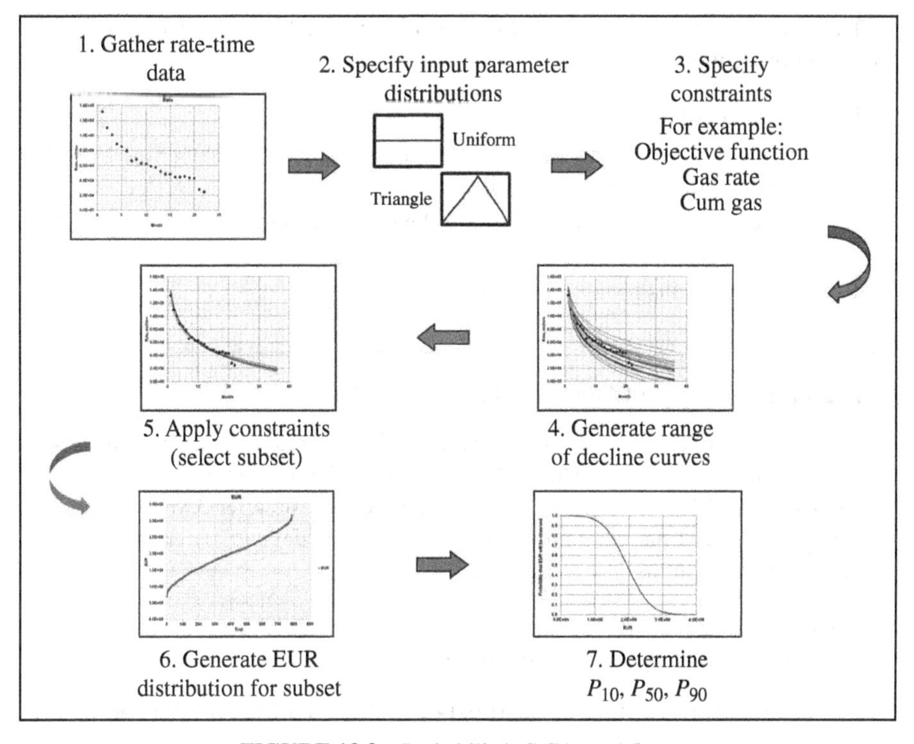

FIGURE 13.2 Probabilistic DCA workflow.

Step 4: Generate decline curve trials

Decline curve trials are obtained by running the DCA model. The number of trials is specified by the analyst.

Step 5: Determine subset of acceptable trials

The trials generated in Step DCA4 are compared to the criteria specified in Step DCA3. Each trial that satisfies the user-specified criteria is included in a subset of acceptable trials.

Step 6: Generate percentile distribution of performance results

The 10th (PC10), 50th (PC50), and 90th (PC90) percentiles are determined from the distribution of EUR values for the subset of acceptable trials.

Step 7: Generate probability distribution of performance results

Each EUR percentile is related to the probability that the amount actually recovered equals or exceeds the estimate using the relationships P_{10}=PC90, P_{50}=PC50, and P_{90}=PC10.

The workflow in Figure 13.2 is a Monte Carlo method because it incorporates the following procedure:

1. Define a set of input parameter distributions.
2. Generate a set of input parameter values by randomly sampling from the associated probability distributions.
3. Use the input parameter values in a deterministic model to calculate a trial result.
4. Gather the results for a set of trials.

Example 13.4 LNDM Model of Shale Gas Decline

The LNDM model was used to model gas production rate for a set of shale gas wells (Fanchi et al., 2013). The LNDM parameters for a high flow rate gas well are $a = -1.0 \times 10^4$ and $b = 1.5 \times 10^5$. Estimate the gas rate after 3 years of production. Express time in months since the LNDM model parameters were calculated using monthly gas production.

Answer
The gas rate at $t=3$ years = 36 months is

$$q = a\ln(t) + b = -1.0 \times 10^4 \times \ln(36) + 1.5 \times 10^5 = 1.14 \times 10^5 \text{MSCF/mo}$$

13.4 OIL RESERVOIR MATERIAL BALANCE

Material balance calculations are based on the law of conservation of mass. They account for material in a system, entering a system, and exiting a system. The reservoir is treated as a large, flexible tank of material. Material can be injected into the tank and produced from the tank. It can also expand and shrink within the tank. The balance can be written in simplified form as

$$\text{Original mass in tank} + \text{mass entering tank} - \text{mass exiting tank} = \text{mass in tank} \quad (13.7)$$

Material balance calculations use measurable quantities to determine the amount of a material in the tank. The measurable quantities include cumulative fluid production volumes and cumulative fluid injection volumes for oil, water, and gas phases; reservoir pressure; and fluid property data from samples of produced fluids. Since reservoir pressure and cumulative volumes are needed in the material balance calculation, it is necessary to have data measurements over a period of time long enough to establish a trend. Consequently, the material balance calculation can yield more accurate results as more data are acquired over time.

Material balance calculations provide an independent method of estimating the original volume of oil, water, and gas in a reservoir. These volume estimates can be compared with volumetric estimates of original volumes. The two methods are independent and provide a comparison between a geologic method that depends on static data and an engineering method that depends on dynamic, time-dependent data.

Different material balance methods are used for oil reservoirs and gas reservoirs. The material balance method for oil reservoirs is discussed here, and the material balance method for gas reservoirs is discussed in the next section.

13.4.1 Undersaturated Oil Reservoir with Water Influx

We illustrate the material balance method for analyzing an oil reservoir by deriving the material balance equation for an undersaturated oil reservoir with water influx. Water influx is water flowing into the reservoir from a natural source such as an aquifer. We simplify the derivation by assuming the effect of compressibility is negligible. In this case, the decrease in oil volume ΔV_o at reservoir conditions that occurs when oil is produced must be matched by an increase in water volume ΔV_w at reservoir conditions so that

$$\Delta V_w = \Delta V_o \tag{13.8}$$

The change in oil volume at reservoir conditions is

$$\Delta V_o = NB_{oi} - \left(N - N_p\right)B_o \tag{13.9}$$

where N is OOIP (STB) and NB_{oi} is initial OIP at reservoir conditions (RB) for initial oil formation volume factor (FVF) B_{oi} (RB/STB). N_p is oil produced (STB), so $(N-N_p)B_o$ is remaining OIP at reservoir conditions.

The change in water volume at reservoir conditions is

$$\Delta V_w = \left(W + W_e - W_p\right)B_w - WB_w = \left(W_e - W_p\right)B_w \tag{13.10}$$

where W is OWIP (STB), W_e is water influx (STB), W_p is cumulative water production (STB), and B_w is water FVF (RB/STB) which is often approximated as 1.0 RB/STB. Water influx refers to water entering the system from a natural source such as an aquifer or by water injection.

The material balance equation for the oil–water system is obtained by substituting Equations 13.9 and 13.10 into Equation 13.8 to give

$$\left(W_e - W_p\right)B_w = NB_{oi} - \left(N - N_p\right)B_o \tag{13.11}$$

This equation is one equation for the two unknowns N and W_e. Solving Equation 13.11 for N gives

$$N = \frac{N_p B_o + \left(W_p - W_e\right)B_w}{B_o - B_{oi}} \tag{13.12}$$

The value of N is determined by using production data and assuming a water influx model for water influx W_e. By contrast, solving Equation 13.11 for W_e gives

$$W_e B_w = N\left(B_{oi} - B_o\right) + N_p B_o + W_p B_w \tag{13.13}$$

The value of W_e can be estimated using production data and a value of N from volumetric analysis.

Example 13.5 Water Influx

Calculate water influx W_e using the material balance equation for an oil–water system: $W_e B_w = N(B_{oi} - B_o) + N_p B_o + W_p B_w$. We have the estimate OOIP $= N \approx N_{vol} = 8.5$ MMSTB where N_{vol} is from volumetric analysis. Initial oil FVF is $B_{oi} = 1.347$ RB/STB. Oil FVF $B_o = 1.348$ RB/STB when 46 MSTB oil has been produced. Water FVF is approximately constant and has the value $B_w = 1.0$ RB/STB. No water production has been reported.

Answer

No reported water production implies $W_p = 0$ so that $W_e B_w = N(B_{oi} - B_o) + N_p B_o$. Substituting values into $W_e B_w = N(B_{oi} - B_o) + N_p B_o$ gives $W_e B_w = -7.65$ MSTB + 62.02 MSTB so that $W_e \approx 54.37$ MSTB.

13.4.2 Schilthuis Material Balance Equation

Schilthuis (1936) presented a general material balance equation that accounted for oil, water, and gas in the system. The general material balance equation is derived by assuming that the system is an isothermal system in pressure equilibrium. It is also based on the assumption that the distribution of oil, water, and gas phases does not affect tank model results. Following the discussion by Fanchi (2010b), we write the general material balance equation as

$$\begin{aligned}
&N\left(B_t - B_{ti}\right) + NmB_{ti}\left(\frac{B_{gc} - B_{gi}}{B_{gi}}\right) + N\frac{B_{ti}S_{wio}}{1 - S_{wio}}\left(\frac{B_{tw} - B_{twi}}{B_{twi}}\right) \\
&+N\frac{mB_{ti}S_{wig}}{1 - S_{wig}}\left(\frac{B_{tw} - B_{twi}}{B_{twi}}\right) + N\left(\frac{1}{1 - S_{wio}} + \frac{m}{1 - S_{wig}}\right)B_{ti}c_f\Delta p \\
&= N_p B_o - N_p R_{so} B_g + \left[G_{ps}B_g + G_{pc}B_{gc} - G_i B_g'\right] - \left(W_e + W_i - W_p\right)B_w
\end{aligned} \tag{13.14}$$

The terms in Equation 13.7 and associated units are specified in Table 13.2. Application of the general material balance equation presumes that fluid property data from fluid samples are representative of reservoir fluids and that production, injection, and pressure data are reliable.

TABLE 13.2 Nomenclature for the General Material Balance Equation

B_g	Gas formation volume factor (FVF) (RB/SCF)
B_{gc}	Gas cap FVF (RB/SCF)
B_g'	Injected gas FVF (RB/SCF)
B_o	Oil FVF (RB/STB)
B_t	$B_o+(R_{si}-R_{so})B_g$ = composite oil FVF (RB/STB)
B_{ti}	Initial value of B_t (RB/STB)
B_{tw}	$B_w+(R_{swi}-R_{sw})B_g$ = composite water FVF (RB/STB)
B_w	Water FVF (RB/STB)
c_f	Formation (rock) compressibility (1/psia)
G	Initial gas in place (SCF)
G_i	Cumulative gas injected (SCF)
G_{pc}	Cumulative gas cap gas produced (SCF)
G_{ps}	Cumulative solution gas produced as evolved gas (SCF)
m	Ratio of gas reservoir volume to oil reservoir volume
N	Initial oil in place (STB)
N_p	Cumulative oil produced (STB)
R_{so}	Solution gas–oil ratio (SCF/STB)
R_{si}	Initial solution gas–oil ratio (SCF/STB)
R_{sw}	Solution gas–water ratio (SCF/STB)
R_{swi}	Initial solution gas–water ratio (SCF/STB)
S_g	Gas saturation (fraction)
S_o	Oil saturation (fraction)
S_w	Water saturation (fraction)
S_{wi}	Initial water saturation (fraction)
S_{wig}	Initial water saturation in gas cap (fraction)
S_{wio}	Initial water saturation in oil zone (fraction)
W_e	Cumulative water influx (STB)
W_i	Cumulative water injected (STB)
W_p	Cumulative water produced (STB)
Δp	p_i-p = reservoir pressure change (psia)
p_i	Initial reservoir pressure (psia)
p	Reservoir pressure corresponding to a cumulative fluid value (psia)

The physical significance of the terms in the general material balance equation is displayed by first defining the terms

$$
\begin{aligned}
D_o &= B_t - B_{ti}, \\
D_{go} &= mB_{ti}\left(\frac{B_{gc}-B_{gi}}{B_{gi}}\right), \\
D_w &= \frac{B_{ti}S_{wio}}{1-S_{wio}}\left(\frac{B_{tw}-B_{twi}}{B_{twi}}\right), \\
D_{gw} &= \frac{mB_{ti}S_{wig}}{1-S_{wig}}\left(\frac{B_{tw}-B_{twi}}{B_{twi}}\right), \\
D_r &= \left(\frac{1}{1-S_{wio}}+\frac{m}{1-S_{wig}}\right)B_{ti}c_f\Delta p
\end{aligned}
\tag{13.15}
$$

Given these definitions, the general material balance equation has the form

$$N\left[D_o + D_{go} + D_w + D_{gw} + D_r\right] = N_p B_o - N_p R_{so} B_g + \left[G_{ps} B_g + G_{pc} B_{gc} - G_i B_g'\right] - \left(W_e + W_i - W_p\right) B_w \tag{13.16}$$

Changes in fluid volume in the reservoir are represented by the terms on the left-hand side of Equation 13.16. Fluid production and injection are represented by the terms on the right-hand side. The physical significance of each term is summarized in Table 13.3.

The relative importance of different drive mechanisms can be estimated by rearranging the general material balance equation. Indices for different drive mechanisms are shown in Table 13.4 relative to the hydrocarbon production D_{HC} defined by

$$D_{HC} = N_p B_o + G_{pc} B_{gc} + \left[G_{ps} - N_p R_{so}\right] B_g \tag{13.17}$$

The sum of the drive indices shown in Table 13.4 equals one, thus

$$I_{sg} + I_{gc} + I_w + I_i + I_e = 1 \tag{13.18}$$

Equation 13.18 can be derived by rearranging Equation 13.14. A comparison of the magnitudes of the drive indices indicates which drive is dominating the performance of the reservoir.

We can illustrate the value of the drive indices by considering an undersaturated oil reservoir that is attached to an aquifer. The reservoir is in the primary production stage, that is, the reservoir is being produced and no fluids are being injected. What drive mechanisms do we need to consider? The undersaturated oil reservoir does not have a gas cap, so m = 0. Furthermore, no fluids are being injected.

TABLE 13.3 Physical Significance of Material Balance Terms

Term	Physical Significance
ND_o	Change in volume of initial oil and associated gas
ND_{go}	Change in volume of free gas
$N(D_w + D_{gw})$	Change in volume of initial connate water
ND_r	Change in formation pore volume
$N_p B_o$	Cumulative oil production
$N_p R_{so} B_g$	Cumulative gas produced in solution with oil
$G_{ps} B_g$	Cumulative solution gas produced as evolved gas
$G_{pc} B_{gc}$	Cumulative gas cap gas production
$G_i B_g'$	Cumulative gas injection
$W_e B_w$	Cumulative water influx
$W_i B_w$	Cumulative water injection
$W_p B_w$	Cumulative water production

TABLE 13.4 Drive Indices for the General Material Balance Equation

Drive	Index
Solution gas	$I_{sg} = \dfrac{ND_o}{D_{HC}}$
Gas cap	$I_{gc} = \dfrac{ND_{go}}{D_{HC}}$
Water	$I_w = \dfrac{\left[(W_e - W_p)B_w\right]}{D_{HC}}$
Injected fluids	$I_i = \dfrac{\left[W_i B_w + G_i B'_g\right]}{D_{HC}}$
Connate water and rock expansion	$I_e = \dfrac{\left[N(D_w + D_{gw}) + ND_r\right]}{D_{HC}}$

Therefore, the remaining possible drive mechanisms are solution gas drive, water drive from aquifer influx, and connate water and rock expansion drive. More information is needed to determine which drive is dominating flow. In addition, it is possible that the relative importance of the drive mechanisms will change during the life of the reservoir.

13.5 GAS RESERVOIR MATERIAL BALANCE

Material balance in a gas reservoir can be derived from the general material balance equation as follows. Original gas in place G is expressed in terms of the variables in Table 13.2 as

$$GB_{gi} = NmB_{ti} \tag{13.19}$$

Combining Equation 13.19 with the general material balance equation gives

$$\begin{aligned} &N(B_t - B_{ti}) + GB_{gi}\left(\frac{B_{gc} - B_{gi}}{B_{gi}}\right) + N\frac{B_{ti}S_{wio}}{1-S_{wio}}\left(\frac{B_{tw} - B_{twi}}{B_{twi}}\right) \\ &+G\frac{B_{gi}S_{wig}}{1-S_{wig}}\left(\frac{B_{tw} - B_{twi}}{B_{twi}}\right) + \left(\frac{NB_{ti}}{1-S_{wio}} + \frac{GB_{gi}}{1-S_{wig}}\right)c_f\Delta p \\ &= N_pB_o + \left[G_{ps}B_g + G_{pc}B_{gc} - G_iB_g'\right] - N_pR_{so}B_g - (W_e + W_i - W_p)B_w \end{aligned} \tag{13.20}$$

A gas reservoir does not include an original oil phase, so $N=0$ and $N_p=0$. Substituting $N=0$ and $N_p=0$ into Equation 13.20 gives

$$GB_{gi}\left(\frac{B_{gc}-B_{gi}}{B_{gi}}\right)+G\frac{B_{gi}S_{wig}}{1-S_{wig}}\left(\frac{B_{tw}-B_{twi}}{B_{twi}}\right)+\left(\frac{GB_{gi}}{1-S_{wig}}\right)c_f\Delta p$$
$$=\left[G_{pc}B_{gc}-G_iB_g'\right]-\left(W_e+W_i-W_p\right)B_w \quad (13.21)$$

We simplify Equation 13.21 further by recognizing that water compressibility and formation compressibility are relatively small compared to gas compressibility. Neglecting water and formation compressibility terms gives

$$G\left(B_g-B_{gi}\right)=\left[G_pB_g-G_iB_g'\right]-\left(W_e+W_i-W_p\right)B_w \quad (13.22)$$

The subscript c denoting gas cap has been dropped since there is no oil in the reservoir.

13.5.1 Depletion Drive Gas Reservoir

An important but relatively simple application of gas reservoir material balance is the analysis of production from a gas reservoir containing only gas and irreducible water. If we produce the gas reservoir without injection, the water influx, injection, and production terms are zero, formation compressibility is negligible, and there is no gas injection. The resulting material balance equation for the depletion drive gas reservoir is

$$G\left(B_g-B_{gi}\right)=G_pB_g \quad (13.23)$$

where G is original free gas in place and G_p is cumulative free gas produced. Substituting the real gas law into gas FVF and rearranging gives

$$G_p=\left[1-\frac{(p/Z)_t}{(p/Z)_i}\right]G \quad (13.24)$$

where p is reservoir pressure and Z is the real gas compressibility factor. Subscript t indicates that p/Z should be calculated at time t corresponding to G_p and subscript i indicates that p/Z should be calculated at initial conditions. We can write Equation 13.24 in the form

$$\left(\frac{p}{Z}\right)_t=\left(\frac{p}{Z}\right)_i-\left[\left(\frac{p}{Z}\right)_i\frac{1}{G}\right]G_p \quad (13.25)$$

Equation 13.25 is a linear equation for $(p/Z)_t$ when plotted versus G_p. The original gas in place G is the value of G_p when $(p/Z)_t$ is zero.

Example 13.6 Depletion Drive Gas Reservoir

Early data from a depletion drive gas reservoir are shown in the following. Estimate OGIP.

G_p (Bscf)	P (psia)	Z	P/Z (psia)
0.015	1946	0.813	2393
0.123	1934	0.813	2378

Answer

The equation for a straight line is $y = mx + b$ where m is the slope and b is the intercept. Comparing the equation for a straight line with Equation 13.25 gives $x = G_p$, $y = (p/Z)_t$. The slope and intercept are $m = -(p/Z)_i \times (1/G) = -138.9 \frac{\text{psia}}{\text{Bscf}}$ and $b = y - mx = 2395$ psia. G is the value of G_p when $(p/Z)_t$ is zero. This corresponds to $y = mx + b = 0$ or $G = x$ at $y = 0$ so $G = -(b/m) = \left(\frac{2395 \text{ psia}}{138.9\,(\text{psia/Bscf})}\right) \approx 17.2\text{ Bscf}$. The actual value of G based on additional data from a longer production period is 16.1 Bscf. Additional data can provide a more accurate straight line for performing the OGIP analysis.

13.6 DEPLETION DRIVE MECHANISMS AND RECOVERY EFFICIENCIES

Depletion drive mechanisms can be identified by examining the production performance of properties such as reservoir pressure and GOR. Reservoir depletion occurs when production occurs without injection. The removal of fluids from the reservoir without replacement results in reservoir pressure decline. Figure 13.3 presents production profiles for three depletion drives during primary production of an oil reservoir.

The solution gas drive shows a significant increase in GOR followed by a decline in GOR as available gas is produced. An undersaturated oil reservoir consists of oil and immobile water. When it is produced, only single-phase oil flows into the wellbore. The production of oil reduces pressure in the reservoir, and the oil in the reservoir expands. The rate of pressure decline when reservoir pressure is above bubble point pressure depends on the compressibility of the oil and the formation. If production continues, eventually reservoir pressure will drop below bubble point pressure. Gas dissolved in the oil comes out of solution and forms a free gas phase as reservoir pressure declines. Bubbles of free gas expand and help displace oil to the well. If enough free gas is present, the gas bubbles coalesce and increase gas saturation. When gas saturation exceeds critical gas saturation, typically 3–5% of pore volume, the gas forms a flow path for free gas flow. Production GOR is constant initially until

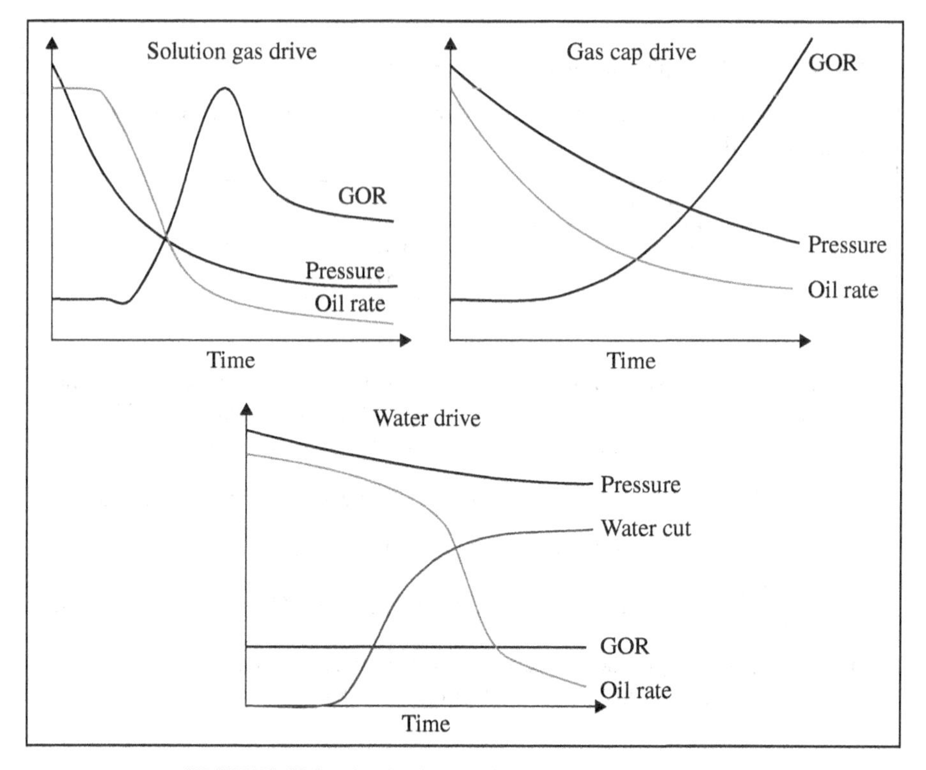

FIGURE 13.3 Production profiles of drive mechanisms.

a dip occurs in the GOR curve. The GOR dip shows the formation of critical gas saturation when reservoir pressure drops below bubble point pressure. When gas saturation exceeds critical gas saturation, free gas flows into the well and the GOR increases.

A gas cap is present in a reservoir when the volume of oil is not large enough to dissolve all of the gas at original reservoir temperature and pressure. Over geologic time, free gas migrates to the crest of the reservoir and forms a gas cap. A well completed in the oil zone will produce single-phase oil. Gas cap expansion displaces oil to the well. Production GOR is constant initially until the gas–oil contact reaches the well. Increasing GOR shows that free gas is being produced.

Water drive is the most efficient drive mechanism for producing oil. Pressure support and displacement of oil by water helps reduce the rate of pressure depletion. Recovery efficiencies for the most common depletion drive mechanisms for primary production are shown in Table 13.5.

In some cases, reservoir dip is so large that gravity drainage is an important natural drive mechanism during primary depletion of an undersaturated oil reservoir. The East Texas oil field is an example of a gravity drainage oil field. The giant oil field was discovered in 1930, and oil production to date exceeds five billion barrels of oil. The East Texas oil field is the part of the Woodbine sand that is wedged between the

TABLE 13.5 Recovery Efficiencies for Different Depletion Drive Mechanisms

Depletion Drive Mechanisms	Recovery Efficiency (% OOIP)
Water drive	35–75
Gas cap drive	20–40
Solution gas drive	5–30

Source: Data from Ahmed (2000).

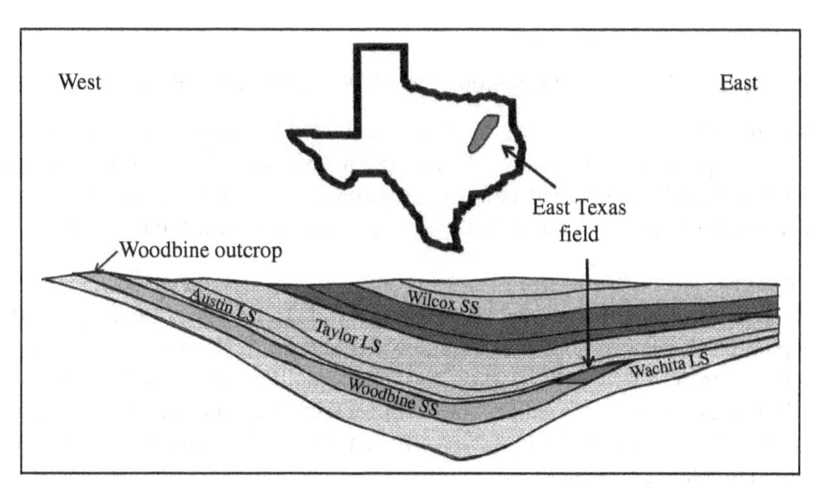

FIGURE 13.4 Cross section of the East Texas Basin. (Source: Adapted from Halbouty (2000).)

Austin and Wichita limestones sketched in Figure 13.4. The section of the Woodbine sand from the outcrop to the East Texas oil field is an aquifer that provided pressure support to the oil field. The outcrop made it possible to recharge the aquifer when rain and snow fell. Both gravity drainage and water drive functioned as natural drive mechanisms during primary depletion.

A combination drive is active when two or more natural drive mechanisms are functioning at the same time during production of an oil reservoir. Primary production by gravity drainage and water drive from the East Texas oil field is an example of a combination drive. The relative importance of drive mechanisms can be determined using drive indices calculated from the general material balance equation. The production profile of a combination drive reservoir depends on which drive is dominant at different points during the primary production period.

Natural reservoir energy can be supplemented by injecting fluids into the reservoir. Water and gas are typical injection fluids. The injected fluids reduce the rate of pressure decline that would have occurred during primary depletion. Water drive can be an effective means of displacing oil to production wells regardless of the source of water. Thus, if a reservoir does not have significant aquifer support, injection wells can be used to supplement existing natural resources. Water breakthrough occurs when injected water arrives at a production well. It is observed as a significant increase in produced WOR or produced WCT. Similarly, the breakthrough of injected gas is recognized as a significant increase in production GOR.

The recovery factor for dry gas reservoir depletion can be as high as 80%–90% of original gas in place (Ahmed, 2000). Gas recovery from water-drive gas reservoir depletion ranges from 50% to 70% of original gas in place (Ahmed, 2000). It is typically lower than gas recovery from dry gas reservoir depletion because gas can be trapped by encroaching water.

Example 13.7 Primary Depletion of an Undersaturated Oil Reservoir

An undersaturated oil reservoir is being produced by primary depletion until reservoir pressure is just above bubble point pressure. The reservoir did not have an initial gas cap and is not in communication with any mobile water. There is no water production. What are the possible drive mechanisms?

Answer

Since there is no initial gas cap, we have $I_{gc} = 0$. The reservoir is under primary depletion; therefore there is no mobile water and no fluids are being injected. Consequently, we have $I_w = 0$ and $I_i = 0$. The remaining drive mechanisms are solution gas drive and connate water and rock expansion drive. The resulting drive indices equation is $I_{sg} + I_e = 1$. It can be used to establish a suitable material balance equation.

13.7 INFLOW PERFORMANCE RELATIONSHIPS

Wellbore inflow represents fluid flow from the reservoir into the wellbore. Reservoir fluid flow may be modeled using either analytical methods or numerical methods. Analytical methods rely on models of inflow performance relationships (IPR). Inflow refers to fluid entering the production tubing from the reservoir. By contrast, outflow refers to fluid flowing through the production tubing to surface facilities. An IPR is the functional relationship between reservoir production rate and bottomhole flowing pressure. Darcy's law is a simple example of an IPR for single-phase liquid flow. The gas well back pressure equation is an example of an IPR for single-phase gas flow. Figure 13.5 illustrates the relationship between an IPR curve and a tubing performance curve (TPC). The IPR curve is solid and the TPC curve is dashed. TPC curves represent outflow and can be calculated using models of fluid flow in pipe.

The IPR versus TPC plot is a plot of fluid flow rate q_{fluid} versus bottomhole flowing pressure p_{wf}. Reservoir pressure p_{res} is the pressure at $q_{fluid} = 0$. The intersection of the IPR and TPC curves identifies the flow rate and bottomhole flowing pressure that simultaneously satisfy inflow into the well from the reservoir and outflow from the well. The subscript "op" in the figure designates the operating flow rate $q_{fluid,op}$ at a particular bottomhole flowing pressure $p_{wf,op}$.

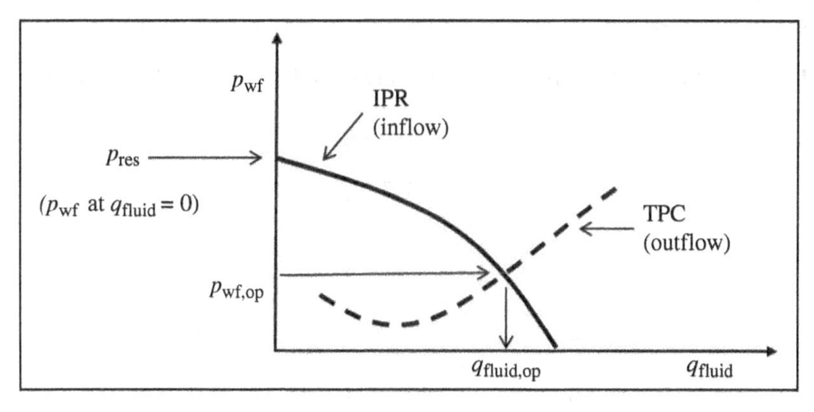

FIGURE 13.5 Illustration of an IPR versus TPC plot.

13.8 ACTIVITIES

13.8.1 Further Reading

For more information about production evaluation techniques, see Economides et al. (2013), Hyne (2012), Satter et al. (2008), and Craft et al. (1991).

13.8.2 True/False

13.1 Original fluids in place can be independently estimated using volumetric analysis and material balance analysis.

13.2 Decline curve analysis is used to fit a curve to production rate data plotted as a function of distance from the wellbore.

13.3 In a decline curve, production rate increases with time.

13.4 The Monte Carlo method can introduce uncertainty into decline curve analysis.

13.5 The sum of material balance drive indices for an oil reservoir equals one.

13.6 Water flooding is a primary production mechanism.

13.7 The radius of the circle in a bubble map indicates the magnitude of the variable.

13.8 The material balance method assumes that the change in reservoir pressure depends on the volume of fluids entering and leaving the reservoir.

13.9 The IPR curve represents fluid flow from the reservoir into the wellbore.

13.10 The gas cap is gas dissolved in the oil phase.

13.8.3 Exercises

13.1 Prepare a bubble map for the following cumulative oil production data:

	Location		Oil
Well	I	J	MMSTB
1	2	2	4.0
2	2	3	2.5
3	3	2	2.0
4	3	3	3.0
5	1	1	1.5
6	4	4	1.0

13.2 The equation $q^{-n} = nat + q_i^{-n}$ is a solution of $dq/dt = -aq^{n+1}$ where a, q_i are constants. What is the value of q at $t = 0$?

13.3 **A.** Use the exponential decline equation $q = q_i e^{-at}$ to plot oil flow rate as a function of time for a well that initially produces 12 000 STB/day and has a decline factor $a = 0.05$ per year. Time should be expressed in years and should range from 0 to 50 years. Use a logarithmic scale that ranges from 100 STB/day to 12 000 STB/day.

B. When does flow rate drop below 1000 STB/day?

13.4 The initial gas rate of a gas well is 1000 MSCF/day. The rate declines to 700 MSCF/day after two years of continuous production. Assume the rate of decline can be approximated by the exponential decline equation $q = q_i e^{-at}$. When will the gas rate be 50 MSCF/day? Express your answer in years after the beginning of production.

13.5 The initial production rate of an oil well is 100 STB/month and has an initial decline factor of 10% per year. Calculate flow rates using Arps' decline model for Arps' n-values of 0, 0.5, 1.0, and 2.0 at the following times: 1, 5, 10, 25, 50, and 100 years. Prepare semilog plots of rate versus time for all cases on the same graph.

13.6 **A.** Well PI = well productivity index = flow rate divided by pressure drawdown. Suppose the well originally produces 1000 STBO/day at a pressure drawdown of 20 psia. What is the well PI?

B. The well PI declined 6% a year for the first 2 years of production. What is the well PI at the end of year 2?

C. The well PI at the beginning of year 3 is 35 STBO/day/psia. A well test showed an increase in skin and the skin is positive. Does the well need a workover (Y or N)?

13.7 The production history for a gas reservoir is the following:

G_p (Bscf)	p (psia)	Z	p/Z
0.015	1946	0.813	2393
0.123	1934	0.813	2378
0.312	1913	0.814	2350
0.652	1873	0.815	2297
1.382	1793	0.819	2190
2.21	1702	0.814	2091
2.973	1617	0.828	1953
3.355	1576	0.83	1898
4.092	1490	0.835	1783
4.447	1454	0.838	1734
4.822	1413	0.841	1679

Plot G_p as a function of p/Z. The x-axis variable should be p/Z on an x–y coordinate system. Fit a trend line to the plot and determine OGIP.

13.8 **A.** Use volumetric analysis to estimate OOIP given the following data:

Bulk reservoir volume	9250 acre-ft
Oil saturation	0.70
Porosity	0.228
Initial pressure P_0	3935 psia
Oil FVF at P_0 is B_{oi}	1.3473 RB/STB
Water FVF is B_w	1.0 RB/STB

B. Use the material balance equation for an undersaturated oil reservoir with water influx to calculate original oil in place. In this case, assume no water influx and use the following production data:

Time (days)	Pressure (psia)	B_o (RB/STB)	N_p (MSTB)
0	3935	1.3473	0
91	3898	1.3482	46
183	3897	1.3482	91
274	3895	1.3482	137
365	3892	1.3483	183

C. Use the material balance equation for an undersaturated oil reservoir with water influx to calculate water influx W_e assuming original oil in place is the value obtained in Part A. The production data is given in Part B.

14

RESERVOIR PERFORMANCE

The commercial success of a reservoir development project is often determined by the performance of a reservoir. In this chapter, we outline workflows used to understand reservoir performance and prepare forecasts using reservoir flow models and then present several examples of reservoir performance.

14.1 RESERVOIR FLOW SIMULATORS

Decline curve analysis and material balance are two methods that are commonly used to understand and predict reservoir performance. A more sophisticated analysis of reservoir performance is provided using reservoir flow simulators. A reservoir flow simulator is a computer program that is designed to solve equations based on the physics and chemistry of fluid flow in porous media. Reservoir flow simulators include computer algorithms that represent reservoir structure, rock properties, fluid properties, interactions between rocks and fluids, and fluid flow in production and injection wells. A reservoir flow model is the data entered into a reservoir flow simulator.

Reservoir flow model studies are important when significant reservoir management options are being considered. By studying a range of scenarios, the reservoir management team can provide decision makers with information that can help them decide how to commit limited resources to activities that can achieve management objectives. These objectives may refer to the planning of a single well

Introduction to Petroleum Engineering, First Edition. John R. Fanchi and Richard L. Christiansen.
© 2017 John Wiley & Sons, Inc. Published 2017 by John Wiley & Sons, Inc.
Companion website: www.wiley.com/go/Fanchi/IntroPetroleumEngineering

or the development of a world-class reservoir. The reservoir flow model must include a representation of the reservoir, which often is based on reservoir characterization using flow units.

14.1.1 Flow Units

Geological characterization of the reservoir begins by defining flow units. A flow unit has been defined as "a volume of rock subdivided according to geological and petrophysical properties that influence the flow of fluids through it" (Ebanks, 1987). Geologic properties include the texture of the rock, mineral content, sedimentary structure, location and type of bedding contacts, and location and distribution of permeability barriers. Petrophysical properties include porosity and permeability distributions, formation compressibility and moduli, and the distribution of fluid saturations. Ebanks and his colleagues refined the definition to state that a flow unit is "a mappable portion of the total reservoir within which geological and petrophysical properties that affect the flow of fluids are consistent and predictably different from the properties of other reservoir rock volumes" (Ebanks et al., 1993, page 282). They identified several characteristics of flow units.

A specified volume of the reservoir contains one or more rock types that are reservoir quality and contain commercial fluids. The specified volume can also include rock types that are not reservoir quality. For example, the reservoir volume may include sandstone and conglomerate formations with layers of embedded shale. Compared to sandstone and conglomerate permeabilities, the permeability of shale is insignificant. Flow unit zones can be correlated and mapped at a scale that is comparable to the distance between wells. Flow units in the specified volume may be in communication with other flow units.

The initial identification of flow units in the reservoir is based on static data. Static data is considered data that does not change significantly with time and includes the structure of the reservoir. By contrast, dynamic data is data that changes with time and includes flow rates and pressure. Flow units identified using static data can be validated by comparing the identified flow units to dynamic data such as actual flow measurements in flow tests. It may be necessary to modify the set of flow units to assure consistency between static and dynamic data.

14.1.2 Reservoir Characterization Using Flow Units

Flow units can be used to characterize the reservoir. The goal is to produce field-wide maps of flow units and geologic regions. A stratigraphic and structural framework provides a 3-D representation of rock layers. Modern maps are prepared using geologic modeling software. The resulting geologic or static reservoir model includes a structure of the reservoir and rock properties distributed throughout the structure.

Rock properties are distributed by subdividing the 3-D volume into many smaller volumes called grid blocks, grid cells, or simply blocks or cells. Reservoir and rock properties include elevations or structure tops, gross thickness, net to gross

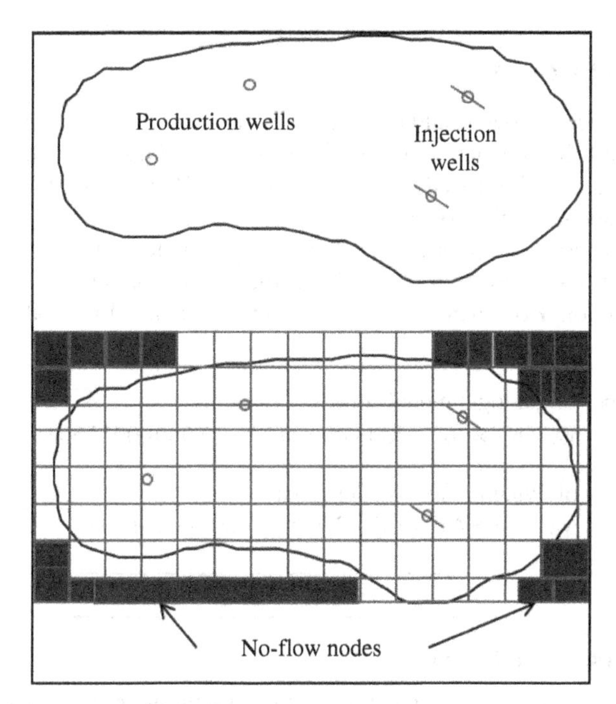

FIGURE 14.1 Overlay of a reservoir grid.

thickness, porosity, permeability, saturations, and numerical representations of geologic features such as faults, fractures, and vugs. Each geologic model of reservoir and rock properties is considered a realization of the reservoir. Other realizations can be prepared by recognizing that the data is uncertain, and it is possible to construct other 3-D models that are consistent with the range of uncertainty of the data.

Reservoir flow simulators are used to model fluid flow in the reservoir, which is based on the geologic model. A numerical representation of the geologic model is used in the reservoir flow simulator. Figure 14.1 illustrates the procedure for digitizing the reservoir. The upper figure is a plan view of a reservoir with four wells. The lower figure shows a grid placed over the plan view. The dark blocks in the lower figure are grid blocks, also known as grid cells, which do not contain pay. The other blocks contain pay. Reservoir and fluid properties are provided for every grid block. The reservoir simulator is designed to model flow between grid cells and into or out of grid cells through wells.

Reservoir management is most effective when as much relevant data as possible is collected and integrated into a reservoir management study. The preparation of a reservoir flow model can help an asset management team coordinate the acquisition of the resources needed to determine the optimum plan for operating a field. If reservoir flow model performance is especially sensitive to a particular parameter, then a plan should be made to minimize or quantify uncertainty in the parameter.

Example 14.1 Grid Size

A. A geologic model is built using a grid with 300 cells in the x-direction and 400 cells in the y-direction based on seismic areal resolution and 100 cells in the z-direction based on well log analysis. How many grid cells are in the static geologic model?

B. A flow model is upscaled from the geologic model using a grid with 150 cells in the x-direction, 200 cells in the y-direction, and 10 cells in the z-direction. How many grid cells are in the dynamic flow model?

Answer

A. Number of geologic model cells:
$NB_{geo} = NX \times NY \times NZ = 300 \times 400 \times 100 = 12\,000\,000$ blocks.

B. Number of dynamic flow model cells:
$NB_{flow} = NX \times NY \times NZ = 150 \times 200 \times 10 = 300\,000$ blocks.

Example 14.2 Grid Cell Size

A. Suppose a well is completed in a grid cell with the following properties: $\Delta x = \Delta y = 100$ ft, net thickness = 16 ft, and porosity = 0.14. What is the pore volume of the grid cell in reservoir barrels? Note: 1 bbl = 5.6146 cu ft.

B. If the well is producing 100 RB/day of fluid, what percentage of the grid cell pore volume is being produced in a 5-day timestep?

Answer

A. Cell pore volume is $V_p = \Delta x \times \Delta y \times \Delta z_{net} \times \phi$,; therefore
$V_p = 100\,\text{ft} \times 100\,\text{ft} \times 16\,\text{ft} \times 0.14 = 22\,400\,\text{ft}^3 = 3990\,\text{RB}$

B. The amount of fluid produced from the cell is $100\,\text{RB/day} \times 5\,\text{days} = 500\,\text{RB}$ which is approximately 12% of cell pore volume produced during the timestep.

14.2 RESERVOIR FLOW MODELING WORKFLOWS

Different workflows exist for designing, implementing, and executing reservoir projects. Modern reservoir flow modeling relies on two types of workflows: green field workflow and brown field workflow. Green fields include discovered, undeveloped fields and fields that have been discovered and delineated but are undeveloped. Brown fields are fields with significant development history. Both workflows are intended to be systematic procedures for quantifying uncertainty.

The workflow for conducting a probabilistic forecast of green field performance using a reservoir flow model is shown in Figure 14.2. The workflow begins by gathering data for the study. Key parameters and associated uncertainties are identified. For example, the permeability distribution and location of fluid contacts may not be well known. Probability

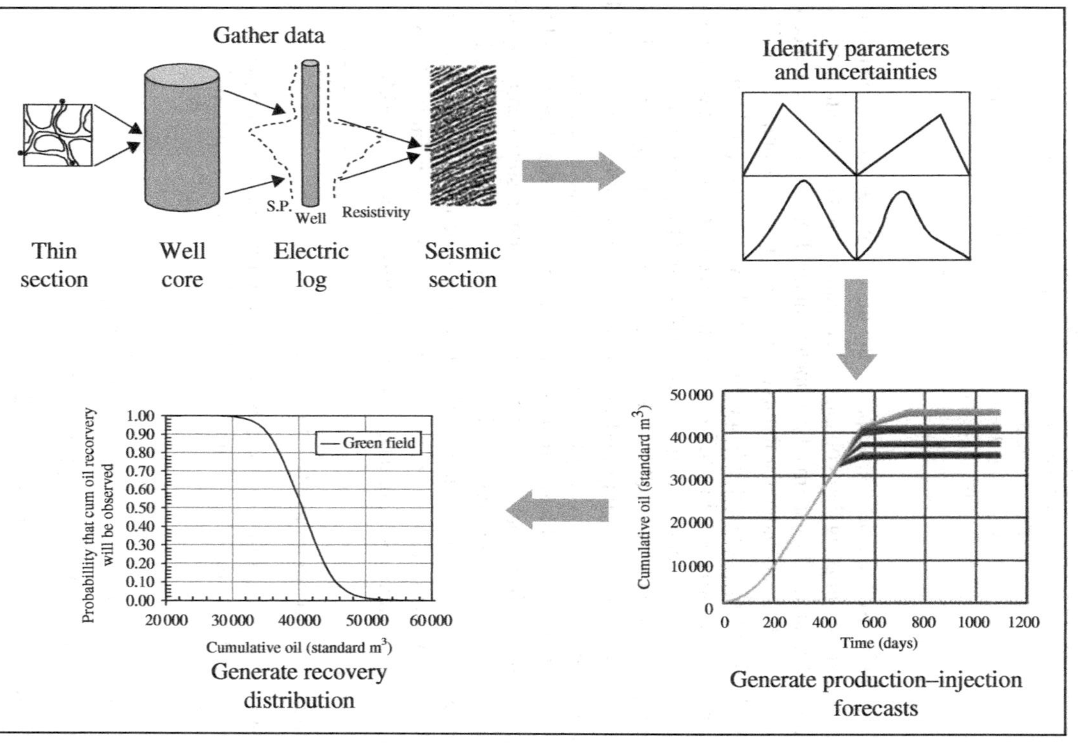

FIGURE 14.2 Green field flow modeling workflow.

TABLE 14.1 Brown Field Flow Modeling Workflow

Step	Task
B1	Gather data
B2	Identify key parameters and associated uncertainties
B3	Identify history match criteria and history match variables
B4	Generate forecast of field performance results
B5	Determine quality of history match
B6	Generate distribution of field performance results
B7	Verify workflow

Source: Fanchi (2010).

distributions can be used to characterize parameter uncertainty. A set of production forecasts is generated by sampling the probability distributions and developing a realization of the reservoir for each set of sampled parameter values. Reservoir performance is calculated for each realization and a distribution of recovery results is prepared. In the case of green fields, the results are relatively unconstrained by historical production.

The workflow for brown fields differs from the green field workflow because historical data is available to constrain the set of results used to generate a distribution of recovery forecasts. Two brown field workflows are currently being used in industry: deterministic reservoir forecasting and probabilistic reservoir forecasting. In deterministic reservoir forecasting, a single reservoir realization is selected and matched to historical performance. The history match is used to calibrate the flow model before the forecast is made. In probabilistic reservoir forecasting, a statistically significant collection, or ensemble, of reservoir realizations is prepared. Dynamic models are run for each possible realization, and the results of the dynamic model runs are then compared to historical performance of the reservoir. The workflow in Table 14.1 presents the steps for conducting a probabilistic brown field flow modeling workflow.

Example 14.3 Brown Field Model Reserves

Estimate P_{10}, P_{50}, P_{90} reserves for a brown field with a normal distribution of reserves. The distribution has a mean of 255 MSTBO and a standard deviation of 25 MSTBO.

Answer

Proved reserves $= P_{90} = \mu - 1.28\sigma = 223\,\text{MSTBO}$

Probable reserves $= P_{50} = \mu = 255\,\text{MSTBO}$

Possible reserves $= P_{10} = \mu + 1.28\sigma = 287\,\text{MSTBO}$

14.3 PERFORMANCE OF CONVENTIONAL OIL AND GAS RESERVOIRS

We have introduced many factors that affect the performance of reservoirs in previous chapters. For example, primary depletion of oil reservoirs depends on the natural drive mechanisms discussed in Chapter 13. In this section we consider examples of reservoir performance of conventional oil and gas reservoirs.

14.3.1 Wilmington Field, California: Immiscible Displacement by Water Flooding

A water flood uses injection wells to inject water into a reservoir. The injected water provides pressure support and can displace oil. A pattern water flood uses injection and production wells in a repeating pattern. The performance of the Wilmington Field in California (see Figure 14.3) illustrates immiscible displacement by water flooding.

The Wilmington Field is in the Los Angeles Basin and was discovered in 1932 as part of the development of the Torrance field. The Wilmington Field was first recognized as a separate field in 1936. The field is a northwest to southeast trending anticline that is approximately 13 mi. long and 3 mi. wide. It is crosscut into fault blocks that are vertically separated by normal faults that are perpendicular to the long axis of the anticline. The southeastern part of the field is beneath Long Beach Harbor and is called the Long Beach Unit (LBU). Otott and Clark (1996) reported that the Wilmington Field contained approximately 8.8 billion barrels of original oil in place (OOIP). The OOIP of the LBU of the Wilmington Field is greater than 3 billion barrels.

Seven stratigraphic zones have been identified between 2000 and 7500 ft true vertical depth subsea (TVDSS). From shallowest to deepest, the zones are the Tar, Ranger, Upper Terminal, Lower Terminal, Union Pacific, Ford, and "237." An unconformity lies above the Tar zone and another unconformity lies below the "237" zone. The zones and some fault blocks are sketched in Figure 14.4 (Fanchi et al., 1983; Clarke and Phillips, 2003).

The upper six zones consist of unconsolidated to poorly consolidated sandstones interbedded with shales and exhibit high compaction characteristics. Surface subsidence up to 29 ft was observed as a result of producing the onshore part of the

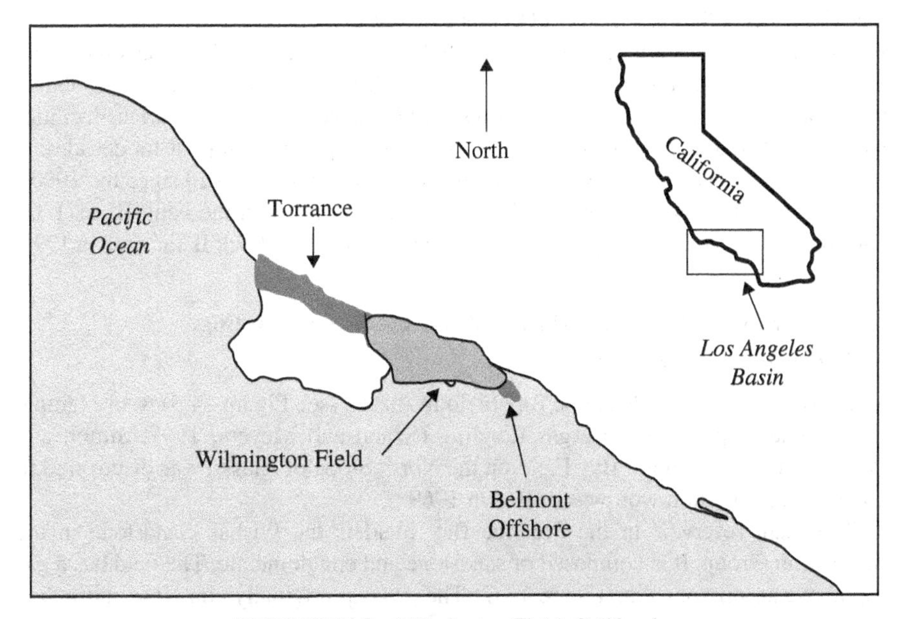

FIGURE 14.3 Wilmington Field, California.

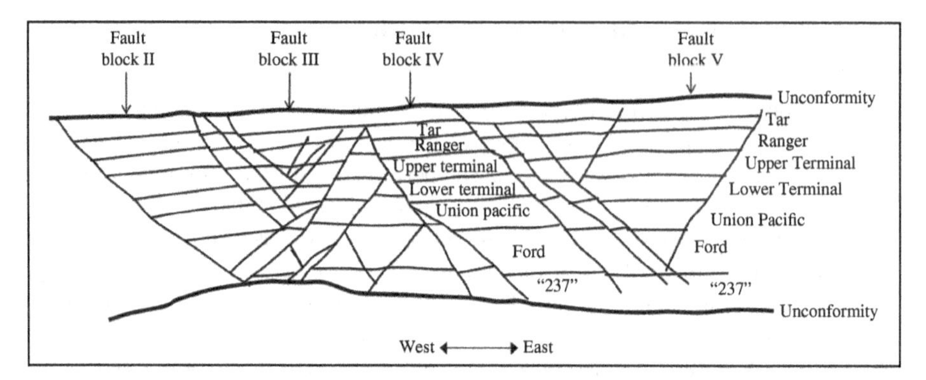

FIGURE 14.4 Illustration of Wilmington Field fault blocks and stratigraphic zones.

Wilmington Field (the non-LBU area) by pressure depletion from 1936 to the 1950s. The principal drive mechanism was solution gas drive. Some onshore areas subsided below sea level but were protected by dikes. As a consequence of subsidence, the LBU area could not be developed until an agreement was reached with governing agencies that would prevent further subsidence. Water flooding was implemented in the LBU area from the outset.

The relatively low API gravity (12–21°API) and high viscosity (15–70 cp) of oil in the Ranger zone implied that the mobility ratio for immiscible displacement of oil by injected water would be unfavorable. A staggered line drive water flood with 10-acre well spacing was implemented in the LBU Ranger zone. Peripheral water flooding was used in other zones.

LBU wells were directionally drilled from nearby Pier J and four artificial islands. The artificial islands were built in 1964 and were named after four astronauts that lost their lives during the early years of US space exploration (Grissom, White, Chaffee, and Freeman). Water injection rate was as high as 1 million barrels per day. Peak oil production rate was 150 000 barrels of oil per day in 1969. Oil production rate decline combined with water production rate increase has resulted in the need to provide surface facilities which have had to handle high water cuts (80–97%) in produced liquids for decades.

Several improved oil recovery techniques have been implemented since the 1990s. For example, the first horizontal well project was conducted in the Fault Block I Tar zone in 1993. A steam flood project was initiated in the Fault Block II Tar zone in 1995.

14.3.2 Prudhoe Bay Field, Alaska: Water Flood, Gas Cycling, and Miscible Gas Injection

The performance of the Prudhoe Bay Field in Alaska (see Figure 14.5) is an example of miscible displacement by gas flooding (Szabo and Meyers, 1993; Simon and Petersen, 1997). Prudhoe Bay Field on the North Slope of Alaska was discovered in 1968. A confirmation well was drilled in 1969.

The main reservoir in the Prudhoe Bay Field is the Ivishak Sandstone in the Sadlerochit Group. It is composed of sandstone and conglomerate. The field has a gas cap above an oil zone (see Figure 14.6). The gas cap originally contained more than

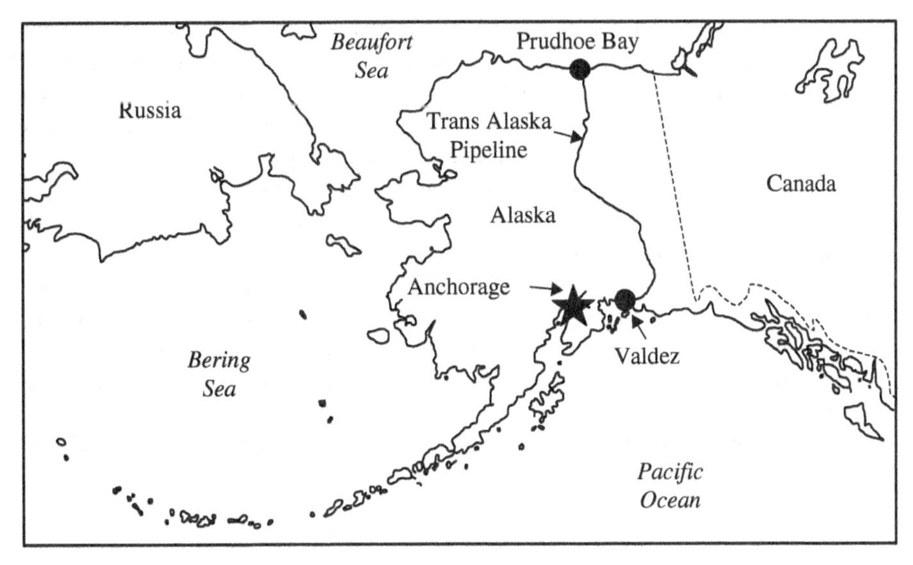

FIGURE 14.5 Prudhoe Bay Field, Alaska.

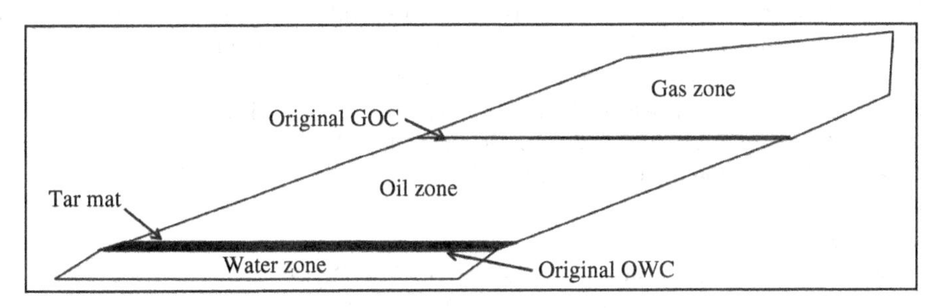

FIGURE 14.6 Schematic cross section of the Prudhoe Bay Field, Alaska.

30 TCF gas and the oil zone originally contained more than 20 billion barrels of oil with an average API gravity of approximately 28°API. A 20- to 60-ft tar mat called the Heavy Oil Tar (HOT) is at the base of the oil zone above the oil–water contact. Tar mats are made up of high molecular weight, high viscosity hydrocarbons. They are often found at the base of an oil column and function as low permeability zones that hinder or prevent liquid flow. The tar mat at the base of the Prudhoe Bay Field oil zone blocked aquifer influx, so gas cap expansion and gravity drainage were the dominant primary recovery mechanisms. Pressure support to the field had to be provided by injection.

Prudhoe Bay Field was brought online in 1977 after the Trans Alaska Pipeline System was built to transport oil from Prudhoe Bay to Valdez (see Figure 14.5). A plateau oil production rate of 1.5 MMSTB/D was reached in 1979 and was maintained until 1988. Produced gas was reinjected into the gas cap to provide pressure support. The production and reinjection of gas is called gas cycling. The reinjected gas contributed to gas cap expansion. A Central Gas Facility was installed to separate natural gas liquid from the produced gas stream in 1986. The natural gas liquid could be mixed with produced oil and transported through the Trans Alaska Pipeline System.

Seawater injection commenced in 1984. The scope of the water flood was determined by the character of the reservoir, aquifer strength in different parts of the reservoir, water sources, production performance, and the timing of gas sales. Water flood infrastructure had to be built and included a seawater treatment plant, water injection plants, and a network of seawater distribution pipelines.

The reinjection of hydrocarbon gas created an opportunity for implementing a miscible gas injection project. Miscibility is achieved by injecting gas at a high enough pressure that the interfacial tension between injected gas and oil is significantly reduced so that the gas and oil phases combine into a single phase. Miscibility occurs when reservoir pressure is greater than minimum miscibility pressure (MMP) of the system, which can be measured in the laboratory. Injected gas mixes with *in situ* oil and swells the oil in a miscible process. A small-scale pilot program was conducted in 1982 to confirm the viability of miscible gas injection as an EOR project. The large-scale Prudhoe Bay Miscible Gas Project began in 1987 and was subsequently expanded.

Example 14.4 Production Stages

A. Primary recovery from an oil reservoir was 100 MMSTBO. A water flood was implemented following primary recovery. Incremental recovery from the water flood was 25% OOIP. Total recovery (primary recovery plus recovery from water flooding) was 50% OOIP. How much oil (in MMSTBO) was recovered by the water flood?

B. What was the OOIP (in MMSTBO)?

Answer

A. Total recovery (primary + water flood) = 50% OOIP
Incremental water flood recovery = 25% OOIP
Therefore primary recovery = total recovery − water flood recovery = 25% OOIP
Water flood recovery = primary recovery = 100 MMSTBO

B. Primary recovery plus water flood recovery = 200 MMSTBO
Total recovery = 200 MMSTBO = 0.5 × OOIP
So OOIP = 200 MMSTBO/0.5 = 400 MMSTBO

14.4 PERFORMANCE OF AN UNCONVENTIONAL RESERVOIR

In this section we consider examples of unconventional reservoir performance.

14.4.1 Barnett Shale, Texas: Shale Gas Production

Geologists discovered a shale outcrop near the Barnett Stream in San Saba County, Texas, in the early twentieth century. The stream was named after John W. Barnett who settled a large tract of land in central Texas. The thick, black, organic-rich shale

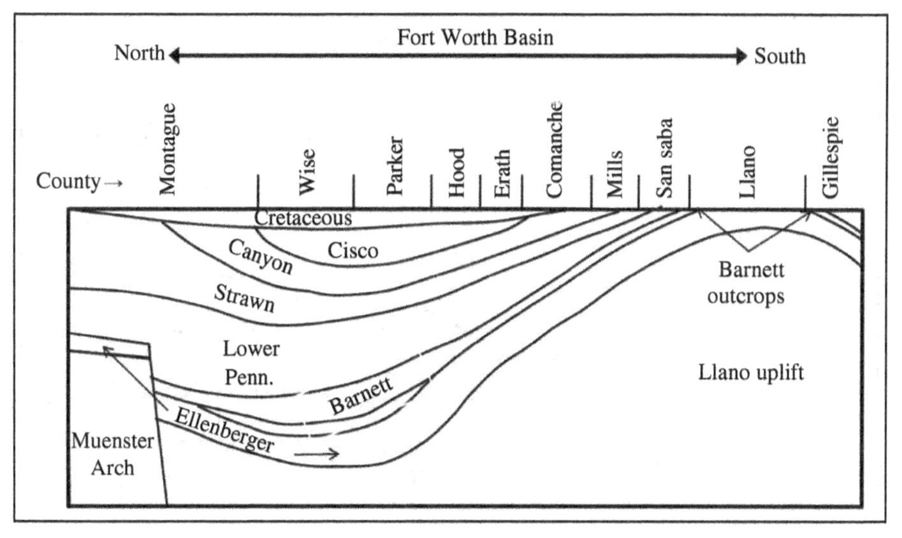

FIGURE 14.7 Cross section of the Fort Worth Basin.

was named the Barnett Shale. The Barnett Shale outcrop in San Saba County is shown in Figure 14.7.

The Barnett Shale is part of the Fort Worth Basin. The shale is a very low permeability mud rock that is several hundred feet thick at depths of 5000–8000 ft. Commercial quantities of gas are being produced from the shale in several north central Texas counties (Figure 14.8). Production was not economically feasible until Mitchell Energy and Development Company (MEDC), led by George P. Mitchell, combined directional drilling and hydraulic fracturing (Steward, 2013).

MEDC drilled the discovery well for the Newark East Field in the Barnett Shale in 1981. Vertical wells drilled and completed in the shale produce gas at relatively low flow rates because shale permeability can be as small as 0.01 md to 0.00001 md in unfractured shale (Arthur et al., 2009). By 1986, MEDC had shown that shale gas production depended on establishing a large pressure difference between shale matrix and fractures and the amount of source rock contacted by induced fractures. The use of hydraulic fracturing in the completion process was the first major technological breakthrough to increase flow rate. Hydraulic fracturing in a vertical well tends to create horizontal fractures and increases shale gas production rate. Directional drilling of horizontal wells was the second major technological breakthrough to increase flow rate. A horizontal test well was drilled in the Barnett Shale in 1981 by MEDC and the Gas Research Institute (GRI). GRI administered research funding provided by a surcharge on natural gas shipments through interstate pipelines. Directional drilling made it possible to drill and complete thousands of feet of shale. Hydraulic fracturing along segments of the horizontal well created vertical fractures in the shale. The combination of directional drilling and hydraulic fracturing resulted in commercial shale gas production rates. MEDC merged with Devon Energy in 2002.

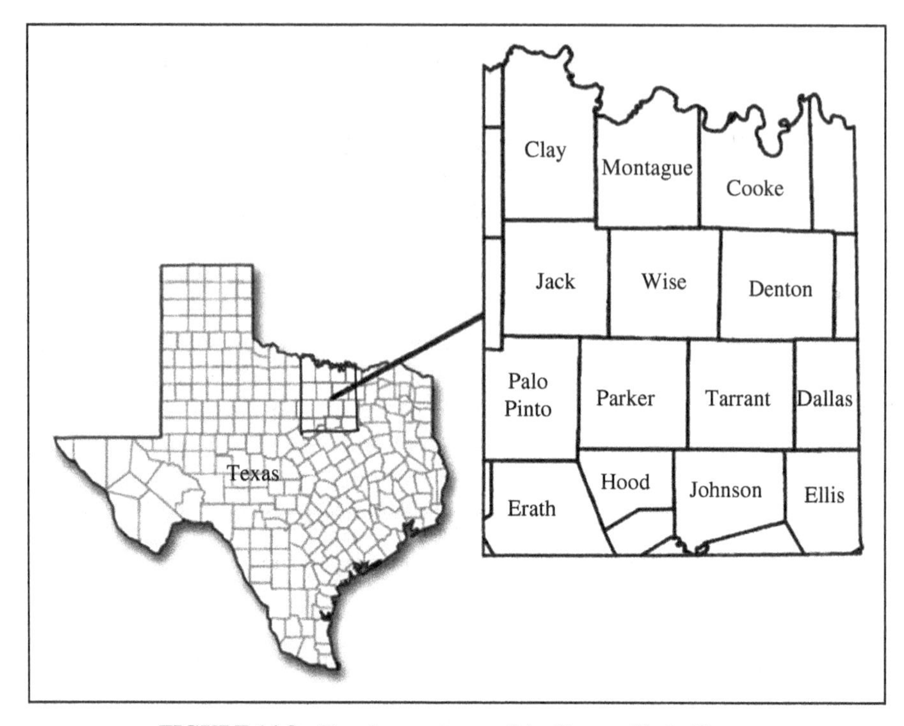

FIGURE 14.8 Development area of the Barnett Shale, Texas.

Hydraulic fracturing is conducted in several stages (Arthur et al., 2009). Each stage requires tens of thousands of barrels of water and has a length ranging from 1000 to 5000 ft. Over 90% of the injected volume is water, approximately 9% of the injected volume is proppant such as sand, and the remaining fraction of 1% is composed of chemicals. The most common hydraulic fracturing technique used in the Barnett Shale is "slickwater" hydraulic fracturing. "Slickwater" refers to the combination of water and a friction-reducing chemical additive that allows the water-based fluid to be pumped into the formation faster.

Drilling and completion issues were not the only challenges faced by operators in the Barnett Shale. Some of the most productive areas of the Barnett Shale were beneath the Dallas–Fort Worth Metroplex. Barnett Shale development required drilling in urban areas. The American Petroleum Institute issued ANSI–API Bulletin 100–3 on Community Engagement. Operators were advised to prepare communities for exploration activities in their neighborhoods and minimize disruption to communities in addition to managing resources (Donnelly, 2014).

The drilling process followed several steps: pad site development, rig setup, drilling, completion and hydraulic fracturing, gas gathering, production, and abandonment. Site preparation required acquiring right of way to pad sites in areas ranging from residential to industrial. Road access was controlled because some roads could not handle the weight of the equipment used in operations, while others handled too much day-to-day traffic to allow field operations to proceed during regular working hours.

FIGURE 14.9 Drilling rig on a shale development well pad.

The pad site had to be as small as possible in the urban area yet large enough to accommodate operations that included rig setup, drilling, hydraulic fracturing, storage tanks for produced liquids, and pipelines. Some sites included gathering centers, gas compressors, and gas treatment facilities. Preparation included clearing and grading the pad site. Barriers were erected to mitigate noise pollution and restrict the view of operations. Figure 14.9 shows a drilling rig on a well pad and a barrier designed to block the view of operations and reduce noise emanating from the well pad. More aesthetically pleasing walls were erected on some pad sites where large-scale operations were completed.

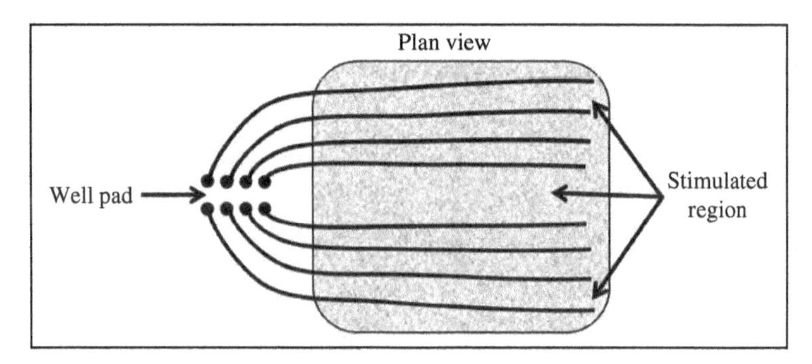

FIGURE 14.10 Sketch of wellbore trajectories drilled from a shale development well pad.

FIGURE 14.11 Surface equipment at a shale gas well pad.

Many wells were drilled at each pad site (Figure 14.10). Wellbore trajectories were designed to contact the largest volume of stimulated rock. The size of the stimulated shale region can be estimated using measurements of microseismic events associated with rock breaking as a result of hydraulic fracturing.

Produced gas from many Barnett Shale gas wells requires relatively little treatment because it is dry and predominantly methane. A separator may be needed to separate gas and liquids at some sites. Produced liquids are stored in storage tanks until they are transported by truck away from the site. Produced gas is compressed at compressor stations and sent through flow lines to gas processor facilities. Figure 14.11 shows compressors, storage tanks, and noise mitigation barrier at a shale gas well pad in Fort Worth area. Eventually the operation will be completed and well sites will be abandoned. At that point, reclamation will be necessary.

Shale gas operators must also manage environmental issues that include produced water handling, gas emissions, and injection-induced seismicity (IIS). Drilling and hydraulic fracturing operations typically use approximately three to five million gallons of water per well. Water used for shale gas production operations in the Fort Worth area has been estimated to be about 2% of total water use. Over a third of the water injected into the well during hydraulic fracturing operations is returned with produced gas as flowback water. Produced water is recycled or injected into the water-bearing Ellenberger limestone. The Ellenberger limestone is below the Barnett Shale at a depth of approximately 10 000–12 000 ft. The disposal wells are classified as Class II disposal wells by the US Environmental Protection Agency. Class II disposal wells inject produced water from oil and gas operation into subsurface sedimentary formations.

Wastewater disposal has been associated with earthquakes and is an example of IIS. IIS is earthquake activity resulting from human activity. A significant increase in low magnitude earthquake activity has been observed in regions where hydraulic fracture operations in shale have been conducted. Microseismic events are associated with hydraulic fractures. Larger magnitude earthquakes on the order of magnitude 3 appear to be correlated to wastewater injection into disposal wells at high flow rates (Hornbach et al., 2015).

Gas emissions from shale gas production wells have been reported (King, 2012; Jacobs, 2014). Gas leaks are associated with faulty well casing and poor cement jobs. Cement may not uniformly fill the annulus of wells, especially directionally drilled or horizontal wells. The result can be a flow path from the formation to shallower formations, such as water-bearing formations.

14.5 PERFORMANCE OF GEOTHERMAL RESERVOIRS

Exploration, drilling, and reservoir technology developed in the oil and gas industry can be applied to the discovery, development, and management of geothermal reservoirs. In this section we use reservoir management principles to understand the relationship between a geothermal reservoir and a geothermal power plant on the Big Island of Hawaii.

Geothermal reservoirs in the crust of the Earth are typically heated by magma close to the surface. For example, the Hawaiian Islands were formed by the movement of the Pacific plate over a fixed hotspot in the mantle. The hotspot is at the interface between magma convection cells in the mantle. The Big Island of Hawaii is the closest Hawaiian island to the hotspot and is the youngest of the Hawaiian Islands. Kauai is farthest from the hotspot and is the oldest Hawaiian island. Fractures in the crust carry magma from the hotspot to the surface and are sources of heat for water-bearing formations.

The Puna Geothermal Venture (PGV) uses geothermal power plants to extract heat from geothermal reservoirs. PGV is located in the Puna district of the Big Island (Figure 14.12). The town of Puna is southeast of the city of Hilo and east of the active crater of the Kilauea Volcano. The dashed lines in Figure 14.12 bound the Kilauea rim. Lava flows from Kilauea tend to move downstructure from the Kilauea crater toward the coast along the Kilauea rim. PGV power plants are located at the easterly base of the Kilauea rim.

PGV began commercial operation in 1993. Electrical power from the PGV power plants is sold to Hawaii Electric Light Company. Two types of air-cooled power plants with a combined generating capacity of 38 MW are installed at PGV: a combined cycle system and a binary system.

Figure 14.13 illustrates key components of one of the power plants: an air-cooled binary geothermal power plant. Binary refers to the use of two fluids and air-cooled

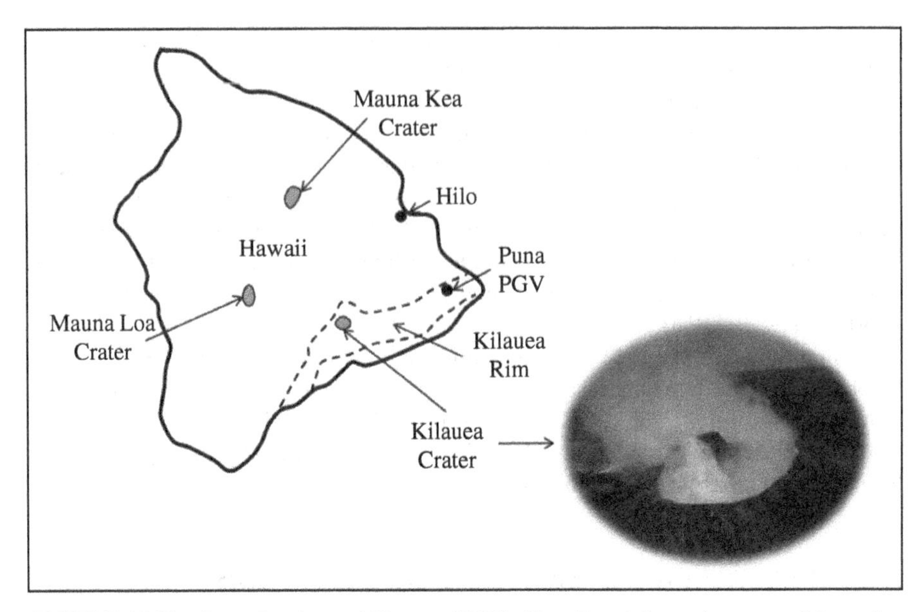

FIGURE 14.12 Puna Geothermal Venture (PGV), Hawaii and the main crater of the active Kilauea Volcano (after Fanchi and Fanchi, 2016).

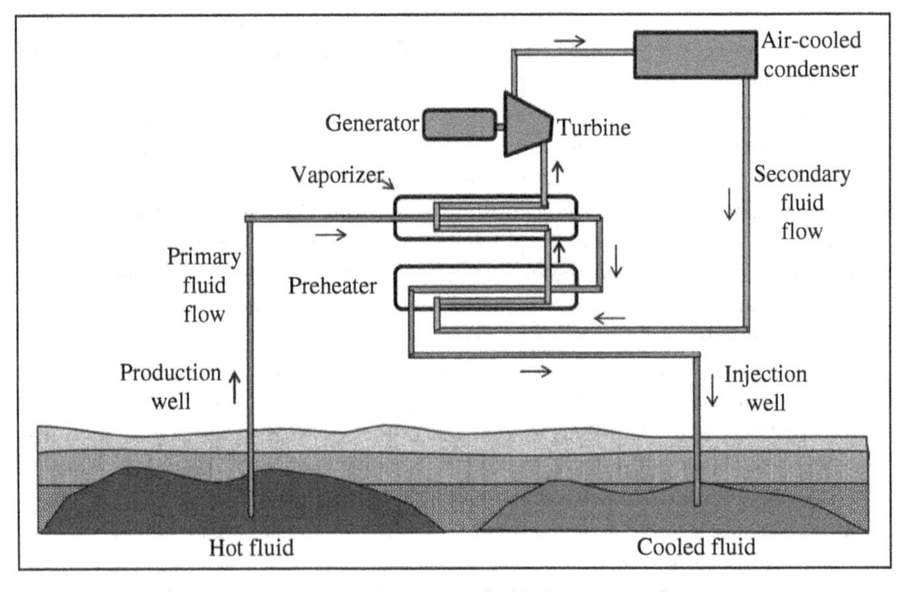

FIGURE 14.13 Schematic of an air-cooled binary geothermal power plant.

refers to the fluid used in the condenser. The production well produces hot fluid from the geothermal reservoir. The produced fluid is the primary fluid entering the power plant and a secondary fluid is the working fluid used to generate electricity. The working fluid is a fluid with a lower boiling point than the primary fluid. Examples of binary fluids include organic fluids such as isobutane, pentane, or a refrigerant.

The working fluid is circulated through the preheater, vaporizer, turbine, and condenser. The working fluid is warmed up in the preheater and vaporized in the vaporizer. The high-pressure working fluid vapor drives a turbine. Mechanical energy from the turbine is transformed into electrical energy by the generator. The working fluid vapor is condensed by the air-cooled condenser and recycled through the preheater.

Heat is transferred from the primary produced fluid to the working fluid in the vaporizer and again in the preheater. The cooler primary fluid is injected into a formation. If the formation is the geothermal reservoir, the cooler injected fluid can reduce the temperature of the geothermal reservoir over time. A heat source like the Hawaii hotspot provides geothermal heat to overcome the cooling effect of reinjected fluid.

The PGV facility is located near residential communities. During the early days of the venture, well failures and gas emissions resulted in serious health concerns. Geothermal fluids are highly toxic and corrosive. Well failures can be caused by corrosion of tubing and casing and exposure to excessive heat by encountering magma or lava. A well failure can result in the emission of produced gases into the environment. Emissions from geothermal water and steam include benzene, hydrogen sulfide, ammonia, mercury vapor, methane and other hydrocarbons, carbon dioxide, arsenic, radon, and radioactive materials that emit alpha and beta emissions. Several of these chemicals can adversely affect human health. For example, hydrogen sulfide can be fatal at 700 ppm, and the PGV wells contained from 750 to 1100 ppm.

Several practices have been implemented to protect the environment in response to community concerns. The footprint, or areal size, of the facility has been minimized and the facility is surrounded by noise reduction enclosures. A closed system is designed to reinject 100% of the produced geothermal fluid, and the entire system is designed to achieve near-zero emissions. Continuous monitoring systems have been built to detect undesired emissions.

14.6 ACTIVITIES

14.6.1 Further Reading

For more information about reservoir flow modeling and reservoir performance, see Gilman and Ozgen (2013), Fanchi (2010b), Satter et al. (2008), Carlson (2003), and Towler (2002).

14.6.2 True/False

14.1 Geostatistics is a method for spatially distributing reservoir parameters.

14.2 Data preparation for a flow model study is usually the quickest and least important step in the process.

14.3 A simulation study can help coordinate activities as a modeling team gathers the resources it needs to determine the optimum plan for operating a field.

14.4 A reservoir flow model is a dynamic model.

14.5 History matching calibrates the reservoir flow model.

14.6 A reservoir flow model should not be used in the reservoir management process if a field has no production history.

14.7 Depletion of the Wilmington Field caused subsidence in the Long Beach area.

14.8 History matching is an essential part of green field flow modeling.

14.9 A realization is one representation of reservoir geology.

14.10 From a business perspective, the objectives of a reservoir model study should yield a solution to an economically important problem.

14.6.3 Exercises

14.1 A grid block in a reservoir flow model is 275 ft wide in the *x*-direction, is 275 ft wide in the *y*-direction, and has a gross thickness of 20 ft. What is the bulk volume of the grid block in cubic feet?

14.2 **A.** Suppose a well is completed in a grid block with the following properties: length = width = 100 ft, net thickness = 15 ft, and porosity = 0.25. What is the pore volume of the grid block in reservoir barrels? Note: 1 bbl = 5.6146 ft^3.

B. If the well is producing 500 RB/day of fluid, what percentage of the grid block pore volume is being produced each day?

14.3 **A.** A reservoir is 10 mi. long and 6 mi. wide. Define a grid with $\Delta x = \Delta y = 1/10$ mi. What is the number of grid blocks needed to cover the areal extent of the reservoir?

B. If five model layers are used, what is the total number of grid blocks in the model?

14.4 **A.** If initial oil saturation is 0.7, residual oil saturation is 0.25, and pore volume is 961 800 RB in a grid block, what is the volume of mobile oil in the grid block? Express your answer in RB.

B. If a well produces 500 RB oil/day from the block, how long does it take to produce all of the mobile oil in the grid block?

14.5 **A.** A geologic model is built using a grid with 300 blocks in the *x*-direction and 400 blocks in the *y*-direction based on seismic areal resolution and 35 blocks in the *z*-direction based on well log analysis. How many grid blocks are in the static geologic model?

B. A flow model is upscaled from the geologic model using a grid with 150 blocks in the *x*-direction, 200 blocks in the *y*-direction, and 10 blocks in the *z*-direction. How many grid blocks are in the dynamic flow model?

14.6 The standard deviation of a population σ_{pop} with mean μ and N values $\{X_i\}$ is

$\sigma_{pop} = \sqrt{\frac{1}{N}\sum_{i=1}^{N}(X_i - \mu)^2}$. The standard deviation σ_{sample} of N' values in a subset of values sampled from the population is $\sigma_{sample} = \sqrt{\frac{1}{N'-1}\sum_{i=1}^{N'}(X_i - \mu)^2}$.

A. Estimate P_{10}, P_{50}, P_{90} reserves for a brown field model study. Assume a normal distribution of reserves. The distribution has a mean of 303.1 MMSTBO and a population standard deviation $\sigma_{pop} = 47.8$ MMSTBO.

B. Estimate P_{10}, P_{50}, P_{90} reserves for a brown field model study. Assume a normal distribution of reserves. The distribution has a mean of 303.1 MMSTBO and a sample standard deviation $\sigma_{sample} = 50.3$ MMSTBO.

15

MIDSTREAM AND DOWNSTREAM OPERATIONS

The oil and gas industry can be separated into three sectors: upstream, midstream, and downstream. The upstream sector includes the subsurface resource, its production to the surface, and the basic facilities at the well location (such as wellhead, separator, and storage tanks). The midstream sector connects the upstream and downstream sectors, and it encompasses the transportation and storage of oil and gas between upstream production operations and downstream refining and processing operations. An alternative classification to the three-sector system is to view the oil and gas industry as two sectors: upstream and downstream. In the case of the two-sector system, the midstream sector is part of the downstream sector. We discussed the upstream sector previously. The midstream and downstream sectors are discussed in this chapter. We then present a case study to illustrate the effort that can be required to create the infrastructure needed for midstream and downstream operations.

15.1 THE MIDSTREAM SECTOR

Midstream operations are designed to transport hydrocarbons from upstream production operations to downstream refining and processing operations. Many modes of transportation are used to transport oil and gas. They include pipelines, tanker trucks and ships, and trains. The mode of transport depends on such factors as safety, distance, and state of the fluid. For example, pipelines are often the first choice

Introduction to Petroleum Engineering, First Edition. John R. Fanchi and Richard L. Christiansen.
© 2017 John Wiley & Sons, Inc. Published 2017 by John Wiley & Sons, Inc.
Companion website: www.wiley.com/go/Fanchi/IntroPetroleumEngineering

FIGURE 15.1 Installing onshore pipelines. (Source: © energy.fanchi.com (2010).)

for transportation. But there are many obstacles to building a network of pipelines capable of transporting produced hydrocarbons from upstream to downstream operations.

Laying pipelines is a construction project that requires specialized heavy equipment, as shown in Figure 15.1. Pipelines are used to transport fluids both onshore and offshore. The transport distance can be a few miles to thousands of miles. A right of way is required to lay pipelines, which involves negotiating with stakeholders that control the right of way.

Pipelines that transport fluids over long distances usually need compressor stations or pump stations to keep fluids moving. Gas compression can also maximize use of space in the pipeline by increasing the density of gas flowing through the pipeline. A gas compressor is shown in Figure 15.2. The metal wall helps mitigate the amount of noise that reaches neighborhoods on the other side of the wall. The number and location of compressor stations and pump stations depends on distance traveled and terrain. Pipelines are built over several different types of terrain, such as plains, mountain ranges, deserts, swamps, frozen tundra, and bodies of water.

The type of terrain impacts construction and maintenance. Hostile environments such as the desert, Arctic tundra, and North Sea increase the difficulty and cost of constructing a pipeline. Maintenance is needed to prevent pipeline leaks. Pipelines can leak if they crack or burst. Hydrocarbon leaks are harmful to the environment and can impact community support for the pipeline.

Pipeline maintenance helps keep fluids moving efficiently. For example, gas flowing in a pipeline can contain water dissolved in the gas phase. Pressure and temperature conditions change over the length of the pipeline. Under some conditions,

FIGURE 15.2 Gas compressor. (Source: © energy.fanchi.com (2015).)

water can drop out of the gas phase and form a separate water phase that affects gas flow along the pipeline. Pistonlike scrubbing devices called "pigs" are run through pipelines to displace liquid phases to exit points along the pipeline route. Pipeline pigs are driven by the flow of fluids in the pipeline. They remove unwanted debris such as water and residual wax inside the pipeline. Intelligent pigs contain sensors that can be used to inspect the internal pipeline walls and identify possible problems such as corrosion.

Pipelines often pass through multiple jurisdictions. The route may be within a nation or cross-national borders. This introduces a geopolitical component to midstream operations. It can also require security in regions of political instability.

The Henry Hub is an interstate pipeline interchange that is located in Vermilion Parish, Louisiana, in the United States. The interchange is the delivery point for some natural gas sales contracts. The price of gas in these contracts is based on the Henry Hub benchmark price.

Other oil and gas transportation modes include trucks, trains, barges, and tankers. Trucking is the most versatile form of transportation on land because trucks can travel to virtually any land-based destination. Transportation by rail can be a cost-effective and efficient mode of transporting large volumes of oil. Barges and tankers transport oil and gas over bodies of water ranging from rivers to oceans.

Transportation requires moving a commodity from one point to another. It is often necessary to provide facilities that gather and store fluids at the upstream site and provide storage facilities at the downstream site. This requires gathering and processing operations, as well as terminal developers and operators. Gathering lines are smaller diameter pipelines that connect wells to larger diameter trunk lines. Large volumes of gas and oil can be stored in spherical and cylindrical storage tanks, salt caverns, and depleted reservoirs.

Stored gas is considered either base gas or working gas. Base gas is the amount of gas that remains in the storage facility to maintain a safe operating pressure. Working gas is the amount of gas that can be withdrawn from the storage facility for use. In some cases, gas may need to be parked, that is, stored temporarily. Parking may be used by a customer until a better price can be obtained for the customer's gas. Demand for gas is seasonal, which is another reason to store gas temporarily until demand increases.

15.2 THE DOWNSTREAM SECTOR: REFINERIES

The downstream sector is closest to the consumer and encompasses natural gas processing, oil refining, and distribution of products. Processing begins at the well site where the produced wellstream is separated into oil, water, and gas phases. Further processing at natural gas plants and oil refineries such as the refinery shown in Figure 15.3 separates the hydrocarbon fluid into marketable products.

Products associated with refining and processing are shown in Table 15.1. An oil refinery converts a typical barrel of crude oil into gasoline, diesel, jet fuel, liquefied petroleum gas (LPG), heavy fuel oil, and other products. A natural gas processing plant purifies natural gas and converts it into products such as LPG, liquefied natural gas (LNG), and fuel gas for residential, commercial, and industrial use. The operation of refineries is discussed in this section, and natural gas processing plants are discussed in the next section.

FIGURE 15.3 South Texas refinery. (Source: © energy.fanchi.com (2002).)

TABLE 15.1 Downstream Sector Products

Liquefied petroleum gas (LPG)	Asphalt
Liquefied natural gas (LNG)	Synthetic rubber
Propane	Plastics
Gasoline	Lubricants
Diesel oil	Pharmaceuticals
Jet fuel	Fertilizers
Heating oil	Pesticides
Other fuel oils	Antifreeze

FIGURE 15.4 Distillation towers at a Texas refinery. (Source: © energy.fanchi.com (2015).)

15.2.1 Separation

Refining changes crude oil into finished products using three major processes: separation, conversion, and purification processes. The first step in refining is separation with one or more distillation towers. A distillation tower, which is also known as an atmospheric crude fractionator, separates crude oil into mixtures of components based on the boiling points of the mixtures. Figure 15.4 shows some distillation towers at a refinery.

Figure 15.5 is a schematic of a distillation tower system. A furnace is used as a heat source to heat and vaporize liquid crude oil. The liquid phase that remains in the lowermost section of the tower after heating is a mixture of components with the highest boiling points, while the vapor phase that rises in the tower is a mixture of components with lower boiling points. A table of boiling point (B.P.) ranges and hydrocarbon components for several fractions is shown in Figure 15.5 (Olsen, 2014). Each fraction is a mixture of hydrocarbon components that can be used to make different products. For example, naphtha is used to make gasoline for vehicles and kerosene is used to make jet fuel.

The temperature in the distillation tower is hottest in the lowermost section because of the furnace. The hot vapor cools from approximately 700°F in the

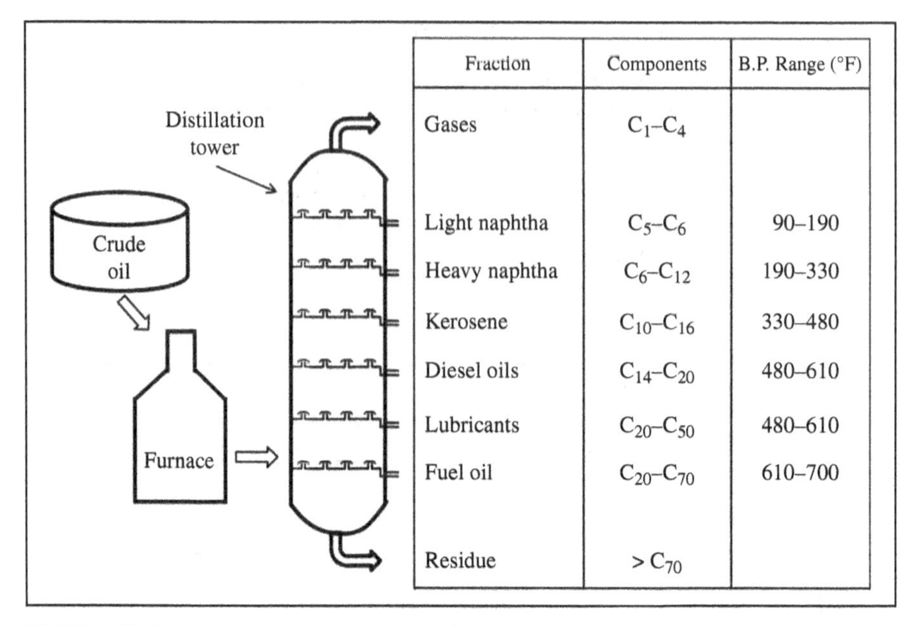

Fraction	Components	B.P. Range (°F)
Gases	C_1–C_4	
Light naphtha	C_5–C_6	90–190
Heavy naphtha	C_6–C_{12}	190–330
Kerosene	C_{10}–C_{16}	330–480
Diesel oils	C_{14}–C_{20}	480–610
Lubricants	C_{20}–C_{50}	480–610
Fuel oil	C_{20}–C_{70}	610–700
Residue	$> C_{70}$	

FIGURE 15.5 Typical distillation tower fractions and components. (Source: © energy.fanchi.com (2015).)

lowermost section of the tower to approximately 90°F in the uppermost section of the tower as the vapor rises in the distillation tower. Each type of hydrocarbon component condenses from the vapor to the liquid state in a temperature range that depends on the boiling point of the molecule.

The boiling point for a hydrocarbon molecule typically increases as the number of carbons in the molecule increases. The normal boiling points (measured at pressure equal to 1.00 atm) for straight chain, or normal, alkanes are shown in Figure 15.6 starting with methane and ending with normal triacontane, n-$C_{30}H_{62}$. Crude distillation towers typically operate near 30 psig, so the normal boiling points in Figure 15.6 do not precisely reflect volatility for the conditions in the tower, but they do reflect relative volatility.

The decline in temperature from the lowermost section of the tower to the uppermost section of the tower establishes sections of the tower with temperature ranges that are suitable for collecting hydrocarbons that condense in the temperature range of a particular section. The liquid mixture collected at each section is the product stream for that section. Product streams vary from high molecular weight, viscous liquids at the base of the tower to low molecular weight gases at the top of the tower.

We can model the separation of components that occurs in each section of the tower with the flash calculation explained in Chapter 11. The flash calculation starts with specification of feed composition and k-values. The next step is the two-phase check, which determines if the feed will separate to gas and liquid phases for the given k-values. The final step is the flash calculation. These three steps are presented for a five-component system in the following examples.

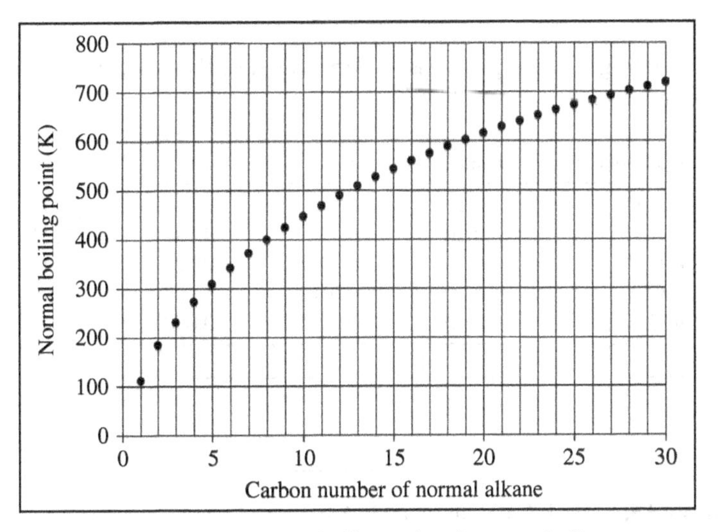

FIGURE 15.6 Normal boiling points for normal alkanes.

Example 15.1 Hydrocarbon *k*-Values

Calculate the *k*-values for $n\text{-}C_5H_{12}$, $n\text{-}C_8H_{18}$, $n\text{-}C_{12}H_{26}$, $n\text{-}C_{16}H_{34}$, and $n\text{-}C_{20}H_{42}$ at 45 psia and 330°F.

Answer

To estimate *k*-values, critical temperature, critical pressure, and acentric factor are needed for each component. Equation 11.4 gives satisfactory estimates for *k*-values as listed in the following table. Verification of these *k*-values will be left as an exercise at the end of the chapter.

Component	T_c (R)	P_c (psia)	Acentric Factor (ω)	*k*-Value
$n\text{-}C_5H_{12}$	845.5	488.1	0.250	6.77
$n\text{-}C_8H_{18}$	1024.0	360.5	0.376	0.898
$n\text{-}C_{12}H_{26}$	1184.6	263.5	0.515	0.101
$n\text{-}C_{16}H_{34}$	1299.8	207.6	0.631	0.0162
$n\text{-}C_{20}H_{42}$	1388.5	173.7	0.731	0.00337

Example 15.2 Two-phase Check

Complete the two-phase check for the five-component system of the previous example at 45 psia and 330°F with feed composition in the following table.

Answer

Use Equations 11.10 and 11.11 for the two-phase check. Results are listed in the following table. The sums for z_ik_i and z_i/k_i both exceed unity, so this feed will split into gas and liquid phases. Verification of these checks is left as an exercise at the end of the chapter.

Component	Feed (z_i)	*k*-Value	$z_i k_i$	z_i/k_i
n-C_5H_{12}	0.20	6.77	1.35	0.03
n-C_8H_{18}	0.20	0.898	0.18	0.22
n-$C_{12}H_{26}$	0.20	0.101	0.02	2.0
n-$C_{16}H_{34}$	0.20	0.0162	0.003	12.3
n-$C_{20}H_{42}$	0.20	0.00337	0.0007	59.3

Example 15.3 Flash Calculation at 330°F

Complete the flash calculation for the five-component system of the previous examples at 45 psia and 330°F.

Answer

Use the kernel of the sum in Equation 11.9 to calculate mole fractions for the liquid phase for a chosen value of G. Vary G until the sum of liquid mole fractions equals unity. Results are listed in the following table for $G = 0.123$. Use Equation 11.5 to calculate the gas phase mole fractions. Verification of these results is left as an exercise at the end of the chapter.

Component	Feed (z_i)	*k*-Value	x_i	y_i
n-C_5H_{12}	0.20	6.77	0.117	0.791
n-C_8H_{18}	0.20	0.898	0.202	0.182
n-$C_{12}H_{26}$	0.20	0.101	0.225	0.022
n-$C_{16}H_{34}$	0.20	0.0162	0.228	0.004
n-$C_{20}H_{42}$	0.20	0.00337	0.228	0.001

Example 15.4 Flash Calculation at 600°F

Perform a flash calculation for the five-component system of the previous examples at 45 psia and 600°F.

Answer

Although the steps of both Examples 15.1 and 15.2 are needed, just the results corresponding to Example 15.3 are shown in the following for a chosen value of $G = 0.841$. Verification of these results is left as an exercise at the end of the chapter. Noticc that the fraction of feed moles in gas *G* is greater at 600°F than it was in Example 15.3 which had the same feed at 330°F.

Component	Feed (z_i)	*k*-Value	x_i	y_i
$n\text{-}C_5H_{12}$	0.20	42.2	0.006	0.237
$n\text{-}C_8H_{18}$	0.20	10.3	0.023	0.234
$n\text{-}C_{12}H_{26}$	0.20	2.25	0.097	0.219
$n\text{-}C_{16}H_{34}$	0.20	0.636	0.288	0.183
$n\text{-}C_{20}H_{42}$	0.20	0.216	0.586	0.127

Additional distillation of the bottom stream, or residue, from a crude distillation unit as shown in Figure 15.5 is achieved by recognizing that the boiling point of a hydrocarbon component decreases as pressure is lowered. Feeding the residue stream into a vacuum distillation tower that operates at a much lower pressure allows separation of components with lower boiling points.

15.2.2 Conversion

The separation process separates the crude oil into product streams that need to be changed into mixtures that are suitable for the consumer. The process of converting low-value, high molecular weight hydrocarbon mixtures into high-value, lower molecular weight hydrocarbon products is known as the conversion process. Conversion is achieved by breaking the hydrocarbon chains of higher molecular weight molecules to produce hydrocarbon molecules with lower molecular weights. High molecular weight hydrocarbon chains are broken in conversion units like the fluidized catalytic cracker (FCC), the hydrocracker, and the delayed coker. The FCC uses a catalyst to convert the high molecular weight product stream into LPG, gasoline, and diesel. The hydrocracker also uses a catalyst at high temperature, but the reactions are performed in the presence of high concentrations of hydrogen. The input product stream flows over a fixed position catalyst in a hydrogen environment. Product streams from the hydrocracker have low levels of sulfur.

The heaviest product from the distillation tower is Vacuum Tower Bottoms (VTB), which is also known as "resid" or residue. Resid would become a solid if allowed to cool to ambient temperature. It is sometimes used as a blend component for the paving asphalt market. It is possible to convert VTB to more commercially valuable products by breaking the high molecular weight hydrocarbon chains. This cannot be achieved in the FCC because resid is too heavy and typically contains too many contaminants. The delayed coker is a conversion unit that uses high temperature to break long carbon chains and convert resid to more valuable products.

Another function of the conversion process is to change the way carbon chains are put together. For example, butane molecules are by-products of some conversion units. They can be combined to form larger, more commercially valuable, hydrocarbon molecules in alkylation units.

The last stage of the conversion process is catalytic reforming. The reformer generates hydrogen and increases the octane number of components used in a gasoline blend. Octane is a saturated hydrocarbon with eight carbon atoms. By contrast, octane number is a measure of the compression a fuel can withstand before it ignites. A fuel with high octane number can be compressed more than a fuel with low octane number before ignition. Fuels with high octane number are used in high-performance engines. Straight chain molecules (paraffins) have a relatively low octane number, while hydrocarbon molecules with rings (aromatics) have a relatively high octane number. The catalytic reformer uses a catalyst in the presence of hydrogen and high temperature to reform paraffin molecules into aromatic molecules.

15.2.3 Purification

Purification is the last step in the refining process following separation and conversion. The primary purpose of purification is to remove sulfur in a process called hydrotreating. Unfinished products are fed into a hydrotreater that contains a catalyst and hydrogen in a high-temperature, high-pressure chamber. The catalyst increases the chemical reaction rate for a reaction that removes sulfur from molecules in the input stream. The primary product of the reaction is hydrogen sulfide, which is removed by extraction in other units. The hydrogen sulfide is separated into elemental sulfur and hydrogen using a desulfurization process. The recovered sulfur can be sold as a refinery by-product and the hydrogen can be used in a hydrocracker or hydrotreater.

15.2.4 Refinery Maintenance

Refinery operations can wear out equipment and consume chemical catalysts. The catalysts need to be replaced and equipment needs to be repaired or replaced. New equipment may have to be installed. Operators shut down the refinery twice a year to allow time to maintain the refinery. Typically, maintenance periods are scheduled during periods when demand for petroleum products is lowest. As a rule, summer driving season and winter heating result in periods of high petroleum demand during the summer and winter. That leaves spring and fall for refinery maintenance. Maintenance periods can require shutting down a refinery for a few days to a few weeks.

15.3 THE DOWNSTREAM SECTOR: NATURAL GAS PROCESSING PLANTS

The typical components of natural gas and natural gas products from natural gas processing plants are presented in Table 15.2. Natural gas consists of hydrocarbon and nonhydrocarbon gases. It also contains water vapor. Water and other impurities must be removed from natural gas before it can be used. Natural gasoline is a mixture of pentane plus smaller amounts of hydrocarbon molecules with more than five carbon atoms. The hydrocarbon molecules in natural gasoline usually do not have more than 10 carbon atoms.

Countries or operators with excess natural gas production may choose to ship gas to distant markets as LNG. LNG consists mostly of methane plus very small amounts of ethane, propane, and nitrogen. LNG is shipped in specialized tankers as a liquid at

TABLE 15.2 Typical Composition of Natural Gas Products

Component	Number of C Atoms	Natural Gas	LNG	Residential Gas	LPG	NGL
Methane	1	X	X	X		
Ethane	2	X	X	X		X
Propane	3	X	X		X	X
Butanes	4	X			X	X
Pentanes	5	X				X
Natural gasoline	5+	X				X
Nonhydrocarbon gases (CO_2, H_2O, H_2S, N_2, etc.)		X				
Water		X				

TABLE 15.3 Typical Recovery of Components at Different NGL Plants

Component	"Lean Oil" (%)	Refrigeration (%)	Cryogenic (%)
Ethane	15–30	80–85	85–90
Propane	65–75	100	100
Butane and isobutane	99	100	100
Pentanes	99	100	100
Natural gasoline	99	100	100

its normal boiling point of −259°F (−162°C). At its destination, LNG is vaporized and heated to appropriate temperature for pipeline transport to industrial and residential markets.

LPG consists mostly of propane or butane or mixtures of the two. LPGs are often used as transportation fuel and as heating fuel in remote locations. LPG must be transported or stored in pressure vessels. Depending on composition, the required pressure varies from about 40 psi to almost 200 psi at ambient temperatures.

Processing plant technology determines the quality of natural gas liquids (NGL) recovered from the input gas stream. Table 15.3 shows the typical recoveries of gas components at different NGL plants. A "lean oil" plant recovers 99% of butanes and components with five or more carbons but is less efficient at recovering lower molecular weight components (ethane and propane). Refrigeration plants use propane to cool the input gas and recover more of the lower molecular weight components. A cryogenic plant is a very low-temperature facility and recovers more ethane than the refrigeration plant.

15.4 SAKHALIN-2 PROJECT, SAKHALIN ISLAND, RUSSIA

The Sakhalin-2 project is operated by Sakhalin Energy and is dedicated to developing two world-class offshore fields: Lunskoye and Piltun-Astokhskoye (P-A). The infrastructure for Sakhalin-2 has largely been built in the twenty-first century. An understanding of the history of Sakhalin Island provides an important cultural perspective on the modern midstream and downstream Sakhalin-2 sectors. We summarize the history of Sakhalin Island (Vysokov, 1996) and then describe the Sakhalin-2 infrastructure.

15.4.1 History of Sakhalin Island

The earliest inhabitants of southern Sakhalin Island, the Kuril Islands, and the northernmost Japanese island of Hokkaido were people known as the Okhotsk culture (Figure 15.7). Their economy was based on fishing, hunting, and shore gathering. The earliest mention of the Okhotsk peoples was in an ancient Chinese geography book called the Sengai-kyo from the Khan dynasty sometime between 206 B.C. and A.D. 220. The Sengai-kyo reported that the frontier of Japan in A.D. 0 extended as far north as the Amur River.

The Nihon shoki, an ancient Japanese history book, described a battle that was fought between a Japanese army and the local population at a big river that was believed to be the Amur River in A.D. 658. Contact between the people of Sakhalin Island and their Asian neighbors was not always hostile. Buddhism was brought to Sakhalin Island in the thirteenth century by a Japanese monk. Matsumae, the Lord of the northern frontier of Japan, sent explorers to Sakhalin Island in A.D. 1635. The explorers mapped Sakhalin Island, the Kuril Islands, and Kamchatka.

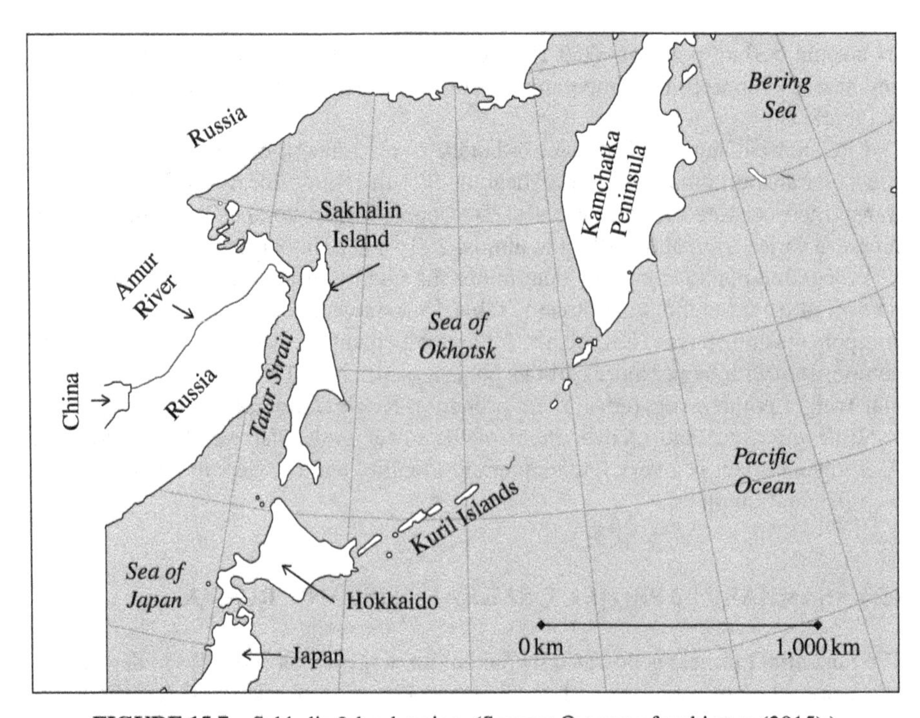

FIGURE 15.7 Sakhalin Island region. (Source: © energy.fanchi.com (2015).)

FIGURE 15.8 Sakhalin Island. (Source: © energy.fanchi.com (2015).)

The Dutch sailor M.G. de Vries was the first European to explore Sakhalin Island and the Kurils in 1643. His voyage put southern Sakhalin Island and the Kuril Islands on European maps. Many of these early maps mistakenly connected Sakhalin Island to the mainland by a narrow land bridge. This mistake was not corrected until the mid-1800s. A modern map of Sakhalin Island is shown in Figure 15.8.

Matsumae established the seaport of Ootomari on southern Sakhalin Island in 1679. Ootomari, which is now called Korsakov, was built as a trading post to control commerce. It became the largest Japanese trading post on Sakhalin Island.

The first Russian expedition to sight Sakhalin Island was led by V.D. Poyarkov. Poyarkov's expedition reached the Amur River estuary in the autumn of 1644 and spent the winter there. The next spring Poyarkov saw the western coast of Sakhalin Island, but he never landed on Sakhalin Island.

Manchurian expansion prevented Russia from conquering Sakhalin Island in the seventeenth and early eighteenth centuries. The people of Sakhalin Island were trading and paying tribute to the Manchurian Emperor during this period. The Chinese were enamored with furs from Sakhalin Island, especially sable, otter, and fox. In exchange, Sakhalin Island received goods that required a warmer climate or industrial processing.

The first armed confrontation between Russia and Japan was a conflict over Sakhalin Island. Russians first visited Sakhalin Island in 1805. The Russian vessel *Nadezhda* entered Aniva Bay at the southern end of Sakhalin Island. When the Russians landed, they were met by the Japanese. A Russian diplomat, N.P. Rezanov, failed to convince the Japanese to sign a trade treaty with Russia. Incensed by his failure, he decided to force the Japanese off Sakhalin Island. A year later, in 1806 the Russian frigate *Yunona* attacked the Ootomari trading post, looted the warehouses, and left them burning. Several Japanese residents were taken to Kamchatka as prisoners. The Japanese returned to Sakhalin Island with a strengthened military presence. In 1808, members of a Japanese expedition realized that Sakhalin Island was an island.

Even though Sakhalin Island is one of the largest islands in the world, it only had a few thousand people living on it when Russian colonization began in the nineteenth century. A couple of thousand Ghilaks were hunting in the north, and a few hundred Oroks led a nomadic life in the mountains. The largest race on Sakhalin Island was the Ainu—a bearded people who also settled in the Kuril Islands.

The Ainu were scattered in a handful of southern settlements. They served a few Japanese merchants who supplied them with corn, salt, and other necessities. In exchange, the merchants took all the fish the Ainu could catch in the bays and river mouths. The Ainu were left with just enough to survive. This servitude was typical under both Chinese and Japanese domination.

Russians became a permanent part of Sakhalin Island population in the middle of the nineteenth century. Russia established a military outpost while the Japanese built fishing and trading villages. The two countries negotiated a treaty that gave Russia the northern Kuril Islands and Japan the southern Kuril Islands. But the treaty did not settle the issue of ownership of Sakhalin Island.

The Russian explorer G.I. Nevelskoy explored the eastern coast of Sakhalin Island and the Amur Firth in 1849. The Amur Firth is a navigable strait between Sakhalin Island and the Russian mainland. It is adjacent to the northern waters of the Tatar Strait and was already well known to the Chinese and Japanese.

Russia attempted to consolidate its hold on Sakhalin Island by building military posts and settlements. The early Russian settlers were either soldiers or exiles. The first exiles were convicts sent to a penal colony in 1858. The Russian government declared Sakhalin Island a place of penal servitude and exile in 1869. By the end of the nineteenth century, Sakhalin Island had become home for many settlements of Russian convicts that had served their sentences and migrants that had volunteered to settle on Sakhalin Island.

Some were shipped two thousand miles down the Amur River from the Kara gold mine. Those from Russia had to travel forty-seven hundred miles to reach the eastern mouth of the Amur. Most of the survivors of the journey suffered from scurvy. The mortality rate in the first few years of colonization was ten percent. Conditions were so bad that it caused a scandal in the press and the route was changed.

Settlers were sent to Odessa and through the Suez Canal. Even that route was risky since travelers were exposed to disease like smallpox.

Americans began their expansion into the Far East in the 1840s and 1850s. The US government sought to negotiate a treaty with the Japanese and failed. As a result, an American military expedition was sent to Japan. This prompted the Russian government to increase its military presence in southern Sakhalin Island.

The Russian military presence on Sakhalin Island grew significantly after the Crimean War (1853–1856), in part as a response to the presence of an Anglo-French squadron patrolling Far Eastern waters. Russia was beginning to view Sakhalin Island as a strategic asset on the Pacific. In addition, colonizing Sakhalin Island also helped Russia colonize Siberia. This gave Russia access to silver and gold in Siberia and coal in Sakhalin Island.

Jurisdiction over Sakhalin Island was decided by the 1875 Treaty of Saint Petersburg when Russia traded the northern Kurils to Japan in exchange for the rights to Sakhalin Island. Hundreds of Ainu preferred Japanese rule and were allowed to move to Hokkaido.

The Russian government needed people to work the mines. Convicts were promised twenty acres of land once they finished their sentences in the mines at Alexandrovsk, which is the northwest region of Sakhalin Island across from the Amur River. Many lives were lost mining coal in the hostile environment, and the coal on Sakhalin Island was inferior to coal from places like Newcastle and Cardiff.

Freed prisoners were expected to raise all the food they would need, but the land on most of Sakhalin Island required cultivation. The Russian government believed that valleys would be better suited for farming than heavily forested mountains. The Tym river valley in the northeast of Sakhalin was considered a promising locale for farming. The Tym river valley extends to the Sea of Okhotsk and has marshy soil during the summer. The summer is too short for oats to ripen, and only barley was grown successfully. The subpolar climate and cold and fog from the Sea of Okhotsk shortened the growing season. The discovery of oil and gas in the northeast provided the only resource that could sustain a modern population.

Sakhalin Island continued to be a source of contention between Russia and Japan. The Japanese took southern Sakhalin away from Russia in 1905 when they defeated Russia in the Russo-Japanese war. The Treaty of Portsmouth divided Sakhalin Island into two regions at the fiftieth parallel. Sakhalin Island stopped being used as a penal colony in 1906, partly because of resistance by Japan, and partly because it was not considered an industrial or agricultural success.

The population of northern Sakhalin Island was less than ten thousand people in the early 1900s. Immigration to Sakhalin Island became voluntary in 1906 when the Russian Council of Ministers repealed penal servitude and exile in Sakhalin Island. It became very difficult to get people to move to Sakhalin Island. The discovery of oil and gas gave the region a natural resource with value.

Japanese firms began to show interest in Sakhalin Island around 1919, especially in the coal and oil deposits. The October Revolution in 1917 did not affect Sakhalin Island until 1918. Soviets replaced Bolsheviks in seats of power by 1920. The Soviets sought to prevent foreign economic supremacy by resisting free competition of capital practiced by the Japanese, Americans, and British. Within a month, a Japanese military detachment occupied Alexandrovsk and took control of northern Sakhalin Island.

Japanese businessmen had a special interest in Sakhalin Island oil. Economic assimilation began as soon as the Japanese military established control. Both the Soviets and Americans protested the occupation, but northern Sakhalin Island remained under Japanese rule for five years. Finally, in January 1925, Russia and Japan signed a treaty that saw the Japanese army withdraw from northern Sakhalin Island in exchange for petroleum concessions.

The Russian–Japanese border on Sakhalin Island did not change again until the end of World War II in 1945. Southern Sakhalin Island and the Kuril Islands were controlled by Japan at the beginning of World War II. Japan was able to control the main sea-lanes that connected the southern part of the Soviet Far East with the rest of the world. The Kuril Island of Iturup was a staging point for the Japanese aircraft carrier fleet that attacked Pearl Harbor in December 1941.

Russia and Japan had a nonaggression pact during most of World War II. The United States dropped the first atomic bomb on Hiroshima on August 6 and another on Nagasaki on August 9. Russia rejected the nonaggression pact and declared war on Japan on August 8, 1945.

The Japanese accepted the Potsdam Declaration that defined terms for Japanese surrender to end World War II on August 15, but hostilities between Japan and Russia continued for several days. The Soviet Red Army crossed the fiftieth parallel on Sakhalin Island and Russian marines landed on several of the Kuril Islands. Hostilities ended on August 20, 1945.

Russia controlled Sakhalin Island and the Kuril Islands by the beginning of September 1945. The Presidium of the Supreme Soviet declared Soviet Sovereignty over the entire area on February 2, 1946. The Yuzhno-Sakhalinsk region was formed to govern southern Sakhalin Island and the Kurils in the Russian Far East. The Yuzhno-Sakhalinsk region and Khabarovsk region were included in the Sakhalin region on January 2, 1947.

Soviet entry into the war against Japan was welcomed by the United States and Britain. When the Cold War began, the United States was no longer happy with Soviet possession of these strategic islands. Many Japanese and Koreans on Sakhalin Island were repatriated in the late 1940s, and many Ainu chose to leave. The Soviet government organized a massive transfer of people to Sakhalin Island. By the beginning of the 1950s, Sakhalin Island had a population of seven hundred thousand people, with over eighty percent Russian. Sakhalin Island has been under the jurisdiction of the Soviet Union and now Russia since 1945.

15.4.2 The Sakhalin-2 Project

The development of oil and gas fields and the growth of the Asian economy turned Sakhalin Island into a valuable prize. Onshore and offshore fields were discovered on the northern half of Sakhalin Island. Figure 15.9 shows five major offshore fields and the infrastructure for the Sakhalin-2 project. Three of the five fields (Arkutun Dagi, Chayvo, and Odoptu) were awarded to the Sakhalin-1 project for development, and two fields (Lunskoye, P-A) were awarded to the Sakhalin-2 project for development. Both projects required substantial infrastructure development because the existing infrastructure associated with onshore field production did not have the capacity to handle offshore field production, transport, and processing. We outline the Sakhalin-2 project to illustrate midstream and downstream operations.

The Lunskoye field was discovered in 1984 and the P-A field was discovered in 1986. The Lunskoye field is a gas reservoir with an oil rim. The P-A field is an oil field with two anticlines connected by a syncline. In simple conceptual terms, the

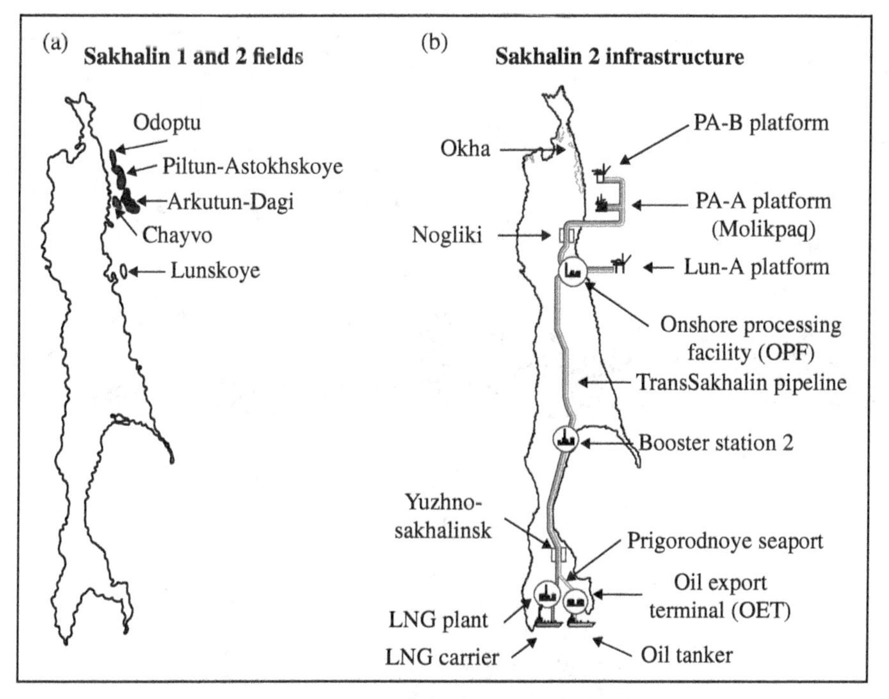

FIGURE 15.9 The Sakhalin 2 project: (a) fields for both Sakhalin 1 and Sakhalin 2 and (b) infrastructure for Sakhalin 2. (Source: © energy.fanchi.com (2015).)

P-A field consists of two hills connected by a valley. The Piltun anticline is in the north, and the Astokh anticline is in the south.

An international tender for the right to conduct a feasibility study of the two fields was announced in 1991. The Russian Federation signed an agreement in 1992 with a consortium of companies (Marathon, McDermott, and Mitsui) to conduct the feasibility study. Royal Dutch Shell and Mitsubishi joined the consortium in 1992. The consortium established the Sakhalin Energy Investment Company Ltd. (Sakhalin Energy) in 1994 after the feasibility study was approved by the Russian Federation in 1993.

A production sharing agreement to develop the Lunskoye and P-A fields was signed in 1994 between the Russian Federation in Moscow, the Sakhalin Oblast Administration in Yuzhno-Sakhalinsk, and Sakhalin Energy. A Law on Production Sharing Agreements was enacted by the Russian Federation in 1996. The Russian Federation then issued licenses to develop the Lunskoye and P-A fields to Sakhalin Energy in 1996. Appraisal work on the two fields began in the summer of 1996.

The consortium changed in 1997 when McDermott sold its shares to the other shareholders. Russian agencies approved the Phase 1 development plan for the P-A field in 1997 and the feasibility study in 1998. Phase 1 development called for installation of the Molikpaq production platform at the P-A field. The Molikpaq platform is called the PA-A platform in Figure 15.9 and is shown in Figure 15.10.

FIGURE 15.10 PA-A (Molikpaq) platform, PA-B platform, and Lun-A platform. (Source: Courtesy of Sakhalin Energy (2016), personal communication.)

The Molikpaq was purchased by Sakhalin Energy and towed to Korea from the Beaufort Sea in 1998. The platform was refurbished in South Korea and then towed to the Amur Shipyard in Russia where it was placed atop a steel base known as a spacer. The Molikpaq was installed as an artificial island in the relatively shallow water off the northeast coast of Sakhalin Island. The steel spacer was set on the seabed, anchored in place, and filled with sand. Oil production from the P-A field to the Molikpaq platform in 1999 was the first offshore oil production in Russia.

Sakhalin Energy moved its corporate headquarters to Yuzhno-Sakhalinsk in 2000. By the end of 2000, Marathon had sold its share of Sakhalin Energy and the remaining shareholders were Royal Dutch Shell (55%), Mitsui (25%), and Mitsubishi (20%).

Full development of the P-A and Lunskoye fields began in 2003. The Lunskoye field provided a gas stream that needed to be liquefied and transported by LNG tanker to market. LNG sales contracts were signed with Japanese and Korean firms in 2003 and 2004. The sale of Russian gas to North American markets was made possible by the sale of Sakhalin Energy gas to Shell Eastern Trading Ltd. in 2004.

FIGURE 15.11 TransSakhalin pipeline system. (Source: Courtesy of Sakhalin Energy (2016), personal communication.)

Sakhalin Energy entered the LNG shipping business by participating in projects to build two LNG tankers.

The shareholders of Sakhalin Energy changed again in 2006 when the Russian company Gazprom acquired majority interest in the Sakhalin-2 project. The shareholders of Sakhalin Energy were now Gazprom (50% plus 1 share), Shell (27.5% minus 1 share), Mitsui (12.5%), and Mitsubishi (10%).

Construction of the TransSakhalin pipeline system officially began in 2004 (Figure 15.11). An oil pipeline and a gas pipeline were built along a single 800 km corridor that runs most of the length of Sakhalin Island. Booster pumps and compressors are used to increase pressure in the oil and gas pipelines. The pipeline system began transporting fluids from north to south of Sakhalin Island in 2008.

Sakhalin-2 platforms are located a few miles offshore Sakhalin Island. Tie in modules and offshore pipelines were installed in 2006 to allow year-round oil and gas production and export from the Molikpaq platform. The Molikpaq platform began year-round operations in 2008.

Concrete gravity base structures were installed for the Lun-A platform and PA-B platform during the summer of 2005. Platform topsides were installed in 2006 for the Lun-A platform and in 2007 for the PA-B platform. The Lun-A platform began natural gas production from the Lunskoye field in January 2009.

Produced fluids from the P-A field and the Lunskoye field make landfall at the Onshore Processing Facility (OPF) in the north near Nogliki. The primary purpose of the OPF is to process the Lunskoye production stream into gas and condensate for transport by the TransSakhalin pipeline system to the LNG plant at Prigorodnoye Production Complex and seaport on Aniva Bay in the south. The OPF began processing fluids in 2008. The LNG plant was commissioned in February 2009. A loading jetty connects the LNG plant to LNG carriers in Aniva Bay (Figure 15.12).

The OPF also prepares oil and gas production from the P-A field for transport by the TransSakhalin pipeline system to the Oil Export Terminal (OET) at Prigorodnoye Production Complex. An offshore pipeline and tanker loading unit connect the OET to oil tankers in Aniva Bay. Year-round oil export from Prigorodnoye Production Complex started in 2008.

FIGURE 15.12 LNG carrier in Aniva Bay. (Source: Courtesy of Sakhalin Energy (2016), personal communication.)

15.5 ACTIVITIES

15.5.1 Further Reading

For more information, see Hyne (2012), Raymond and Leffler (2006), and van Dyke (1997).

15.5.2 True/False

15.1 Gas compression can maximize use of pipeline space by increasing the density of gas flowing through the pipeline.

15.2 Pipeline pigs are driven by the flow of fluids in the pipeline.

15.3 The Henry Hub is an interstate pipeline interchange that is located in Cushing, Oklahoma.

15.4 Working gas is the amount of stored gas that is kept in the storage facility to maintain a safe pressure.

15.5 Gas that is stored temporarily is called parked gas.

15.6 A distillation tower separates crude oil into mixtures of components based on the boiling points of the mixtures.

15.7 Naphtha is used to make gasoline for jet fuel.

15.8 The conversion process changes high molecular weight hydrocarbon mixtures into lower molecular weight hydrocarbon products.

15.9 The lightest product from the distillation tower is Vacuum Tower Bottoms (VTB), or “resid."

15.10 Refrigeration and cryogenics are liquefaction processes.

15.5.3 Exercises

15.1 The density of oil is 900 kg/m^3. What is the pressure (in Pa) at the bottom of a cylindrical storage tank containing a column of oil that is 10 m high? Assume the pressure of air at the top of the oil is 1 atm and the acceleration of gravity is 9.8 m/s^2. Recall that 1 atm = 1.01×10^5 Pa.

15.2 **A.** The height h of the liquid level in a cylindrical tank with a 50 ft radius above the midpoint of a circular valve with radius 0.25 ft is 50 ft. What is the speed of the liquid flowing from the tank (in ft/s)? Assume the acceleration of gravity is 32 ft/s^2 and that the potential energy at an elevation of 50 ft equals the kinetic energy of liquid flowing from the tank.

B. How fast does the liquid level change in the tank compared to flow from the tank?

15.3 **A.** A cylindrical steel pipe with 24-in. inner diameter is 0.10 in. thick. The pipe carries oil under an internal pressure $p = 60$ psi. The circumferential (hoop) stress in the steel is $S = pd/2t$, where p is the internal pressure, d is the inner diameter of pipe, and t is the pipe thickness. Calculate circumferential stress S in the steel.

B. If the pressure is raised to 100 psi in the 24-in. inner diameter steel pipe, what thickness of steel would be required for an allowable circumferential stress $S = 10000$ psi?
Hint: solve $S = pd/2t$ for t.

C. Do you need to replace the steel pipe?

15.4 Verify the k-value calculation in Example 15.1.

15.5 Verify the two-phase check in Example 15.2.

15.6 Verify the flash calculation in Example 15.3.

15.7 Verify the flash calculation in Example 15.4.

APPENDIX

UNIT CONVERSION FACTORS[1]

TIME

1 hour = 1 hr = 3600 s
1 day = 8.64×10^4 s
1 year = 1 yr = 3.1536×10^7 s

LENGTH

1 foot = 1 ft = 0.3048 m
1 kilometer = 1 km = 1000 m
1 mile = 1 mi = 1.609 km
1 cm = 10^{-2} m

VELOCITY

1 foot per second = 0.3048 m/s
1 kilometer per hour = 1 kph = 1000 m/hr = 0.278 m/s
1 mile per hour = 1 mph = 1.609 km/hr = 1609 m/hr = 0.447 m/s

[1] Adapted from Fanchi (2010), Appendix A .

Introduction to Petroleum Engineering, First Edition. John R. Fanchi and Richard L. Christiansen.
© 2017 John Wiley & Sons, Inc. Published 2017 by John Wiley & Sons, Inc.
Companion website: www.wiley.com/go/Fanchi/IntroPetroleumEngineering

AREA

1 square foot = 1 ft^2 = 0.0929 m^2
1 square mile = 1 mi^2 = 2.589 km^2 = 2.589×10^6 m^2
1 square mile = 1 mi^2 = 640 acres
1 acre = 1 ac = 4047 m^2
1 hectare = 1 ha = 1.0×10^4 m^2
1 millidarcy = 1 md = 0.986923×10^{-15} m^2
1 darcy = 1000 md = 0.986923×10^{-12} m^2
1 barn = 1.0×10^{-24} cm^2 = 1.0×10^{-28} m^2

VOLUME

1 liter = 1 L = 0.001 m^3
1 cubic foot = 1 ft^3 = 2.83×10^{-2} m^3
1 standard cubic foot = 1 SCF = 1 ft^3 at standard conditions
1 acre-foot = 1 ac-ft = 1233.5 m^3
1 barrel = 1 bbl = 0.1589 m^3
1 gallon (US liquid) = 1 gal = 3.785×10^{-3} m^3
1 barrel = 42 gallons = 0.1589 m^3
1 barrel = 5.6148 ft^3
1 gallon = 3.788 liters
1 cm^3 = 1 cc = 10^{-6} m^3
1 cm^3 = 3.534×10^{-5} ft^3

MASS

1 gram = 1 g = 0.001 kg
1 pound (avoirdupois) = 1 lb (avdp) = 1 lbm = 0.453592 kg
1 tonne = 1000 kg

MASS DENSITY

1 g/cm^3 = 1000 kg/m^3

FORCE

1 pound-force = 1 lbf = 4.4482 N

PRESSURE

1 pascal = 1 Pa = 1 N/m^2 = 1 kg/m·s^2
1 megapascal = 1 MPa = 10^6 Pa
1 gigapascal = 1 GPa = 10^9 Pa

1 pound-force per square inch = 1 psi = 6894.8 Pa
1 atmosphere = 1 atm = 1.01325×10^5 Pa
1 atmosphere = 1 atm = 14.7 psi
1 bar = 14.5 psi = 0.1 MPa
1 MPa = 145 psi

ENERGY

1 megajoule = 1 MJ = 1.0×10^6 J
1 gigajoule = 1 GJ = 1.0×10^9 J
1 exajoule = 1 EJ = 1.0×10^{18} J
1 eV = 1.6022×10^{-19} J
1 MeV = 10^6 eV = 1.6022×10^{-13} J
1 erg = 10^{-7} J
1 BTU = 1055 J
1 ft·lbf = 0.7376 J
1 calorie (thermochemical) = 1 cal = 4.184 J
1 kilocalorie = 1 kcal = 1000 calories = 4.184×10^3 J
1 Calorie = 1000 calories = 4.184×10^3 J
1 kilowatt-hour = 1 kW·h = 1 kW·1 hr = 3.6×10^6 J
1 quad = 1 quadrillion BTU = 1.0×10^{15} BTU = 1.055×10^{18} J
1 quad = 2.93×10^{11} kW·h = 1.055×10^{12} MJ
1 quad = 1.055 exajoule = 1.055 EJ
1 barrel of oil equivalent = 1 BOE = 5.8×10^6 BTU = 6.12×10^9 J
1 quad = 1.72×10^8 BOE = 172×10^6 BOE

ENERGY DENSITY

1 BTU/lbm = 2326 J/kg
1 BTU/SCF = 3.73×10^4 J/m^3

POWER

1 watt = 1 W = 1 J/s
1 megawatt = 10^6 W = 10^6 J/s
1 kilowatt-hour per year = 1 kW·h/yr = 0.114 W = 0.114 J/s
1 horsepower = 1 hp = 745.7 W
1 hp = 33 000 ft·lbf/min

VISCOSITY

1 centipoise = 1 cp = 0.001 Pa·s
1 mPa·s = 0.001 Pa·s = 1 cp = 10^{-3} Pa·s
1 poise = 100 cp = 0.1 Pa·s

RADIOACTIVITY

1 curie = 1 Ci = 3.7×10^{10} decays/s
1 roentgen = 1 R = 2.58×10^{-4} C/kg
1 radiation absorbed dose = 1 rad = 100 erg/g = 0.01 J/kg
1 gray = 1 Gy = 1 J/kg
100 rems = 1 sievert = 1 Sv

TEMPERATURE

Kelvin to centigrade: °C = °K − 273.15
Centigrade to fahrenheit: °F = (9/5)°C + 32
Rankine to fahrenheit: °F = °R − 460
Rankine to kelvin: °K = (5/9)°R

VOLUMETRIC FLOW RATE

1 cm^3/s = 1 cc/s = 10^{-6} m^3/s
1 cm^3/s = 3.053 ft^3/day = 0.5437 bbl/day
1 bbl/day = 1.839 cm^3/s

REFERENCES

Ahmed, T. (2000): *Reservoir Engineering Handbook*, Houston, TX: Gulf Publishing.

Alvarez, W. and C. Zimmer (1997): *T. Rex and the Crater of Doom*, Princeton, NJ: Princeton University Press.

Amyx, J.W., D.H. Bass, and R.L. Whiting (1960): *Petroleum Reservoir Engineering*, New York: McGraw-Hill.

Arps, J.J. (1945): "Analysis of Decline Curves," *Transactions of AIME*, 160, 228–247.

Arthur, J.D., B. Bohm, and D. Cornue (2009): *Evaluating Implications of Hydraulic Fracturing in Shale Gas Reservoirs*, Paper SPE 122931, Richardson, TX: Society of Petroleum Engineers.

Asquith, G. and D. Krygowski (2004): *Basic Well Log Analysis*, Second Edition, Tulsa, OK: American Association of Petroleum Geologists.

Barree, R.D. and M.W. Conway (March 2005): "Beyond Beta Factors: A Model for Darcy, Forchheimer, and Trans-Forchheimer Flow in Porous Media," *Journal of Petroleum Technology*, 57 (3), 43–45.

Bassiouni, Z. (1994): *Theory, Measurement, and Interpretation of Well Logs*, SPE Textbook Series 4, Richardson, TX: Society of Petroleum Engineers.

Batycky, R.P., M.R. Thiele, K.H. Coats, A. Grindheim, D. Ponting, J.E. Killough, T. Settari, L.K. Thomas, J. Wallis, J.W. Watts, and C.H. Whitson (2007): "Chapter 17: Reservoir Simulation," *Petroleum Engineering Handbook, Volume 5: Reservoir Engineering and Petrophysics*, Edited by E.D. Holstein, Richardson, TX: Society of Petroleum Engineers; Editor-In-Chief L.W. Lake.

Introduction to Petroleum Engineering, First Edition. John R. Fanchi and Richard L. Christiansen.
© 2017 John Wiley & Sons, Inc. Published 2017 by John Wiley & Sons, Inc.
Companion website: www.wiley.com/go/Fanchi/IntroPetroleumEngineering

Batzle, M. (2006): "Chapter 13: Rock Properties," *Petroleum Engineering Handbook, Volume 1: General Engineering*, Edited by J.R. Fanchi, Richardson, TX: Society of Petroleum Engineers; Editor-In-Chief L.W. Lake.

Bear, J. (1972): *Dynamics of Fluids in Porous Media*, New York: Elsevier.

Beggs, H.D. (1984): *Gas Production Operations*, Tulsa, OK: OGCI Publications.

Beggs, H.D. (1991): *Production Optimization Using Nodal Analysis*, Tulsa, OK: OGCI Publications.

Beggs, H.D. and J.P. Brill (May 1973): "A Study of Two-Phase Flow in Inclined Pipes," *Journal of Petroleum Technology*, 255, 607–617.

Bjørlykke, K. (2010): *Petroleum Geoscience*, Berlin: Springer-Verlag.

Boswell, R. (August 2009): "Is Gas Hydrate Energy Within Reach?" *Science*, 325 (5943), 957–958.

Bourgoyne, A.T. Jr., K.K. Milheim, M.E. Chenevert, and F.S. Young, Jr. (1991): *Applied Drilling Engineering*, Richardson, TX: Society of Petroleum Engineers.

BP (2015): BP Statistical Review of World Energy 2015, http://www.bp.com/en/global/corporate/energy-economics/statistical-review-of-world-energy.html, accessed August 17, 2015.

Brill, J.P. (January 1987): "Multiphase Flow in Wells," *Journal of Petroleum Technology*, 39 (1), 15–21.

Brill, J.P. and S.J. Arirachakaran (May 1992): "State of the Art in Multiphase Flow," *Journal of Petroleum Technology*, 44 (5), 538–541.

Brill, J.P. and H. Mukherjee (1999): *Multiphase Flow in Wells*, SPE Monograph #17, Richardson, TX: Society of Petroleum Engineers.

Brooks, J.E. (1997): "*A Simple Method for Estimating Well Productivity*," Paper SPE 38148, Richardson, TX, Society of Petroleum Engineers.

Brundtland, G. (1987): *Our Common Future*, Oxford: Oxford University Press.

Buckley, S.E. and M.C. Leverett (1942): "Mechanisms of Fluid Displacement in Sands," *Transactions of the AIME*, 146, 107–116.

de Buyl, M., T. Guidish, and F. Bell (1988): "Reservoir Description from Seismic Lithologic Parameter Estimation," *Journal of Petroleum Technology*, 40 (4), 475–482.

Canadian Energy Resources Conservation Board (1975): *Theory and Practice of the Testing of Gas Wells*, Third Edition, Calgary: Energy Resources Conservation Board.

Carlson, M.R. (2003): *Practical Reservoir Simulation*, Tulsa, OK: PennWell.

Castagna, J.P., M.L. Batzle, and R.L. Eastwood (1985): "Relationships between Compressional-wave and Shear-wave Velocities in Clastic Silicate Rocks," *Geophysics*, 50, 571–581.

Christiansen, R.L. (2006): "Chapter 15: Relative Permeability and Capillary Pressure," *Petroleum Engineering Handbook, Volume 1: General Engineering*, Edited by J.R. Fanchi, Richardson, TX: Society of Petroleum Engineers; Editor-In-Chief L.W. Lake.

Clarke, D.D. and C.C. Phillips (2003): "Chapter 3: Three-dimensional Geologic Modeling and Horizontal Drilling Bring More Oil Out of the Wilmington Oil Field of Southern California," *Horizontal Wells: Focus on the Reservoir*, AAPG Methods in Exploration, No. 14, Edited by T.R. Carr, E.P. Mason, and C.T. Feazel, Tulsa, OK: American Association of Petroleum Geology, pp. 27–47.

Cohen, R. (2000): *History of Life*, Third Edition, Malden, MA: Blackwell Scientific.

Cohen, K.M., S. Finney, and P.L. Gibbard (2015): International Chronostratigraphic Chart, International Commission on Stratigraphy, www.stratigraphy.org, accessed December 7, 2015.

Craft, B.C. and M.F. Hawkins, revised by R.E. Terry (1991): *Applied Petroleum Reservoir Engineering*, Second Edition, Englewood Cliffs, NJ: Prentice-Hall.

Dahlberg, E.C. (1975): "Relative Effectiveness of Geologists and Computers in Mapping Potential Hydrocarbon Exploration Targets," *Mathematical Geology*, 7, 373–394.

Dake, L.P. (2001): *The Practice of Reservoir Engineering*, Revised Edition, Amsterdam: Elsevier.

Dandekar, A.Y. (2013): *Petroleum Reservoir Rock and Fluid Properties*, Second Edition, Boca Raton, FL: CRC Press.

Denehy, D. (2011): *Fundamentals of Petroleum*, Fifth Edition, Austin, TX: Petroleum Extension Service (PETEX).

Donnely, J. (September 2014): "Community Engagement," *Journal of Petroleum Technology*, 66 (9), 18.

Dorn, G.A. (September 1998): "Modern 3-D Seismic Interpretation," *The Leading Edge*, 17, Tulsa, OK: Society of Exploration Geophysicists, 1262–1283.

van Dyke, K. (1997): *Fundamentals of Petroleum*, Fourth Edition, Austin, TX: Petroleum Extension Service (PETEX).

Earlougher, R.C., Jr. (1977): *Advances in Well Test Analysis*, Richardson, TX: Society of Petroleum Engineers.

Ebanks, W.J., Jr. (1987): "Flow Unit Concept – Integrated Approach to Reservoir Description for Engineering Projects," paper presented at the AAPG Annual Meeting, Los Angeles, CA, USA, *AAPG Bulletin*, 71 (5), 551–552.

Ebanks, W.J., Jr., M.H. Scheihing, and C.D. Atkinson (1993): "Flow Units for Reservoir Characterization," *Development Geology Reference Manual*, Edited by D. Morton-Thompson and A.M. Woods, AAPG Methods in Exploration Series, Number 10, Tulsa, OK: American Association of Petroleum Geologists, pp. 282–285.

Economides, M.J., A.D. Hill, C. Ehlig-Economides, and D. Zhu (2013): *Petroleum Production Systems*, Second Edition, Upper Saddle River, NJ: Prentice-Hall.

Ertekin, T., J.H. Abou-Kassem, and G.R. King (2001): *Basic Applied Reservoir Simulation*, Richardson, TX: Society of Petroleum Engineers.

Fancher, G.H. and J.A. Lewis (1933): "Flow of Simple Fluids through Porous Materials," *Industrial & Engineering Chemistry*, 25, 1139–1147.

Fanchi, J.R. (1990): "Calculation of Parachors for Compositional Simulation: An Update," *Society of Petroleum Engineers Reservoir Engineering*, 5 (3), 433–436.

Fanchi, J.R. (2004): *Energy: Technology and Directions for the Future*, Boston, MA: Elsevier-Academic Press.

Fanchi, J.R. (2006): *Math Refresher for Scientists and Engineers*, Third Edition, New York: John Wiley & Sons.

Fanchi, J.R. (2009): "Embedding a Petroelastic Model in a Multipurpose Flow Simulator to Enhance the Value of 4D Seismic," SPE Reservoir Simulation Symposium, the Woodlands, TX, USA, February 2–4, 2009, subsequently published in *SPE Reservoir Evaluation and Engineering Journal*, 13 (February 2010), 37–43.

Fanchi, J.R. (2010): *Integrated Reservoir Asset Management*, Burlington, MA: Elsevier-Gulf Professional Publishing.

Fanchi, J.R. (2011a): "Flow Modeling Workflow: I. Green Fields," *Journal of Petroleum Science and Engineering* 79, 54–57.

Fanchi, J.R. (2011b): "Flow Modeling Workflow: II. Brown Fields," *Journal of Petroleum Science and Engineering* 79, 58–63.

Fanchi, J.R. and C.J. Fanchi (2015): *"The Role of Oil and Natural Gas in the Global Energy Mix," Volume 3: Oil and Natural Gas, in Compendium of Energy Science and Technology (12 Volumes)*, Edited by R. Prasad, S. Kumar, and U.C. Sharma. New Delhi: Studium Press.

Fanchi, J.R. and C.J. Fanchi (2016): *Energy in the 21st Century*, Fourth Edition; Singapore: World Scientific.

Fanchi, J.R., N.C. Duane, C.J. Hill, and H.B. Carroll, Jr. (1983): "An Evaluation of the Wilmington Field Micellar-Polymer Project," Report DOE/BC/10033-8, Bartlesville, OK: U.S. Department of Energy.

Fanchi, J.R., M.J. Cooksey, K.M. Lehman, A. Smith, A.C. Fanchi, and C.J. Fanchi (September 2013):"Probabilistic Decline Curve Analysis of Barnett, Fayetteville, Haynesville, and Woodford Gas Shales," *Journal of Petroleum Science and Engineering* 109, 308–311.

Gaddy, D.E. (April 1999): "Coalbed Methane Production Shows Wide Range of Variability," *Oil & Gas Journal*, 97 (17), 41–42.

Gassmann, F. (1951): "Elastic Waves through a Packing of Spheres," *Geophysics*, 16, 673–685.

Gilman, J.R. and C. Ozgen (2013): *Reservoir Simulation: History Matching and Forecasting*, Richardson, TX: Society of Petroleum Engineers.

Gluyas, J. and R. Swarbrick (2004): *Petroleum Geoscience*, Malden, MA: Blackwell Publishing.

Govier, G.W., Editor (1978): *Theory and Practice of the Testing of Gas Wells*, Calgary: Energy Resources Conservation Board.

Gradstein, F.M., J.G. Ogg, M.D. Schmitz, (2012): The Geologic Time Scale 2012, Boston, MA: Elsevier.

Gunter, G.W., J.M. Finneran, D.J. Hartmann, and J.D. Miller (1997): *Early Determination of Reservoir Flow Units Using an Integrated Petrophysical Method*, Paper SPE 38679, Richardson, TX: Society of Petroleum Engineers.

Halbouty, M.T. (2000): "Exploration into the New Milennium," presented to the Second Wallace E. Pratt Memorial Conference on Petroleum Provinces of the 21st Century, San Diego, CA, USA, January 12–15, 2000; Tulsa, OK: American Association of Petroleum Geologists, http://archives.datapages.com/data/bulletns/1999/12dec/2033/images/halbouty.pdf, accessed May 12, 2016.

Haldorsen, H.H. and L.W. Lake (1989): "*A New Approach to Shale Management in Field-Scale Models*," Reservoir Characterization 2, SPE Reprint Series #27, Richardson, TX: Society of Petroleum Engineers.

Hanks, T.C. and H. Kanamori (1978): "A Moment Magnitude Scale," *Journal of Geophysical Research*, 84, 2348–2350.

Hibbeler, R.C. (2011): *Mechanics of Materials*, Eighth Edition, Boston, MA: Prentice Hall.

Hirsche, K., J. Porter-Hirsche, L. Mewhort, and R. Davis (March 1997): "The Use and Abuse of Geostatistics," *The Leading Edge*, 16 (3), 253–260.

Holditch, S.A. (2007): "Chapter 7: Tight Gas Reservoirs," *Petroleum Engineering Handbook, Volume 6: Emerging and Peripheral Technologies*, Edited by H.R. Warner, Jr., Richardson, TX: Society of Petroleum Engineers; Editor-In-Chief L.W. Lake.

Holditch, S.A. (2013): "Unconventional Oil and Gas Resource Development – Let's Do It Right," *Journal of Unconventional Oil and Gas Resources*, 1–2, 2–8.

Hornbach, M.J., H.R. DeShon, W.L. Ellsworth, B.W. Stump, C. Hayward, C. Frohlich, H.R. Oldham, J.E. Olson, M.B. Magnani, C. Brokaw, and J.H. Luetgert (2015): "Causal Factors for Seismicity near Azle, Texas," *Nature Communications*, 6, 6728.

Horne, R.N. (1995): *Modern Well Test Analysis*, Palo Alto, CA: Petroway.

Hubbert, M.K. (1956): "Nuclear Energy and the Fossil Fuels," American Petroleum Institute Drilling and Production Practice, Proceedings of the Spring Meeting, San Antonio, pp. 7–25.

Hyne, N.J. (2012): *Nontechnical Guide to Petroleum Geology, Exploration, Drilling & Production*, Third Edition, Tulsa, OK: PennWell.

IER Levelized Costs, 2011, "Levelized Cost of New Electricity Generating Technologies," *Annual Energy Outlook*, http://instituteforenergyresearch.org/analysis/levelized-cost-of-new-electricity-generating-technologies/, accessed June 4, 2012.

Jacobs, T. (2014): "Studying the Source of Methane Migration into Groundwater," *Journal of Petroleum Technology*, 66 (12), 42–51.

Jenkins, C., D. Freyder, J. Smith, and G. Starley (2007): "Chapter 6: Coalbed Methane" *Petroleum Engineering Handbook, Volume 6: Emerging and Peripheral Technologies*, Edited by H.R. Warner, Jr., Richardson, TX: Society of Petroleum Engineers; Editor-In-Chief L.W. Lake.

Johnston, D.C. (2006): "Stretched Exponential Relaxation Arising from a Continuous Sum of Exponential Decays," *Physical Review B*, 74, 184430.

Johnston, D.H., Editor (2010): *Methods and Applications in Reservoir Geophysics*, Tulsa, OK: Society of Exploration Geophysics.

Kelkar, M. (2000): "Application of Geostatistics for Reservoir Characterization – Accomplish ments and Challenges," *Journal of Canadian Petroleum Technology*, 59, 25–29.

King, G.R. (2012): "Hydraulic Fracturing 101: What Every Representative, Environmentalist, Regulator, Reporter, Investor, University Researcher, Neighbor and Engineer Should Know About Estimating Frac Risk and Improving Frac Performance in Unconventional Gas and Oil Wells," Paper SPE 152596, first presented at the SPE Hydraulic Fracturing Technology Conference, February 6–8, 2012, The Woodlands, TX, USA.

King, G.R., W. David, T. Tokar, W. Pope, S.K. Newton, J. Wadowsky, M.A. Williams, R. Murdoch, and M. Humphrey (April 2002): "Takula Field: Data Acquisition, Interpretation, and Integration for Improved Simulation and Reservoir Management," *SPE Reservoir Evaluation & Engineering*, 5 (2), 135–145.

Kuuskraa, V.A. and C.F. Brandenburg (October 1989): "Coalbed Methane Sparks a New Energy Industry," *Oil & Gas Journal*, 87 (41), 49.

Kuuskraa, V.A. and G.C. Bank (December 2003): "Gas from Tight Sands, Shales a Growing Share of U.S. Supply," *Oil & Gas Journal*, 101 (47), 34–43.

Lake, L.W. (1989): *Enhanced Oil Recovery*, Englewood Cliffs, NJ: Prentice Hall.

Lake, L.W. and J.L. Jensen (1989): "*A Review of Heterogeneity Measures Used in Reservoir Characterization*," Paper SPE 20156, Richardson, TX: Society of Petroleum Engineers.

Lee, J. (2007): "Chapter 8: Fluid Flow through Permeable Media," *Petroleum Engineering Handbook, Volume 5: Reservoir Engineering and Petrophysics*, Edited by E.D. Holstein, Richardson, TX: Society of Petroleum Engineers; Editor-In-Chief L.W. Lake.

Levin, H.L. (1991): *The Earth Through Time*, Fourth Edition, New York: Saunders College Publishing.

Makogon, Y.F., W.A. Dunlap, and S.A. Holditch (1997): "Recent Research on Properties of Gas Hydrates," Paper OTC 8299, presented at the 1997 Offshore Technology Conference, Houston, TX, USA, May 5–8, Richardson, TX: Society of Petroleum Engineers.

Masters, J.A. (1979): "Deep Basin Gas Trap, Western Canada," *The American Association of Petroleum Geologists Bulletin*, 63 (2), 152–181.

Mattax, C.C. and R.L. Dalton (1990): *Reservoir Simulation*, SPE Monograph #13, Richardson, TX: Society of Petroleum Engineers.

Matthews, C.S. and D.G. Russell (1967): *Pressure Buildup and Flow Tests in Wells*, Richardson, TX: Society of Petroleum Engineers.

Mavko, G., T. Mukerji, and J. Dvorkin (2009): *The Rock Physics Handbook*, Second Edition, Cambridge, UK: Cambridge University Press.

Mavor, M., T. Pratt, and R. DeBruyn (April 1999): "Study Quantifies Powder River Coal Seam Properties," *Oil & Gas Journal*, 97 (17), 35–40.

McCain, W.D., Jr. (1990): *The Properties of Petroleum Fluids*, Second Edition, Tulsa, OK: PennWell.

McGlade, V.A., J. Speirs, and S. Sorrell (2013): "Unconventional Gas – A Review of Regional and Global Resource," *Energy*, 55, 571–584.

McGuire, W.J. and V.J. Sikora, (October 1960): "The Effect of Vertical Fractures on Well Productivity," *Journal of Petroleum Technology*, 12 (10), 72–74.

Mitchell, R.F. and S.Z. Miska (2010): *Fundamentals of Drilling Engineering*, Richardson, TX: Society of Petroleum Engineers.

Offshore Staff (September 1998): "Refurbishment/Abandonment," *Offshore* Magazine, pp. 148 and 186.

Olsen, T. (May 2014): "An Oil Refinery Walk-Through," *Chemical Engineering Progress*, 34–40, http://www.aiche.org/resources/publications/cep/2014/may/oil-refinery-walk-through, accessed May 12, 2016.

Otott, G.E. Jr. and D.D. Clark (1996): "History of the Wilmington Field – 1986–1996," in *Old Oil Fields and New Life: A Visit to the Giants of the Los Angeles Basin*, Bakersfield, CA: AAPG Pacific Section (2007), pp. 17–22, http://www.searchanddiscovery.com/documents/2007/07014priority/fields%20of%20la%20basin/17.pdf, accessed September 28, 2015.

Peaceman, D.W. (June 1978): "Interpretation of Well-Block Pressures in Numerical Reservoir Simulation," *Society of Petroleum Engineering Journal*, 18 (3), 183–194.

Peaceman, D.W. (June 1983): "Interpretation of Well-Block Pressures in Numerical Reservoir Simulation with Non-square Grid Blocks and Anisotropic Permeability," *Society of Petroleum Engineering Journal*, 23, 531–543.

Pedersen, K.S., A. Fredenslund, and P. Thomassen (1989): *Properties of Oil and Natural Gases*, Houston, TX: Gulf Publishing.

Pennington, W.D. (2001): "Reservoir Geophysics," *Geophysics*, 66, 25–30.

Pennington, W.D. (2007): "Chapter 1: Reservoir Geophysics," *Petroleum Engineering Handbook, Volume 6: Emerging and Peripheral Technologies*, Edited by H.R. Warner, Jr., Richardson, TX: Society of Petroleum Engineers; Editor-In-Chief L.W. Lake.

Petalas, N. and K. Aziz (June 2000): "A Mechanistic Model for Multiphase Flow in Pipes," *Journal of Canadian Petroleum Technology*, 23 (3), 43–55.

Phillips, J.C. (1996): "Stretched Exponential Relaxation in Molecular and Electronic Glasses," *Reports of Progress in Physics*, 59, 1133–1207.

Pletcher, J.L. (February 2002): "Improvements to Reservoir Material Balance Methods," *SPE Reservoir Evaluation & Engineering*, 5 (1), 49–59.

Press, F. and R. Siever (2001): *Understanding Earth*, Third Edition, San Francisco, CA: W.H. Freeman and Company.

Raymond, M.S. and W.L. Leffler (2006): *Oil and Gas Production in Nontechnical Language*, Tulsa, OK: PennWell.

Raza, S.H. (1992): "Data Acquisition and Analysis for Efficient Reservoir Management," *Journal of Petroleum Technology*, 44 (4), 466–468.

Reynolds, S.J., J.K. Johnson, M.M. Kelly, P.J. Morin, and C.M. Carter (2008): *Exploring Geology*, Boston, MA: McGraw-Hill.

Richardson, J.G., J.B. Sangree, and R.M. Sneider (1987): "Applications of Geophysics to Geologic Models and to Reservoir Descriptions," *Journal of Petroleum Technology*, 39 (7), 753–755.

Richter, C.F. (January 1935): "An Instrumental Earthquake Magnitude Scale," *Bulletin of the Seismological Society of America*, 25 (1), 1–32.

Ridley, M. (1996): *Evolution*, Second Edition, Cambridge, MA: Blackwell Science.

Rubinstein, J.L. and A.B. Mahani (July/August 2015): "Myths and Facts on Wastewater Injection, Hydraulic Fracturing, Enhanced Oil Recovery, and Induced Seismicity," *Seismological Research Letters*, 86 (4), 1–8

Ruppel, C. D. (2011): "Methane Hydrates and Contemporary Climate Change," *Nature Education Knowledge*, 3 (10), 29.

Saleri, N.G. (2002): "'Learning' Reservoirs: Adapting to Disruptive Technologies," *Journal of Petroleum Technology*, 54 (3), 57–60.

Saleri, N.G. (2005): "Reservoir Management Tenets: Why They Matter to Sustainable Supplies," *Journal of Petroleum Technology*, 57 (1), 28–30.

Satter, A., G.M. Iqbal, and J.L. Buchwalter (2008): *Practical Enhanced Reservoir Engineering*, Tulsa, OK: PennWell.

Schecter, R.S. (1992): "Chapter 3: Chemical and Mechanical Properties of Injected Fluids," *Oil Well Stimulation*, Englewood Cliffs, NJ, Prentice Hall.

Schilthuis, R.D. (1936): "Active Oil and Reservoir Energy," *Transactions of the AIME*, 118, 33 ff.

Schön, J.H. (1996): *Physical Properties of Rocks: Fundamentals and Principles of Petrophysics*, Volume 18, New York: Elsevier.

Seitz, T. and K. Yanosek (March2015): "Management: Navigating in Deep Water: Greater Rewards through Narrower Focus", *Journal of Petroleum Technology*, 67 (3), 82.

Selley, R.C. (1998): *Elements of Petroleum Geology*, Second Edition, San Diego, CA: Academic Press.

Selley, R.C. and S.A. Sonnenberg (2015): *Elements of Petroleum Geology*, Third Edition, San Diego, CA: Academic Press.

Sheriff, R.E. (1992): *Reservoir Geophysics*, Investigations in Geophysics, Number 7, Tulsa, OK: Society of Exploration Geophysicists.

Simon, A.D. and E.J. Petersen (1997): *Reservoir Management of the Prudhoe Bay Field*, Paper SPE 38847, Richardson, TX: Society of Petroleum Engineers.

Sloan, E.D. (2006): "Chapter 11: Phase Behavior of H_2O + Hydrocarbon Systems," *Petroleum Engineering Handbook, Volume 1: General Engineering*, Edited by J.R. Fanchi, Richardson, TX: Society of Petroleum Engineers; Editor-In-Chief L.W. Lake.

Sloan, E.D. (2007): "Chapter 11: Hydrate Emerging Technologies," *Petroleum Engineering Handbook, Volume 6: Emerging and Peripheral Technologies*, Edited by H.R. Warner, Jr., Richardson, TX: Society of Petroleum Engineers; Editor-In-Chief L.W. Lake.

Snyder, D.J. and R. Seale (2011): "*Optimization of Completions in Unconventional Reservoirs for Ultimate Recovery Case Studies*," Paper SPE 143066, Richardson, TX: Society of Petroleum Engineers.

SPE-PRMS (2011): "SPE/WPC/AAPG/SPEE Petroleum Resources Management System," Richardson, TX: Society of Petroleum Engineers, http://www.spe.org/spe-app/spe/industry/reserves/accessed July 15, 2015.

Steward, D.B. (2013): "George P. Mitchell and the Barnett Shale," *Journal of Petroleum Technology*, 65 (11), 58–68.

Stone, H.L. (1973): "Estimation of Three-Phase Relative Permeability and Residual Oil Data," *Journal of Canadian Petroleum Technology*, 12 (4), 53ff.

Sutton, R.P. (2006): "Chapter 6: Oil System Correlations," *Petroleum Engineering Handbook, Volume 1: General Engineering*, Edited by J.R. Fanchi, Richardson, TX: Society of Petroleum Engineers; Editor-In-Chief L.W. Lake.

Szabo, D.J. and K.O. Meyers (1993): *Prudhoe Bay: Development History and Future Potential*, Paper SPE 25053, Richardson, TX: Society of Petroleum Engineers.

Tatham, R.H. and M.D. McCormack (1991): *Multicomponent Seismology in Petroleum Exploration*, Tulsa, OK: Society of Exploration Geophysicists.

Tearpock, D.J. and J.C. Brenneke (December 2001): "Multidisciplinary Teams, Integrated Software for Shared Earth Modeling Key E&P Success," *Oil and Gas Journal*, 99 (50), 84–88.

Telford, W.M., L.P. Geldart, and R.E. Sheriff (1990): *Applied Geophysics*, Second Edition, New York: Cambridge University Press.

Terry, R.E., M.J. Michnick, and S. Vossoughi (1981): *Manual for Tracer Test Design and Evaluation*, Institute of Mineral Resources Research, University of Kansas, Lawrence, KS.

Tiab, D. and E.C. Donaldson (2011): *Petrophysics*, Third Edition, Boston, MA: Elsevier.

Toronyi, R.M. and N.G. Saleri (1989): "*Engineering Control in Reservoir Simulation: Part II*," Paper SPE 17937, Richardson, TX: Society of Petroleum Engineers.

Towler, B.F. (2002): *Fundamental Principles of Reservoir Engineering* SPE Textbook Series, Volume 8, Richardson, TX: Society of Petroleum Engineers.

Towler, B.F. (2006): "Chapter 5: Gas Properties," *Petroleum Engineering Handbook, Volume 1: General Engineering*, Edited by J.R. Fanchi, Richardson, TX: Society of Petroleum Engineers; Editor-In-Chief L.W. Lake.

United States Geological Survey (USGS) (2013): Historical, USGS, http://pubs.usgs.gov/gip/dynamic/historical.html, accessed April 23, 2016.

US EIA (2001): Annual Energy Review, Appendix F, United States Energy Information Agency, http://www.eia.doe.gov/, accessed April 8, 2016.

US EIA (2011): Shale Gas Map, United States Energy Information Agency, http://www.eia.gov/oil_gas/rpd/shale_gas.jpg, accessed August 18, 2015.

US EIA (2013): Petroleum: United States Energy Information Agency, http://www.eia.gov/cfapps/ipdbproject/iedindex3.cfm?tid=5&pid=5&aid=2&cid=regions&syid=1980&eyid=2010&unit=TBPD, accessed August 17, 2015.

US EIA (2015): US Annual Energy Overview, United States Energy Information Agency, http://www.eia.gov/beta/MER/?tbl=T01.01#/?f=A&start=1949&end=2014&charted=4-6-7-14, accessed August 18, 2015.

Valkó, P.P. and W.J. Lee (2010): *A Better Way to Forecast Production from Unconventional Gas Wells*, Paper SPE 134231, Richardson, TX: Society of Petroleum Engineers.

Vysokov, M.S. (1996): *A Brief History of Sakhalin and the Kurils*, Yuzhno-Sakhalinsk: Sakhalin Book Publishing House/LIK.

Walker, G.J. and H.S. Lane (2007): *Assessing the Accuracy of History Match Predictions and the Impact of Time-Lapse Seismic Data: A Case Study for the Harding Reservoir*, Paper SPE 106019, Richardson, TX: Society of Petroleum Engineers.

Wang, Z. (2000): "Dynamic versus Static Elastic Properties of Reservoir Rocks," *Seismic and Acoustic Velocities in Reservoir Rocks, Volume 3, Recent Developments*, Edited by Z. Wang and A. Nur. Tulsa, OK: Society of Exploration Geophysicists, pp. 145–158.

Warpinski, N.R., J. Du, and U. Zimmer (2012): "Measurements of Hydraulic-Fracture-Induced Seismicity in Gas Shales," *SPE Production & Operations Journal*, 27, 240–252.

Weingarten, M., S. Ge, J.W. Godt, B.A. Bekins, and J.L. Rubinstein (June 2015): "High-rate Injection is Associated with the Increase in U.S. Mid-continent Seismicity," *Science*, 348 (6241), 1336–1340.

Welge, H. (1952): "A Simplified Method for Computing Oil Recoveries by Gas or Water Drive," *Transactions American Institute of Mechanical Engineers*, 195, 91–98.

Westlake, D.W.S. (1999): "Bioremediation, Regulatory Agencies, and Public Acceptance of this Technology," *Journal of Canadian Petroleum Technology*, 38, 48–50.

Wigley, T.M.L., R. Richels, and J.A. Edmonds (January 1996): "Economic and Environmental Choices in the Stabilization of Atmospheric CO_2 Concentrations," *Nature*, 379, 240–243.

Wilkinson, A.J. (September 1997): "Improving Risk-Based Communications and Decision Making," *Journal of Petroleum Technology*, 49 (9), 936–943.

Williams, M.A., J.F. Keating, and M.F. Barghouty (April 1998): "The Stratigraphic Method: A Structured Approach to History-Matching Complex Simulation Models," *SPE Reservoir Evaluation and Engineering Journal*, 1 (2), 169–176.

Williams, M.C., V.L. Leighton, A.A. Vassilou, H. Tan, T. Nemeth, V.D. Cox, and D.L. Howlett (1997): "Crosswell Seismic Imaging: A Technology Whose Time Has Come?" *The Leading Edge*, 16 (3), 285–291.

Wilson, M.J. and J.D. Frederick (1999): *Environmental Engineering for Exploration and Production Activities*, SPE Monograph Series, Richardson, TX: Society of Petroleum Engineers.

Wittick, T. (July 2000): "Exploration vs. Development Geophysics: Why is Development Geophysics so much more Quantitative?" *Oil and Gas Journal*, 98 (31), 29–32.

Wood, H.O. and J.A. Anderson (1925): "Description and Theory of the Torsion Seismometer," *Bulletin of the Seismological Society of America*, 15, 1–14.

Yellig, W.F. and R.S. Metcalfe (1980) "Determination and Prediction of CO_2 Minimum Miscibility Pressures," *Society of Petroleum Engineers Journal*, 32(1): 160–168.

Yergin, D. (1992): *The Prize*, New York: Simon and Schuster.

Yergin, D. (2011): *The Quest: Energy, Security, and the Remaking of the Modern World*, New York: Penguin.

INDEX

Introduction to Petroleum Engineering, First Edition. John R. Fanchi and Richard L. Christiansen.
© 2017 John Wiley & Sons, Inc. Published 2017 by John Wiley & Sons, Inc.
Companion website: www.wiley.com/go/Fanchi/IntroPetroleumEngineering